ANNALS *of* THE NEW YORK ACADEMY OF SCIENCES

The New York Academy of Sciences
7 World Trade Center
250 Greenwich Street, 40th Floor
New York, NY 10007-2157

annals@nyas.org
www.nyas.org/annals

Published by Blackwell Publishing
On behalf of the New York Academy of Sciences

Boston, Massachusetts
2011

ANNALS *of* THE NEW YORK ACADEMY OF SCIENCES

VOLUME 1215

ISSUE

Resveratrol and Health

ISSUE EDITORS

Ole Vang and Dipak K. Das

This volume presents manuscripts stemming from the conference entitled "Resveratrol 2010: First International Conference of Resveratrol and Health," held September 13–16, 2010 in Copenhagen, Denmark.

TABLE OF CONTENTS

1 Resveratrol and cellular mechanisms of cancer prevention
Yogeshwer Shukla and Richa Singh

9 Bioavailability of resveratrol
Thomas Walle

16 Resveratrol: a cardioprotective substance
Joseph M. Wu and Tze-chen Hsieh

22 Resveratrol in cardiovascular health and disease
Goran Petrovski, Narasimman Gurusamy, and Dipak K. Das

34 Anti-diabetic effects of resveratrol
Tomasz Szkudelski and Katarzyna Szkudelska

40 Effect of resveratrol on fat mobilization
Clifton A. Baile, Jeong-Yeh Yang, Srujana Rayalam, Diane L. Hartzell, Ching-Yi Lai, Charlotte Andersen, and Mary Anne Della-Fera

48 Transport, stability, and biological activity of resveratrol
Dominique Delmas, Virginie Aires, Emeric Limagne, Patrick Dutartre, Frédéric Mazué, François Ghiringhelli, and Norbert Latruffe

60 Chemoprevention in experimental animals
Nalini Namasivayam

72 Resveratrol modulates astroglial functions: neuroprotective hypothesis
André Quincozes-Santos and Carmem Gottfried

79 Resveratrol and apoptosis
Hung-Yun Lin, Heng-Yuan Tang, Faith B. Davis, and Paul J. Davis

89 Chemopreventive effects of resveratrol and resveratrol derivatives
Thomas Szekeres, Philipp Saiko, Monika Fritzer-Szekeres, Bob Djavan, and Walter Jäger

96 The beneficial effect of resveratrol on severe acute pancreatitis
Qingyong Ma, Min Zhang, Zheng Wang, Zhenhua Ma, and Huanchen Sha

103 Neuroprotective properties of resveratrol and derivatives
Tristan Richard, Alison D. Pawlus, Marie-Laure Iglésias, Eric Pedrot, Pierre Waffo-Teguo, Jean-Michel Mérillon, and Jean-Pierre Monti

109 MicroRNA signatures of resveratrol in the ischemic heart
Partha Mukhopadhyay, Pal Pacher, and Dipak K. Das

117 Anti-inflammatory effects of resveratrol: possible role in prevention of age-related cardiovascular disease
Anna Csiszar

123 The phenomenon of resveratrol: redefining the virtues of promiscuity
John M. Pezzuto

131 Safety of resveratrol with examples for high purity, *trans*-resveratrol, resVida®
J. A. Edwards, M. Beck, C. Riegger, and J. Bausch

138 Resveratrol and life extension
Beamon Agarwal and Joseph A. Baur

144 Resveratrol in cancer management: where are we and where we go from here?
Mary Ndiaye, Raj Kumar, and Nihal Ahmad

150 Chemosensitization of tumors by resveratrol
Subash C. Gupta, Ramaswamy Kannappan, Simone Reuter, Ji Hye Kim and Bharat B. Aggarwal

161 Clinical trials of resveratrol
Ketan R. Patel, Edwina Scott, Victoria A. Brown, Andreas J. Gescher, William P. Steward, and Karen Brown

---

170 Erratum for Ann. N. Y. Acad. Sci. 1201: 1–7

Ann. N.Y. Acad. Sci. ISSN 0077-8923

ANNALS OF THE NEW YORK ACADEMY OF SCIENCES
Issue: *Resveratrol and Health*

Resveratrol and cellular mechanisms of cancer prevention

Yogeshwer Shukla and Richa Singh

Proteomics Laboratory, Indian Institute of Toxicology Research (Council of Scientific & Industrial Research), Lucknow, India

Address for correspondence: Yogeshwer Shukla, Proteomics Laboratory, Indian Institute of Toxicology Research (Council of Scientific & Industrial Research), MG Marg, P.O. Box 80, Lucknow 226001, India. yogeshwer_shukla@hotmail.com; yshukla@iitr.res.in

The use of novel and improved chemopreventive and chemotherapeutic agents for the prevention and treatment of cancer is on the rise. Natural products have always afforded a rich source of such agents. Epidemiological evidence suggests that a higher flavonoid intake is associated with low cancer risk. Accumulating data clearly indicate that the induction of apoptosis is an important component in the chemoprevention of cancer by naturally occurring dietary agents. Resveratrol, a naturally occurring polyphenol, demonstrates pleiotropic health benefits, including antioxidant, anti-inflammatory, antiaging, cardioprotective, and neuroprotective activities. Because of these properties and their wide distribution throughout the plant kingdom, resveratrol is envisioned as a potential chemopreventive/curative agent. Currently, a number of preclinical findings from our lab and elsewhere suggest resveratrol to be a promising natural weapon in the war against cancer. Remarkable progress in elucidating the molecular mechanisms underlying the anticancer properties of resveratrol has been achieved. Here, we focus on some of the myriad pathways that resveratrol targets to exert its chemopreventive role and advocate that resveratrol holds tremendous potential as an efficient anticancer drug of the future.

Keywords: resveratrol; cancer; apoptosis; miRNA; CD95; Wnt

Introduction

Resveratrol (3,4′,5-trihydroxy-*trans*-stilbene) is a dietary polyphenol derived from grapes, berries, peanuts, and other plant sources. During the last decade resveratrol has been shown to possess a fascinating wide spectrum of pharmacologic properties. Multiple biochemical and molecular actions seem to contribute to its effects against precancerous or cancer cells. Resveratrol affects all three discrete stages of carcinogenesis (initiation, promotion, and progression) by modulating signal transduction pathways that control cell division and growth, apoptosis, inflammation, angiogenesis, and metastasis, and hence is considered by some to be a promising anticancer therapy.[3] The anticancer property of resveratrol has been supported by studies indicating that it inhibits proliferation of a wide variety of human tumor cells *in vitro*; these data have led to numerous preclinical animal studies to evaluate the potential of resveratrol for cancer chemoprevention and chemotherapy. The polyphenolic phytoalexin and its analogues have attracted attention of researchers over the past couple of decades because of a number of reports highlighting its benefits *in vitro* and *in vivo* in a variety of human disease models, including cardio- and neuroprotection, immune regulation, and cancer chemoprevention.

Since its discovery resveratrol has been shown to exhibit a plethora of physiological properties that could be useful in human medicine. Even greater interest in resveratrol developed at the beginning of the 1990s, when it was first reported to be in red wine.[4] Resveratrol is present in many plants and fruits, including red grapes, eucalyptus, spruce, blueberries, mulberries, peanuts, red wine, and grape skins.

The use of complementary and alternative medicine (CAM) is increasing rapidly in developed countries and is evident in the use of traditional medicines in various Asian countries, for example the Indian system of medicine called

doi: 10.1111/j.1749-6632.2010.05870.x

Ayurveda. Many plant products are in use as herbal medicines, as food supplements, or as spices in Indian cooking. Some of them have been studied in various experimental models of cancer and have been shown inhibit cell proliferation. Cancer patients are especially interested in exploring the use of CAM because of the high risk of mortality and long-term morbidity associated with surgical procedures of cancer management and high side effects of chemotherapy. Cancer is the second leading cause of death after cardiovascular disorders, and research shows that the chances of developing cancer can be reduced by lifestyle changes and dietary habits. People worldwide use dietary vegetables, medicinal herbs, and plant extracts to prevent or treat cancer.[5]

Cancer cells are known to have alterations in multiple cellular signaling pathways, and because of complexities in communication between multiple signaling networks, the treatment and the cure for most human malignancies remain challenging.[6] Because no set paradigm for the treatment of a particular malignancy can be easily designed due to species response variation, varying lifestyle, and dietary habits, research on such differences and responses is still underway. Nevertheless, in general, epidemiological data suggest that the consumption of fruits and vegetables is associated with reduced risk of several types of cancer.[7] Nutritional supplements, some based on chemicals found in fruits and vegetables, are currently being investigated for their use in preventing, inhibiting, and reversing the progression of cancer. There is also growing evidence for the use of natural products as adjunctive therapy alongside conventional cancer treatments.

Background

In 1997, Jang *et al.* first reported that topical resveratrol applications prevented skin cancer development in mice treated with a carcinogen.[8] There have since been dozens of studies of the anticancer activity of resveratrol in animal models.[9] However, no results of human clinical trials for cancer have been reported.[10] Clinical trials to investigate the effects of resveratrol on colon cancer and melanoma are currently recruiting patients.

In vitro data have shown that resveratrol interacts with multiple molecular targets and damaged cells of breast, skin, gastric, colon, esophageal, prostate, and pancreatic cancer, as well as leukemia.[9] However, the study of pharmacokinetics of resveratrol in humans concluded that even high doses of resveratrol might be insufficient to achieve resveratrol concentrations *in vivo* required for the systemic prevention of cancer.[11] This is consistent with the results from animal cancer models, which indicate that the *in vivo* effectiveness of resveratrol is limited by its poor systemic bioavailability.[10,12,13] The strongest evidence of anticancer action of resveratrol exists for tumors it can come into direct contact with, such as skin and gastrointestinal tract tumors. For other cancers, the evidence is uncertain, even if massive doses of resveratrol are used.[10]

Topical application of resveratrol in mice, both before and after the UVB exposure, inhibited skin damage and decreased skin cancer incidence; however, oral resveratrol was ineffective in treating mice inoculated with melanoma cells.[10] Resveratrol given orally also had no effect on leukemia and lung cancer;[10,14] however, injected intraperitoneally, 2.5 or 10 mg/kg of resveratrol slowed the growth of metastatic Lewis lung carcinomas in mice.[10,15] Resveratrol (1 mg/kg orally) reduced the number and size of the esophageal tumors in rats treated with a carcinogen.[16] In several studies, small doses (0.02–8 mg/kg) of resveratrol, given prophylactically, reduced or prevented the development of intestinal and colon tumors in rats given different carcinogens.[10]

Resveratrol treatment appeared to prevent the development of mammary tumors in animal models; however, it had no effect on the growth of existing tumors. Paradoxically, treatment of prepubertal mice with high doses of resveratrol enhanced formation of tumors. Yet, injected in high doses into mice, resveratrol slowed the growth of neuroblastomas.[10]

By regulating multiple important cellular signaling pathways, including NF-κB, Akt, MAPK, Wnt, etc., some natural products can activate cell death signals and induce apoptosis in precancerous or cancer cells without adversely modulating the activity of normal cells. Therefore, nontoxic "natural agents" harvested from the bounties of nature could be useful either alone or in combination with conventional therapeutics for the prevention of tumor progression and/or treatment of human malignancies.

A major challenge is to understand the biological processes and molecular pathways by which resveratrol induces beneficial effects. Below we will summarize some of the studies on resveratrol's affects on key signaling pathways.

CD95 signaling pathway

The Fas receptor (FasR) is a death receptor on the surface of cells that leads to programmed cell death (PCD; also called apoptosis); it is one of two apoptosis pathways, the other being the mitochondrial pathway. FasR is also known as CD95, Apo-1, and tumor necrosis factor receptor superfamily member 6 (TNFRSf6). FasR is located on chromosome 10 in humans and 19 in mice. Similar sequences related by evolution (orthologs) are found in most mammals. Fas forms the death-inducing signaling complex (DISC) upon ligand binding. Membrane-anchored Fas ligand trimer on the surface of an adjacent cell causes trimerization of Fas receptor.[17] This event is also mimicked by binding of agonistic Fas antibody, though some evidence suggests that the apoptotic signal induced by the antibody is unreliable in the study of Fas signaling.

Resveratrol, like other anticancer drugs, induces tumor cell death by targeting pathways through modulating the levels of Fas and FasL.[10–13,18] A recent study on resveratrol-induced apoptosis in multiple myeloma and T cell leukemia cells highlighted the role of recruitment of Fas/CD95 signaling in lipid rafts in antimyeloma and antileukemia chemotherapy through coclustering of Fas/CD95 death receptor and lipid rafts, whereas normal lymphocytes were spared.[10] Earlier reports have also documented this effect in leukemia cell lines,[11,12] colon,[13] and breast carcinoma cells.[18]

Apoptotic pathway

Apoptosis is the process of PCD that occurs in multicellular organisms. Biochemical events lead to characteristic cell changes and death; these changes include blebbing, loss of cell membrane asymmetry and attachment, cell shrinkage, nuclear fragmentation, chromatin condensation, and chromosomal DNA fragmentation. Apoptosis may be mediated by either death receptors or signals arising from within the cell, that is, by a mitochondrial mechanism or by generation of reactive oxygen species.[3]

Data amassed from several laboratories point to the fact that resveratrol exerts its antiproliferative effect by targeting members of the apoptotic family in cancers of various tissues, such as prostate, breast, colon, brain, endometrium, blood, rectum, pancreas, skin, lung, liver, ovary, and bladder.[19–21]

Table 1. Potential of resveratrol as an anticancer agent

Site	Pathway involved	Reference
Lymphocytes	CD95	10
Colon	CD95	13
Breast	CD95	14
Skin	Apoptosis	18
Skin	NF-κB	20
Prostate	NF-κB	21
Lung	NF-κB	22
Endometrium	PI3K/Akt	24
Lung	SIRT1	30
Colon	Wnt	34–36

Nuclear factor κB pathway

Naturally occurring polyphenolic compounds, such as curcumin and resveratrol, are potent agents for modulating inflammation. Recently, both compounds were shown to mediate some of their effects by targeting the NF-κB (a protein complex that controls the transcription of DNA) signaling pathway. It was shown that resveratrol modulates the NF-κB pathway by inhibiting the proteasome in human articular chondrocytes. NF-κB is found in almost all animal cell types and is involved in cellular responses to stimuli such as stress, cytokines, free radicals, ultraviolet irradiation, oxidized LDL, and bacterial or viral antigens.[23] NF-κB plays a key role in regulating the immune response to infection. Conversely, incorrect regulation of NF-κB has been linked to cancer, inflammatory and autoimmune diseases, septic shock, viral infection, and improper immune development. NF-κB has also been implicated in processes of synaptic plasticity and memory. NF-κB is widely used by eukaryotic cells as a regulator of genes that control cell proliferation and cell survival.[24] As such, many different types of human tumors have misregulated NF-κB. Active NF-κB turns on the expression of genes that keep the cell proliferating and protect the cell from conditions that would otherwise cause it to die via apoptosis.

Defects in NF-κB result in increased apoptosis because NF-κB regulates antiapoptotic genes, for example TRAF1 and TRAF2, and thereby alters the activities of the caspase family of enzymes that are central to most apoptotic processes.[25]

Resveratrol suppressed NF-κB–regulated gene products involved in inflammation (cyclooxygenase-2, matrix metalloproteinase [MMP]-3, MMP-9, and vascular endothelial growth factor), inhibited apoptosis (Bcl-2, Bcl-xL, and TNF-α receptor-associated factor 1), and prevented activation of caspase-3.[26] In our laboratory, resveratrol and UVB treatment was shown to decrease the phosphorylation of tyrosine 701 of the important transcription factor signal transducer and activator of transcription (STAT1), which in turn inhibited translocation of phospho-STAT1 to the nucleus and metastatic protein LIMK1.[27] Likewise, in another study, NF-κB–mediated transcriptional activity induced by EGF and TNF-α were inhibited by resveratrol in prostate cancer cell lines.[28] The suppression of MMP-2 expression by resveratrol led to an inhibition of A549 cell invasion by inactivating phosphorylation of SAPK/c-Jun N-terminal kinase (JNK) and p38 MAPK signaling pathways. A time-dependent inhibition of protein levels for p65, c-Jun, and c-Fos in the nucleus by MR-3 treatment was also observed.[29] In addition, NF-κB promotes liver regeneration by upregulating IL-6 and other molecules such as hepatocyte growth factor. We observed that resveratrol inhibited IL-1β–induced apoptosis, caspase-3 activation, and PARP cleavage in human articular chondrocytes.[30]

Phosphoinositol 3 kinase/Akt pathway

Phosphatidylinositol 3-kinases (PI3-kinases or PI3Ks) are a family of enzymes involved in cellular functions such as cell growth, proliferation, differentiation, motility, survival, and intracellular trafficking, which in turn are involved in cancer. They have also been linked to an extraordinarily diverse group of cellular functions, including cell growth, proliferation, differentiation, motility, survival, and intracellular trafficking. PI3Ks phosphorylate the 3 position hydroxyl group of the inositol ring of phosphatidylinositol. The pathway, including oncogene PIK3CA and tumor suppressor PTEN (gene), is implicated in insensitivity of cancer tumors to insulin and IGF1 in calorie restriction.[31]

Inhibition of beta-arrestin 2 increases the number of apoptotic cells and caspase-3 activation by reducing Akt/GSK3-β levels in endometrial cancer cells.[32] Resveratrol downregulated cell cycle-related proteins, including the expression of cyclin-dependent kinase (CDK) 2, cyclin E, CDK4, cyclin D1, retinoblastoma (Rb), and proliferative cell nuclear antigen (PCNA), thereby blocking the Akt pathway in rat aortic vascular smooth muscle cell,[33] bladder cancer,[34] and liver cancer cells.[35]

The SIRT1-regulated pathway

It has been established that genes control almost every aspect of human physiology, including longevity and aging processes on the cellular level, which in turn affects the lifespan and aging of the individual. One of the genes that may be associated with cellular longevity and ability to slow down the aging process is sirtuin 1 (SIRT1).[36]

SIRT1 is known to deacetylate histones and nonhistone proteins, including transcription factors, thereby regulating metabolism, stress resistance, cellular survival, cellular senescence, inflammation-immune function, endothelial functions, and circadian rhythms.[37] Resveratrol has been shown to activate SIRT1 directly or indirectly in a variety of models. Activation of SIRT1 by resveratrol may be beneficial for regulation of calorie restriction, oxidative stress, inflammation, cellular senescence, autophagy/apoptosis, autoimmunity, metabolism, adipogenesis, circadian rhythm, skeletal muscle function, mitochondria biogenesis, and endothelial dysfunction.[37] Resveratrol mimics the effects of calorie restriction in lower organisms, and mice fed a high-fat diet demonstrate reduced insulin resistance.[38] SIRT1 also plays an important role in regulating autophagy in response to cigarette smoke.[39] SIRT1 activation by resveratrol may also play a role in treatment strategies for stroke and other neurodegenerative disorders. The goal here is to provide a better understanding of the mode of action of resveratrol and its possible use as a potential therapeutic agent to ameliorate stroke damage as well as other age-related neurodegenerative disorders. Similarly, research on elucidating the beneficial effects of resveratrol during colitis revealed that it may be mediated through several mechanisms.[40] One mechanism may involve the negative regulation of NF-κB activity by SIRT1, as the NF-κB pathway has been shown to contribute to colitis and colon cancer associated with colitis Thus, downregulation of SIRT1 during colitis may induce inflammatory cytokines through activation of NF-κB, which is reversed by resveratrol.[41]

Table 2. Resveratrol health claims: animals versus humans

In vivo studies	Human cell lines	Human
√ Increased longetivity (C57BL/6NIA mice)	√ Inhibition of proliferation in many cancer cell lines	√ None
√ Increased mitochondria and endurance (Sprague-Dawley rats)	√ Improves cardiac muscle cells	
√ Impedes cancer cell growth(BALB/cAnNCr-nu/nu mice)	√ Improves insulin sensitivity	
√ Improves insulin sensitivity(db/db mice)		

miRNAs as molecular targets of resveratrol

In recent years, microRNAs have received greater attention in cancer research. MicroRNAs (miRNAs) are small RNA molecules that regulate gene expression post-transcriptionally. About 3% of human genes encode for miRNAs, and up to 30% of human protein coding genes may be regulated by miRNAs. MiRNAs play a key role in diverse biological processes, including development, cell proliferation, differentiation, and apoptosis. These small, noncoding RNAs inhibit target gene expression by binding to the 3′ untranslated region of target mRNAs, resulting in either mRNA degradation or inhibition of translation. MiRNAs play important roles in many normal biological processes; however, studies have also shown that aberrant miRNA expression is correlated with the development and progression of some cancers.[42] Thus some miRNAs could have oncogenic or tumor suppressor activities. Moreover, specific miRNAs may regulate formation of cancer stem cells and epithelial-mesenchymal transition phenotype of cancer cells, which are typically drug resistant. MiRNAs may also be used as biomarkers for diagnosis and prognosis. Thus miRNAs are emerging as targets for cancer therapy.[43] The manipulation of miRNAs and other genes in animal models can increase or decrease lifespan. Transcriptional and posttranscriptional regulatory mechanisms, some of which involve miRNAs, as well as modifications to chromatin and histones, can influence longevity. A decline in the function of stem cells might also be responsible for some aspects of mammalian aging or senescence which is also a mechanism of inhibiting tumor cell proliferation.[44]

Wnt pathway

One reason why specific inhibitors that target only one pathway have typically failed in cancer treatment is the complexities of the communication between multiple signaling networks. *In vitro* and *in vivo* studies have demonstrated that natural products such as isoflavones, indole-3-carbinol (I3C), 3,3′-diindolylmethane (DIM), curcumin, (–)-epigallocatechin-3-gallate (EGCG), resveratrol, and lycopene have inhibitory effects on human and animal cancers through targeting multiple cellular signaling pathways, and thus these "natural agents" could be classified as multitargeted agents.

The Wnt signaling pathway consists of a network of proteins most well known for its role in embryogenesis and cancer, though this network is also involved in normal physiological processes in adult animals.[45] The canonical Wnt pathway describes a series of events that occur when Wnt proteins bind to cell surface receptors of the frizzled family, causing the receptors to activate the disheveled family of proteins and, ultimately, a change in the amount of β-catenin that reaches the nucleus. Recently it was reported that resveratrol suppresses colon cancer cell proliferation and elevates apoptosis even in the presence of IGF-1 (via suppression of IGF-1R/Akt/Wnt signaling pathways and activation of p53), suggesting its potential role as a chemotherapeutic agent.[46] The effects of low concentrations of resveratrol on the Wnt pathway was evaluated; and in the absence of effects on cell proliferation, resveratrol significantly inhibited Wnt signaling in colon-derived cells, which do not have a basally activated Wnt pathway.[47] This inhibitory effect may be due in part to regulation of intracellular

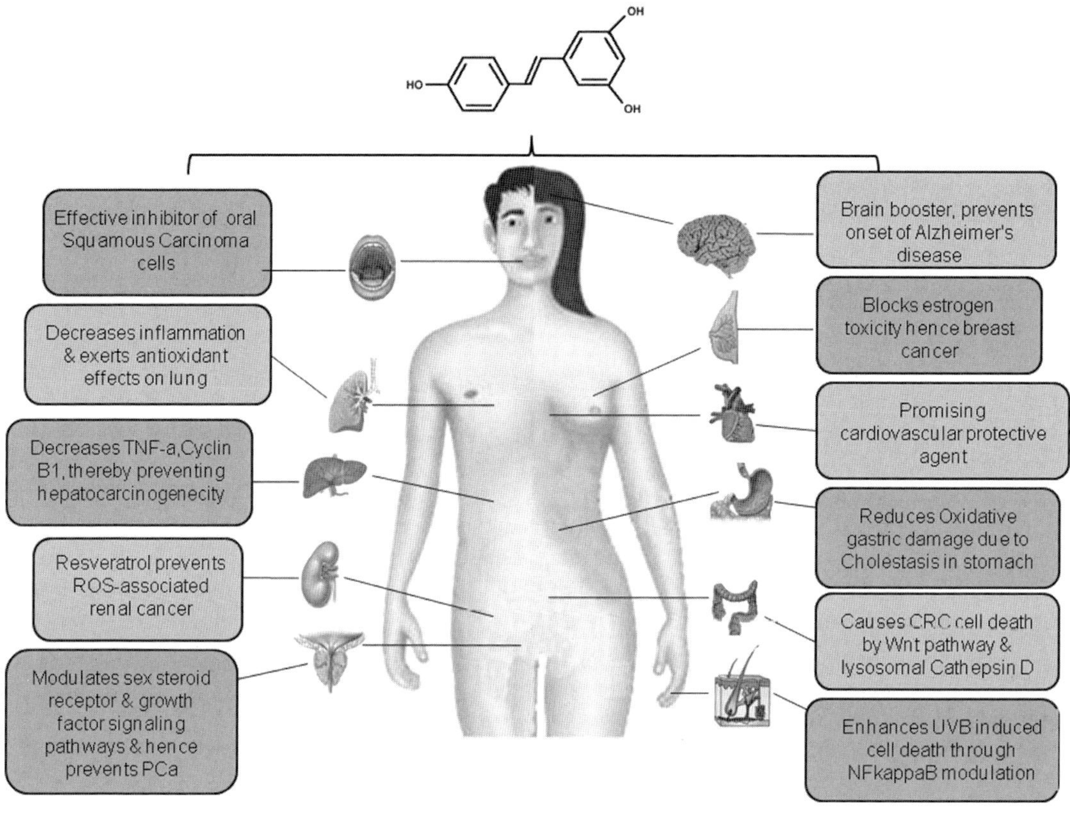

Figure 1. Beneficial health effects of resveratrol.

β-catenin localization.[47] Resveratrol inhibited proliferation and induced apoptosis in colon cancer cells by reducing both of survivin expression and the Wnt/β-catenin signaling pathway.[48]

Conclusion

Resveratrol, a natural nonflavonoid polyphenol found in grapes and red wine, is recognized as a bioactive agent with potential benefits for health. A large number of pharmacological properties, including cardioprotective, antioxidant, and anticancer effects, are thought to be associated with its beneficial effects (Figure 1). Few studies have been performed with resveratrol in humans, and the results of these studies appear fragmentary and sometimes contradictory due to variations in conditions of administration, protocols, and methods of assessment; for example, it appears that the presence of the matrix in which resveratrol is administered (e.g., in alcohol, or other polyphenolic compounds in wine) or feeding conditions (fed versus fasting) results in discrepancies between studies. Although differences in resveratrol administration, doses, and assay methods make data from different studies difficult to compare, these data nevertheless provide important information and raise some interesting questions. First, resveratrol seems to be well tolerated; however, no information is available on long-term administration. For use in chronic diseases such as diabetes, colorectal cancer, or Alzheimer's disease, or for the prevention of cardiovascular disease and antiaging antioxidative care, resveratrol administration would occur over several months/years at doses that have yet to be determined. Although resveratrol is considered a food supplement and a relatively safe natural medication, further investigations are required to determine its long-term effects. Second, resveratrol is rapidly absorbed and metabolized, mainly as sulfo- and gluco-conjugates that are excreted in urine. This high metabolic rate probably allows the transport, the distribution, and the excretion of resveratrol. In rodents, the gut epithelium has been shown to be highly implicated in the metabolic process, resulting in polar resveratrol compounds that require specific transporters to cross cell membranes. Several ATP-binding cassette transporters may be involved in the tissue distribution and subsequent elimination of resveratrol from the body. However, concentrations of free *trans*-resveratrol are very low in plasma, and hence several authors have raised doubts about its efficiency. Because many lacunae in our understanding of resveratrol action and biochemistry remain to be filled, humans clinical trials have understandably lagged behind other animal and *in vitro* studies.

Several studies on elucidating the effects of resveratrol demonstrate that it may have potential beneficial activities against cancer, cardiovascular disease, diabetes, and autoimmune diseases; and it may even interfere with the normal physiological processes of aging. But to fully realize the potential of resveratrol, clinical trials are needed. Its analogues, with improved pharmacokinetic and pharmacodynamics, will also help the field move forward. Safety during long-term administration, combined with its cost and future therapeutic potential, makes resveratrol an ideal agent for both prevention and therapy of chronic illnesses, either alone or in combination with other drugs. Reverse pharmacology, in the case of resveratrol, is likely to prove correct Hippocrates correct, who remarked 25 centuries ago, "Let food be thy medicine and medicine be thy food." Natural products such as resveratrol have gained considerable attention as cancer chemopreventive or cardioprotective agents and as antitumor agents. Among its wide range of biological activities, resveratrol has been reported to interfere with many intracellular signaling pathways that regulate cell survival or apoptosis. Hence, resveratrol may hold promise in the near future as a chemotherapeutic drug in the treatment of cancer and other diseases.

Conflicts of interest

The authors declare no conflict of interest.

References

1. Patel, V.B., S. Misra, B.B. Patel & A.P. Majumdar. 2010. Colorectal cancer: chemopreventive role of curcumin and resveratrol. *Nutr. Cancer*. **62:** 958–967.
2. Gullett, N.P., A.R. Ruhul Amin, S. Bayraktar, *et al.* 2010. Cancer prevention with natural compounds. *Semin. Oncol.* **37:** 258–281.
3. Kraft, T.E., D. Parisotto, C. Schempp & T. Efferth. 2009. Fighting cancer with red wine? Molecular mechanisms of resveratrol. *Crit. Rev. Food Sci. Nutr*. **49:** 782–799.
4. Sadruddin, S. & R.J. Arora. 2009. Resveratrol: biologic and therapeutic implications. *Cardiometab. Syndr*. **4:** 102–106.
5. Granados-Principal, S., J.L. Quiles, C.L. Ramirez-Tortosa, *et al.* 2010. New advances in molecular mechanisms and the prevention of adriamycin toxicity by antioxidant nutrients. *Food Chem. Toxicol.* **48:** 1425–1438.
6. Liu, E.H., L.W. Qi, Q. Wu, *et al.* 2009. Anticancer agents derived from natural products. *Mini. Rev. Med. Chem.* **9:** 1547–1555.
7. Liu, B.L., X. Zhang, W. Zhang & H.N. Zhen. 2007. New enlightenment of French Paradox: resveratrol's potential for cancer chemoprevention and anti-cancer therapy. *Cancer Biol. Ther.* **6:** 1833–1836.
8. Jang, M., L. Cai, G.O. Udeani, *et al.* 1997. Cancer chemopreventive activity of resveratrol, a natural product derived from grapes. *Science* **275:** 218–220.
9. Baur, J.A. & D.A. Sinclair. 2006. Therapeutic potential of resveratrol: the in vivo evidence. *Nat. Rev. Drug Discov.* **5:** 493–506.
10. Athar, M., J.H. Back, X. Tang, *et al.* 2007. Resveratrol: a review of preclinical studies for human cancer prevention. *Toxicol. Appl. Pharmacol.* **224:** 274–283.
11. Boocock, D.J., G.E. Faust, K.R. Patel, *et al.* 2007. Phase I dose escalation pharmacokinetic study in healthy volunteers of resveratrol, a potential cancer chemopreventive agent. *Cancer Epidemiol. Biomarkers Prev.* **16:** 1246–1252.
12. Niles, R.M., C.P. Cook, G.G. Meadows, *et al.* 2006. Resveratrol is rapidly metabolized in athymic (nu/nu) mice and does not inhibit human melanoma xenograft tumor growth. *J. Nutr.* **136:** 2542–2456.
13. Wenzel, E., T. Soldo, H. Erbersdobler & V. Somoza. 2005. Bioactivity and metabolism of trans-resveratrol orally administered to Wistar rats. *Mol. Nutr. Food Res.* **49:** 482–494.
14. Gao, X., Y.X. Xu, G. Divine, *et al.* 2002. Disparate in vitro and in vivo antileukemic effects of resveratrol, a natural polyphenolic compound found in grapes. *J. Nutr.* **132:** 2076–2081.
15. Kimura, Y. & H. Okuda. 2001. Resveratrol isolated from Polygonum cuspidatum root prevents tumor growth and metastasis to lung and tumor-induced neovascularization in Lewis lung carcinoma-bearing mice. *J. Nutr.* **131:** 1844–1849.
16. Li, Z.G., T. Hong, Y. Shimada, *et al.* 2002. Suppression of N-nitrosomethylbenzylamine (NMBA)-induced esophageal tumorigenesis in F344 rats by resveratrol. *Carcinogenesis* **23:** 1531–1536.
17. Falschlehner, C., T.M. Ganten, R. Koschny, *et al.* 2009. TRAIL and other TRAIL receptor agonists as novel cancer therapeutics. *Adv. Exp. Med. Biol.* **647:** 195–206.
18. Pervaiz, S. 2001. Resveratrol: from the bottle to the bedside? *Leuk. Lymphoma* **40:** 491–498.
19. Gatouillat, G., E. Balasse, D. Joseph-Pietras, *et al.* 2010. Resveratrol induces cell-cycle disruption and apoptosis in chemoresistant B16 melanoma. *J. Cell Biochem.* [Epub ahead of print].
20. Brizuela, L., A. Dayon, N. Doumerc, *et al.* 2010. The sphingosine kinase-1 survival pathway is a molecular target for the tumor-suppressive tea and wine polyphenols in prostate cancer. *FASEB J.* [Epub ahead of print].

21. Li, H., W.K. Wu, Z. Zheng, *et al.* 2010. 3,3′,4,5,5′-pentahydroxy-trans-stilbene, a resveratrol derivative, induces apoptosis in colorectal carcinoma cells via oxidative stress. *Eur. J. Pharmacol.* **637:** 55–61.
22. Madan, E., S. Prasad, P. Roy, *et al.* 2008. Regulation of apoptosis by resveratrol through JAK/STAT and mitochondria mediated pathway in human epidermoid carcinoma A431 cells. *Biochem. Biophys. Res. Commun.* **377:** 1232–1237.
23. Staudt, L.M. 2010. Oncogenic activation of NF-kappaB. *Cold Spring Harb. Perspect. Biol.* **2:** a000109.
24. Kucharczak, J., M.J. Simmons, Y. Fan & C. Gélinas. 2003. To be, or not to be: NF-kappaB is the answer–role of Rel/NF-kappaB in the regulation of apoptosis. *Oncogene.* **22:** 8961–8982.
25. Sughra, K., A. Birbach, R. de Martin & J.A. Schmid. 2010. Interaction of the TNFR-receptor associated factor TRAF1 with I-kappa B kinase-2 and TRAF2 indicates a regulatory function for NF-kappa B signaling. *PLoS One* **5:** e12683.
26. Csaki, C., A. Mobasheri & M. Shakibaei. 2009. Synergistic chondroprotective effects of curcumin and resveratrol in human articular chondrocytes: inhibition of IL-1beta-induced NF-kappaB-mediated inflammation and apoptosis. *Arthritis Res. Ther.* **11:** R165.
27. Roy, P., E. Madan, N. Kalra, *et al.* 2009. Resveratrol enhances ultraviolet B-induced cell death through nuclear factor-kappaB pathway in human epidermoid carcinoma A431 cells. Biochem. *Biophys. Res. Commun.* **384:** 215–220.
28. Benitez, D.A., M.A. Hermoso, E. Pozo-Guisado, *et al.* 2009. Regulation of cell survival by resveratrol involves inhibition of NF kappa B-regulated gene expression in prostate cancer cells. *Prostate.* **69:** 1045–1054.
29. Yang, Y.T., C.J. Weng, C.T. Ho & G.C. Yen. 2009. Resveratrol analog-3,5,4′-trimethoxy-trans-stilbene inhibits invasion of human lung adenocarcinoma cells by suppressing the MAPK pathway and decreasing matrix metalloproteinase-2 expression. *Mol. Nutr. Food Res.* **53:** 407–416.
30. Shakibaei, M., C. Csaki, S. Nebrich & A. Mobasheri. 2008. Resveratrol suppresses interleukin-1beta-induced inflammatory signaling and apoptosis in human articular chondrocytes: potential for use as a novel nutraceutical for the treatment of osteoarthritis. *Biochem. Pharmacol.* **76:** 1426–1439.
31. Harikumar, K.B. & B.B. Aggarwal. 2008. Resveratrol: a multitargeted agent for age-associated chronic diseases. *Cell Cycle* **7:** 1020–1035.
32. Sun, X., Y. Zhang, J. Wang, *et al.* 2010. Beta-arrestin 2 modulates resveratrol-induced apoptosis and regulation of Akt/GSK3beta pathways in human endometrial cancer cells. *Biochim. Biophys. Acta.* [Epub ahead of print].
33. Park, E.S., Y. Lim, J.T. Hong, *et al.* 2010. Pterostilbene, a natural dimethylated analog of resveratrol, inhibits rat aortic vascular smooth muscle cell proliferation by blocking Akt-dependent pathway. *Vascul. Pharmacol.* **53:** 61–67.
34. Bai, Y., Q.Q. Mao, J. Qin, *et al.* 2010. Resveratrol induces apoptosis and cell cycle arrest of human T24 bladder cancer cells in vitro and inhibits tumor growth in vivo. *Cancer Sci.* **101:** 488–493.
35. Parekh, P., L. Motiwale, N. Naik & K.V. Rao. 2010. Downregulation of cyclin D1 is associated with decreased levels of p38 MAP kinases, Akt/PKB and Pak1 during chemopreventive effects of resveratrol in liver cancer cells. *Exp. Toxicol. Pathol.* [Epub ahead of print].
36. Queen, B.L. & T.O. Tollefsbol. 2010. Polyphenols and aging. *Curr. Aging Sci.* **3:** 34–42.
37. Chung, S., H. Yao, S. Caito, *et al.* Regulation of SIRT1 in cellular functions: role of polyphenols. *Arch. Biochem. Biophys.* **501:** 79–90.
38. de Kreutzenberg, S.V., G. Ceolotto, I. Papparella, *et al.* 2010. Downregulation of the longevity-associated protein sirtuin 1 in insulin resistance and metabolic syndrome: potential biochemical mechanisms. *Diabetes* **59:** 1006–1015.
39. Hofseth, L.J., U.P. Singh, N.P. Singh, *et al.* 2010. Taming the beast within: resveratrol suppresses colitis and prevents colon cancer. *Aging (Albany NY).* **2:** 183–184.
40. Cui, X., Y. Jin, A.B. Hofseth, *et al.* 2010. Resveratrol suppresses colitis and colon cancer associated with colitis. *Cancer Prev. Res. (Phila).*;**3** : 549–59. Epub 2010 Mar 23.
41. Arunachalam, G., H. Yao, I.K. Sundar, *et al.* 2010. SIRT1 regulates oxidant- and cigarette smoke-induced eNOS acetylation in endothelial cells: role of resveratrol. *Biochem. Biophys. Res. Commun.* **393:** 66–72.
42. Wang, J., Q. Wang, H. Liu, *et al.* 2010. MicroRNA expression and its implication for the diagnosis and therapeutic strategies of gastric cancer. *Cancer Lett.* **297:** 137–143.
43. Li, Y., D. Kong, Z. Wang & F.H. Sarkar. 2010. Regulation of microRNAs by natural agents: an emerging field in chemoprevention and chemotherapy research. *Pharm. Res.* **27:** 1027–1041.
44. Marques, F.Z., M.A. Markus & B.J. Morris. 2010. The molecular basis of longevity, and clinical implications. *Maturitas* **65:** 87–91.
45. Sarkar, F.H., Y. Li, Z. Wang & D. Kong. 2009. Cellular signaling perturbation by natural products. *Cell Signal.* **21:** 1541–1547.
46. Vanamala, J., L. Reddivari, S. Radhakrishnan & C. Tarver. 2010. Resveratrol suppresses IGF-1 induced human colon cancer cell proliferation and elevates apoptosis via suppression of IGF-1R/Wnt and activation of p53 signaling pathways. *BMC Cancer.* **10:** 238.
47. Hope, C., K. Planutis, M. Planutiene, *et al.* 2008. Low concentrations of resveratrol inhibit Wnt signal throughput in colon-derived cells: implications for colon cancer prevention. *Mol. Nutr. Food Res.* **52:** 52–61.
48. Sakoguchi-Okada, N., F. Takahashi-Yanaga, K. Fukada, *et al.* 2007. Celecoxib inhibits the expression of survivin via the suppression of promoter activity in human colon cancer cells. *Biochem. Pharmacol.* **73:** 1318–1329.

Ann. N.Y. Acad. Sci. ISSN 0077-8923

ANNALS OF THE NEW YORK ACADEMY OF SCIENCES
Issue: *Resveratrol and Health*

Bioavailability of resveratrol

Thomas Walle
Department of Pharmacology, Medical University of South Carolina, Charleston, South Carolina

Address for correspondence: Thomas Walle, Department of Pharmacology, Medical University of South Carolina, Charleston, SC 29425. wallet@musc.edu

This paper reviews our current understanding of the absorption, bioavailability, and metabolism of resveratrol, with an emphasis on humans. The oral absorption of resveratrol in humans is about 75% and is thought to occur mainly by transepithelial diffusion. Extensive metabolism in the intestine and liver results in an oral bioavailability considerably less than 1%. Dose escalation and repeated dose administration of resveratrol does not appear to alter this significantly. Metabolic studies, both in plasma and in urine, have revealed major metabolites to be glucuronides and sulfates of resveratrol. However, reduced dihydroresveratrol conjugates, in addition to highly polar unknown products, may account for as much as 50% of an oral resveratrol dose. Although major sites of metabolism include the intestine and liver (as expected), colonic bacterial metabolism may be more important than previously thought. Deconjugation enzymes such as β-glucuronidase and sulfatase, as well as specific tissue accumulation of resveratrol, may enhance resveratrol efficacy at target sites. Resveratrol analogs, such as methylated derivatives with improved bioavailability, may be important in future research.

Keywords: resveratrol; bioavailability; metabolism

Introduction: effects *in vivo* versus *in vitro*

Large numbers of studies have demonstrated that resveratrol (Fig. 1) has protective effects on biological processes, many of which are important in cancer and/or cardiovascular disease, supporting the idea that this dietary chemical may be a useful chemoprotective agent.[1,2] Most of these biologic studies have been conducted *in vitro*, i.e., using cell or tissue models or subcellular fractions. When expanding such studies to *in vivo* animal models of disease, however, it has been difficult to show such effects. Proof of efficacy of resveratrol in humans *in vivo* has been even more elusive; in fact, no convincing studies exist as yet. A major reason for the *in vitro*/*in vivo* discrepancies may be the lack of bioavailability of resveratrol *in vivo*, in particular after oral administration. In short, concentrations of resveratrol at potential tissue or cellular sites of action appear to have thus far been inadequate to demonstrate efficacy in humans.

This paper will review our current understanding of the bioavailability and metabolism of resveratrol, with the emphasis on humans. Areas where further in depth studies should be important will be pointed out.

Absorption and bioavailability, single doses

Two of the initial human studies of the absorption and bioavailability of resveratrol used single 25 mg oral doses,[3,4] which corresponds to a moderate intake of red wine. Despite the use of highly sensitive and molecularly specific analytical methodology, it was difficult to detect unmetabolized resveratrol in the circulating plasma. Rough estimates showed peak concentrations of <10 ng/mL at 0.5–2 h after the oral dose. Estimates of resveratrol-plus-total-metabolite plasma concentrations were considerably higher, around 400–500 ng/mL (≈2 μM),[3,4] indicating very low oral bioavailability for resveratrol. The good agreement between the two studies emphasized the use of radioactive doses and liquid chromatography (LC), as well as unlabeled doses and liquid chromatography/mass spectrometry (LC/MS), or gas chromatography mass spectrometry (GC)/MS, as methods of choice in the

doi: 10.1111/j.1749-6632.2010.05842.x

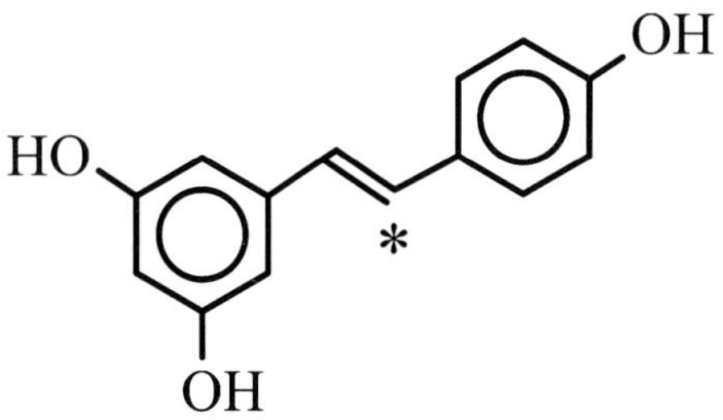

Figure 1. Chemical structure of resveratrol (*trans*-3,5,4′-trihydroxystilbene). *Position of ^{14}C-label in radiolabeled compound.

continuing investigation of the absorption and bioavailability of dietary resveratrol.

Based on the urinary excretion of total metabolites after radiolabeled doses, the oral absorption of resveratrol appeared to be at least 75%. This may also be inferred from a comparison of the plasma total area under the plasma concentration time curve (AUC) data after oral and intravenous doses, if resveratrol is assumed to be metabolized similarly after both routes of administration.[4] This absorption is unusually high for a dietary polyphenol, particularly when considering the poor aqueous solubility of this compound. The high oral absorption of resveratrol is consistent with findings in the human intestinal Caco-2 cell monolayer, which is a good model of human intestinal absorption.[5] Resveratrol transport in this model appeared to be direction-independent, occurring principally by transepithelial diffusion.[6] Active transport might occur as well, but likely only by resveratrol metabolites.[7,8] However, the transport of resveratrol in the Caco-2 cell monolayer was nonlinear with time and, together with extensive metabolite formation, suggested metabolism to be rate-limiting with respect to bioavailability.

Effects of dose escalation and repeated doses

Severely limited availability of resveratrol at potential sites of action may be one major explanation for the lack of chemopreventive effects of resveratrol *in vivo* in humans as well as in animal models of human disease. Several approaches to increase the availability of resveratrol in humans have been used. Dose escalation, a common logical approach, has been examined in two studies using a total dose range of 25–5,000 mg,[9,10] thus covering the wide range used in chemopreventive studies. As seen in Figure 2, there appeared to be a linear increase in plasma resveratrol concentrations with increasing doses, i.e., no evidence of saturation of metabolism or possible efflux transport. Even after the highest dose (5,000 mg), the peak plasma levels only reached ≈500 ng/mL,[9] possibly due to limited solubility. Repeated or chronic dosing might also result in saturation of metabolism, leading to higher plasma and tissue levels of resveratrol. It should be noted that in the few studies where chemoprotective effects of resveratrol have been observed in experimental rat models, these effects have been seen after as long as 15–20 weeks of treatment.[11–13] Only one repeated dose pharmacokinetic study over two days of treatment has so far been published; it was unclear whether an effect on the pharmacokinetics was observed.[10] Both of these approaches should be examined further.

Metabolism by glucuronidation and sulfation

A combination of solid-phase extraction and LC/MS/UV separation and detection techniques was used to identify resveratrol metabolites in human urine (Fig. 3):[4] the top tracing (Fig. 3A) used MS detection of resveratrol (RV), three glucuronide metabolites (M1, M2, M3), and two sulfates (M4, M5); the bottom tracing (Fig. 3B) used UV to detect two glucuronides (M1, M2) and one sulfate metabolite (M4). M3 and M5 were not detectable by UV at 305 nm due to reduction of the aliphatic double bond and loss of the chromofore. The two

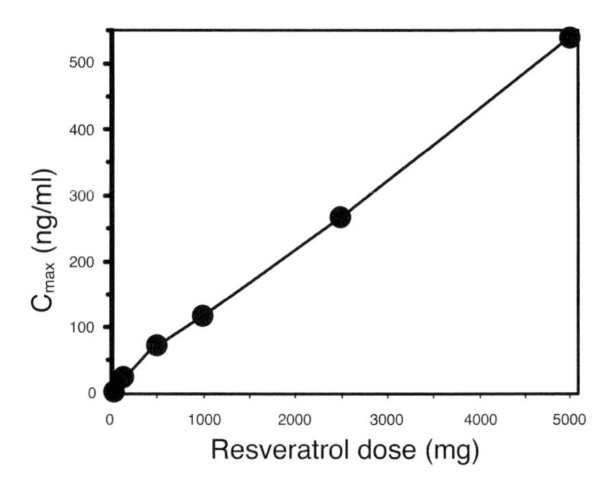

Figure 2. Mean peak plasma resveratrol concentrations (C_{max}) in healthy volunteers who received single oral doses of 25 to 5000 mg. Plotted from data in Refs. 9, 10.

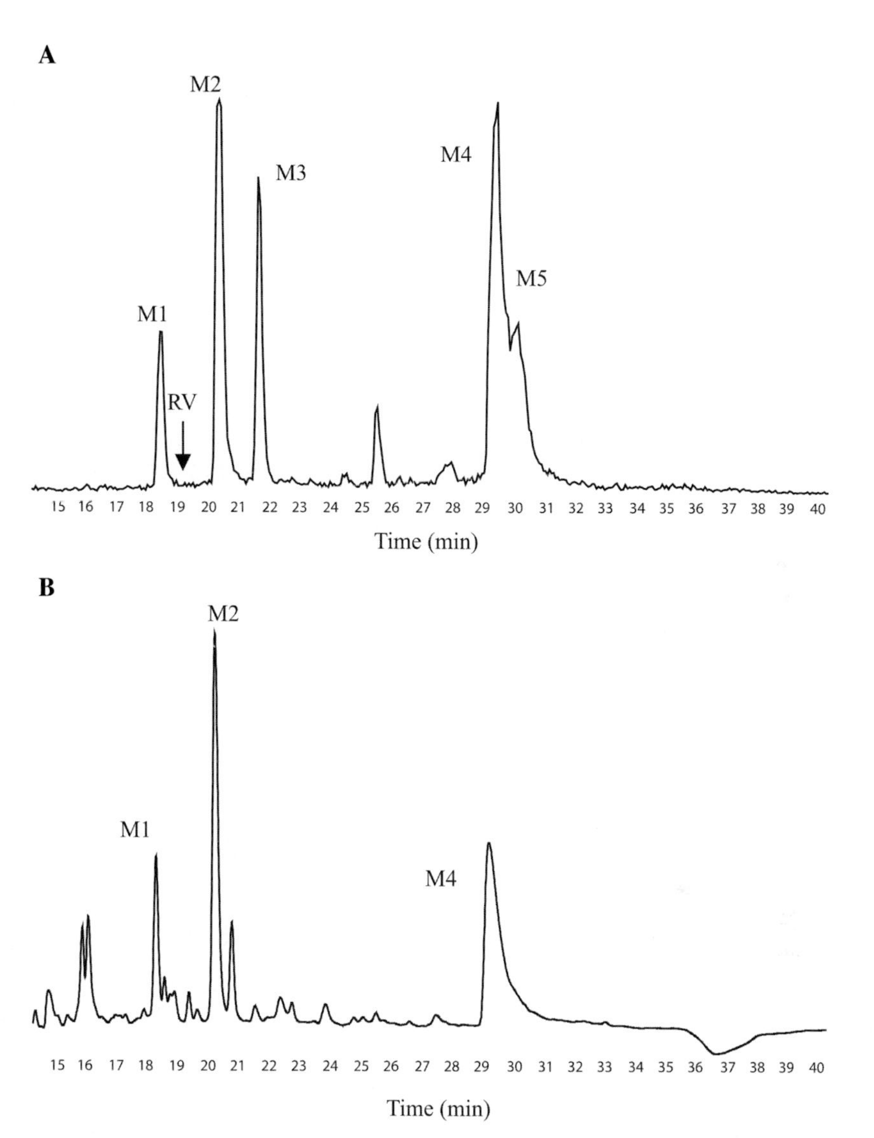

Figure 3. LC/MS tracings of the urinary excretion (0–12 h) of resveratrol and its metabolites (M1–M5) after a 100-mg unlabeled oral dose.[4] (A) MS detection of the $[M-1]^{-1}$ ions for RV (*m*/*z* 227), M1 and M2 (*m*/*z* 403), M3 (*m*/*z* 405), M4 (*m*/*z* 307), and M5 (*m*/*z* 309). (B) UV detection at 305 nm. The sample was run using a gradient with 0.1% acetic acid and acetonitrile. RV denotes the retention time of resveratrol.

dihydroresveratrol metabolites thus identified were one glucuronide (M3) and one sulfate conjugate (M5). There was no evidence of any oxidative metabolites. The preference for conjugation of the hydroxyl groups in the resveratrol molecule was obvious, which is similar to most other phenolic dietary chemicals. The sites of glucuronidation in the resveratrol molecule, i.e., the 3- and 4′-positions, have been firmly established after synthesis of the metabolites.[14] Similarly, the major site of sulfate conjugation, i.e., in the 4′-position, was also established by synthesis.[15–17] Whereas glucuronidation was considered the main conjugation pathway, more recent studies favor sulfation as more important, in particular in humans.[4,9] Sulfate conjugates have been more difficult to characterize and quantify, partly due to their poor chromatographic behavior by LC,[4] with adsorption and tailing on the reversed-phase columns commonly used. Optimizing the conditions for detection of the sulfate conjugates made it possible to detect them also in plasma after both intravenous and oral doses.[4] This

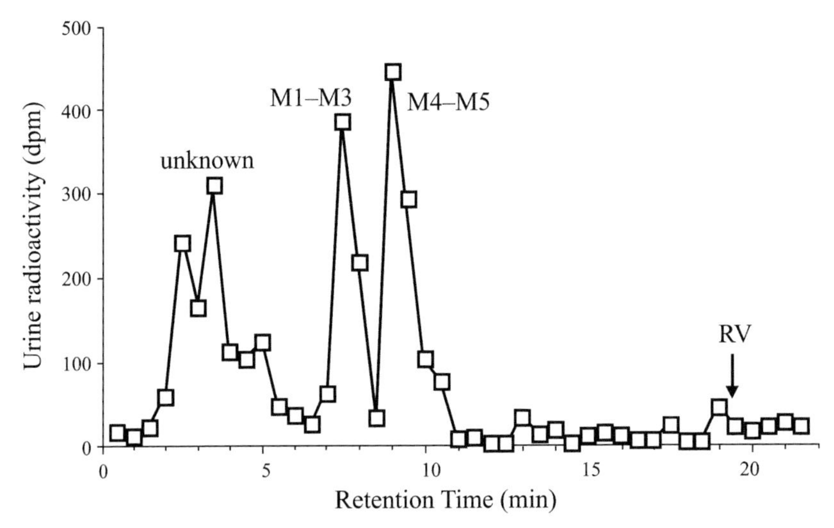

Figure 4. LC/radioactivity tracings with fraction collection of a 0–12-h urine extract after a 25-mg oral dose of ^{14}C-resveratrol.[4] M1–M5 are glucuronides and sulfate conjugates. RV is resveratrol.

led to the finding that sulfate conjugation occurs very rapidly and might be the primary metabolic pathway. Resveratrol sulfation may be of special significance with respect to breast cancer. Thus, resveratrol was efficiently sulfated in the 4′-position by SULT1A1 in breast cancer cell lines, in particular in the hormone-dependent cell line.[18] In contrast, in normal human breast cells, which contain SULT1E1 exclusively rather than SULT1A1, resveratrol was a potent inhibitor of estradiol sulfation, with an IC_{50} value as low as 1 μM.[19]

Reduced metabolic products

In spite of many studies of the metabolism of resveratrol in both animals and humans, it is still not clear how much of administered doses can be accounted for, i.e., what is the mass balance. The finding of two reduced metabolites of resveratrol, i.e., the glucuronic acid and sulfate conjugates of dihydroresveratrol, i.e., M3 and M5, respectively (Fig. 3A),[4] was surprising. First, they were not detectable by UV at 305 nm, explaining why they had not been recognized earlier by this common detection technique. Second, they appeared to account for a substantial fraction of the resveratrol dose. Third, they most likely were derived from intestinal bacterial metabolism. Dihydroresveratrol was later also identified in rat urine,[20] plasma,[21] and, most importantly, from incubates of resveratrol with mammalian fecal bacterial species.[22] Dihydroresveratrol's biological activity is still incompletely understood but may be less than for resveratrol.[23] A very recent study indicates that dihydroresveratrol may in fact have potent proliferative effects on hormone-sensitive breast cancer cells, e.g., MCF-7 cells.[24] The latter is similar to previous observations for quercetin.[25]

Metabolism by bacterial enzymes

When using a radioactive resveratrol dose additional metabolites, labeled "unknown," were detected in the human urine (Fig. 4).[4] These metabolic products appeared very early by reversed-phase LC, ahead of the glucuronic acid and sulfate conjugates M1–M5. They may together account for as much as 25% of an oral 25 mg dose. Because of their high polarity, as deduced from the low LC retention, they might be further conjugated metabolites, or could more likely be bacterial breakdown products from intestinal degradation. Together with the dihydroresveratrol conjugates, these polar metabolites, derived from bacterial degradation followed by enteric absorption and eventually urinary excretion, may account for as much as 50% of the dose. These products may contribute to resveratrol's chemoprotective effects in similar ways as has been shown for other dietary natural products, such as the bacterial degradation of the soy isoflavone daidzein to equol, *O*-desmethylangolensin, and other products,[26] and the degradation of plant lignans by the intestinal microbiota to the mammalian lignans enterodiol and enterolactone.[27] This will require much further investigation for resveratrol. In this regard we believe that the metabolism by Caco-2 cells, i.e., conjugation reactions represented in Figure 3 by M1, M2, and M4, are truly mammalian in nature, whereas M3, M5, and the unknown metabolites shown in Figure 4 may be bacterial.

Deconjugation of major conjugates

As conjugation to glucuronic acid and sulfate seems to be the major metabolic reactions of resveratrol in many studies, it is not surprising that these products have been favored to explain some of resveratrol's unknown effects. Although resveratrol glucuronides have been synthesized,[14] it is not yet clear whether they are biologically active. Glucuronic acid conjugates of several other polyphenols, such as quercetin, indeed have shown activity.[9] For the resveratrol sulfate conjugates, their cytotoxicity toward cancer cells appears to be reduced compared to resveratrol itself.[17,28] In addition, a common hypothesis is that these conjugates, in particular the glucuronic acid conjugates, serve as a pool in the body from which active resveratrol can be released locally by tissue β-glucuronidases. There is, however, currently no experimental evidence that this may occur. Similarly, as previously suggested,[4] the ubiquitously present human tissue sulfatases may convert the sulfate conjugates back to resveratrol.[28]

Tissue accumulation of resveratrol

One additional deficit in our understanding of resveratrol pharmacokinetics relates to its tissue accumulation. After all, its binding to cellular sites, including possible receptors, should be much more important than its concentrations in plasma, urine, and other body fluids. We only have fragmentary knowledge in this area. There is presently no support for the existence of specific resveratrol receptors; however, resveratrol is known to accumulate in tissues as compared with plasma. One common indication of tissue distribution is the volume of distribution (V_d) after intravenous administration. In the only instance when this route of administration was employed for resveratrol, a V_d of about 1.8 L/kg indicated a fair amount of extravascular distribution.[4] This occurred in spite of some binding of resveratrol to serum albumin.[29,30] Preferential accumulation of resveratrol in tissues was seen in rats as early as in 1996,[31] with high concentrations particularly in kidneys and liver. Similar tissue accumulation has been found in later studies in rats.[32,33] Very high resveratrol accumulation has been noted in intestinal tissue in mice,[34,35] although this may be partially related to the oral route of administration with residual nonabsorbed drug. In support of this very high concentrating effect of the intestine is the observation in the human colonic cell line Caco-2, in which resveratrol accumulated almost 40-fold compared to the surrounding buffer,[6] suggesting the colonocyte as a major target for this dietary preventive compound. Another site that may be important for resveratrol activities is the oral tissue. Using a novel bioengineered tissue model,[36] high tissue uptake and facile transport was demonstrated in this complex stratified epithelium. This raises the possibility of protection by resveratrol in oral cancers.

Resveratrol analogs

A logical approach to improving bioavailability, thereby potentially enhancing the chemopreventive effect of resveratrol, has been to make derivatives. This can include synthetic derivatives or testing naturally occurring derivatives, which sometimes are quite abundant in the human diet. The most common choice has been the methylated derivatives of resveratrol. 3,4,5,4′-tetramethoxystilbene showed particularly high accumulation in the intestinal mucosa.[35] Pterostilbene, a dimethyl derivative of resveratrol, was shown to be effective in the prevention of colon cancer in a rat model[37] and to have improved bioavailability.[38] Also a naturally occurring active analog of resveratrol, 3,5,4′-trimethoxy-*trans*-stilbene had greater plasma exposure, longer half-life, and lower clearance than resveratrol in rats.[39] Similarly, 3′,4′,5′,5,7-pentamethoxyflavone had greatly improved metabolic stability compared with the unmethylated flavone *in vivo* in mice.[40] These studies follow the general directions for improving bioavailability of polyphenols by methylation.[41]

Additional factors

The fact that resveratrol exists in nature as two isomers, i.e., *cis*- and *trans*-resveratrol, raises questions about stereoisomeric differences in their disposition and metabolism. Presently, we have very limited information on this topic, although large stereoselective differences have been shown in the glucuronidation pathway.[42] A more complete knowledge in this area awaits further studies.

Once metabolism of resveratrol is completely understood, it may be important to examine the potential for phenotypic and genotypic differences in metabolism. Understanding the importance of the microbiota in the metabolism of resveratrol could be critically important, as is thought to be the case for the soy isoflavone daidzein[26] and the lignans.[27]

Summary: need for further studies

In conclusion, there seems to be multiple areas worthy of further investigation regarding the disposition and metabolism of resveratrol, where explanations regarding its chemoprotective effects may be found. Some of those areas may be specific tissue accumulation as well as hitherto unknown metabolic reactions in relation to chronic resveratrol administration. Of particular significance would be to analyze the role of bacterial metabolism. Novel derivatives of resveratrol, including naturally occurring or synthetic compounds, will be of much further interest.

Conflicts of interest

The author declares no conflicts of interest.

References

1. Baur, J.A. & D.A. Sinclair. 2006. Therapeutic potential of resveratrol: the *in vivo* evidence. *Nat. Rev. Drug Discov.* **5:** 493–506.
2. Gescher, A.J. & W.P. Steward. 2003. Relationship between mechanisms, bioavailability, and preclinical chemopreventive efficacy of resveratrol: a conundrum. *Cancer Epidemiol. Biomark. Prev.* **12:** 953–957.
3. Goldberg, D.M., J. Yan & G.J. Soleas. 2003. Absorption of three wine-related polyphenols in three different matrices by healthy subjects. *Clin. Biochem.* **36:** 79–87.
4. Walle, T., F. Hsieh, M.H. DeLegge, *et al.* 2004. High absorption but very low bioavailability of oral resveratrol in humans. *Drug Metab. Dispos.* **32:** 1377–1382.
5. Artursson, P. & J. Karlsson. 1991. Correlation between oral drug absorption in humans and apparent drug permeability coefficients in human intestinal epithelial (Caco-2) cells. *Biochem. Biophys. Res. Commun.* **175:** 880–885.
6. Kaldas, M.I., U.K. Walle & T. Walle. 2003. Resveratrol transport and metabolism by human intestinal Caco-2 cells. *J. Pharm. Pharmacol.* **55:** 307–312.
7. Maier-Salamon, A., B. Hagenauer, G. Reznicek, *et al.* 2008. Metabolism and disposition of resveratrol in the isolated perfused rat liver: role of Mrp2 in the biliary excretion of glucuronides. *J. Pharm. Sci.* **97:** 1615–1628.
8. van der Wetering, K., A. Burkon, W. Feddema, *et al.* 2009. Intestinal breast cancer resistance protein (BCRP)/Bcpr1 and multidrug resistance protein 3 (MRP3)/Mrp3 are involved in the pharmacokinetics of resveratrol. *Mol. Pharmacol.* **75:** 876–885.
9. Boocock, D.J., G.E.S. Faust, K.R. Patel, *et al.* 2007. Phase I dose escalation pharmacokinetic study in healthy volunteers of resveratrol, a potential cancer chemopreventive agent. *Cancer Epidemiol. Biomark. Prev.* **16:** 1246–1252.
10. Almeida, L., M. Vaz-da-Silva, A. Falcão, *et al.* 2009. Pharmacokinetic and safety profile of *trans*-resveratrol in a rising multiple-dose study in healthy volunteers. *Mol. Nutr. Food Res.* **53:** S7–S15.
11. Tessitore, L., A. Davit, I. Sarotto, *et al.* 2000. Resveratrol depresses the growth of colorectal aberrant crypt foci by affecting *bax* and *p21CIP* expression. *Carcinogenesis* **21:** 1619–1622.
12. Li, Z.G., T. Hong, Y. Shimada, *et al.* 2002. Suppression of *N*-nitrosomethylbenzylamine (NMBA)-induced esophageal tumorigenesis in F344 rats by resveratrol. *Carcinogenesis* **23:** 1531–1536.
13. Banerjee, S., C. Bueso-Ramos & B.B. Aggarwal. 2002. Suppression of 7,12-dimethylbenz(*a*)anthracene-induced mammary carcinogenesis in rats by resveratrol: role of nuclear factor-κB, cyclooxygenase 2, and matrix metalloprotease 9. *Cancer Res.* **62:** 4945–4954.
14. Learmonth, D.A. 2003. A concise synthesis of the 3-*O*-β-D- and 4′-*O*-β-D-glucuronide conjugates of *trans*-resveratrol. *Bioconjugate Chem.* **14:** 262–267.
15. Yu, C., Y.G. Shin, A. Chow, *et al.* 2002. Human, rat, and mouse metabolism of resveratrol. *Pharm. Res.* **19:** 1907–1914.
16. Miksits, M., A. Maier-Salamon, S. Aust, *et al.* 2005. Sulfation of resveratrol in human liver: evidence of a major role for the sulfotransferases SULT1A1 and SULT1E1. *Xenobiotica* **35:** 1101–1119.
17. Hoshino, J., E.-J. Park, T.P. Kondratyuk, *et al.* 2010. Selective synthesis and biological evaluation of sulfate-conjugated resveratrol metabolites. *J. Med. Chem.* **53:** 5033–5043.
18. Murias, M., M. Miksits, S. Aust, *et al.* 2008. Metabolism of resveratrol in breast cancer cell lines: impact of sulftransferase 1A1 expression on cell growth inhibition. *Cancer Lett.* **261:** 172–182.
19. Otake, Y., A.L. Nolan, U.K. Walle, *et al.* 2000. Quercetin and resveratrol potently reduce estrogen sulfotransferase in normal human mammary epithelial cells. *J. Steroid Biochem. Mol. Biol.* **73:** 265–270.
20. Wang, D., T. Hang, C. Wu, *et al.* 2005. Identification of the major metabolites of resveratrol in rat urine by HPLC-MS/MS. *J. Chromatogr. B.* **829:** 97–106.
21. Juan, M.E., I. Alfaras & J.M. Planas. 2010. Determination of dihydroresveratrol in rat plasma by HPLC. *J. Agric. Food Chem.* **58:** 7472–7475.
22. Jung, C.M., T.M. Heinze, L.K. Schnackenberg, *et al.* 2009. Interaction of dietary resveratrol with animal-associated bacteria. *FEMS Microbiol. Lett.* **297:** 266–273.
23. Stivala, L.A., M. Savio, F. Carafoli, *et al.* 2001. Specific structural determinants are responsible for the antioxidant activity and the cell cycle effects of resveratrol. *J. Biol. Chem.* **276:** 22586–22594.

24. Gakh, A.A., N.Y. Anisimova, M.V. Kiselevsky, *et al.* 2010. *Dihydro*-resveratrol—a potent dietary polyphenol. *Bioorg. Med. Chem. Lett.* **20:** 6149–6151.
25. van der Woude, H., A. Gliszczynska-Swiglo, K. Struijs, *et al.* 2003. Biphasic modulation of cell proliferation by quercetin at concentrations physiologically relevant in humans. *Cancer Lett.* **200:** 41–47.
26. Frankenfeld, C.L., A. McTiernan, E.J. Aiello, *et al.* 2004. Mammographic density in relation to daidzein-metabolizing phenotypes in overweight, postmenopausal women. *Cancer Epidemiol. Biomark. Prev.* **13:** 1156–1162.
27. Woting, A., T. Clavel, G. Loh, *et al.* 2010. Bacterial transformation of dietary lignans in gnotobiotic rats. *FEMS Microbiol. Ecol.* **72:** 507–514.
28. Miksits, M., K. Wicek, M. Svoboda, *et al.* 2009. Antitumor activity of resveratrol and its sulfated metabolites against human breast cancer cells. *Planta Med.* **75:** 1227–1230.
29. Burkon, A. & V. Somoza. 2008. Quantitation of free and protein-bound *trans*-resveratrol metabolites and identification of *trans*-resveratrol-C/O-conjugated diglucuronides—two novel resveratrol metabolites in human plasma. *Mol. Nutr. Food Res.* **52:** 549–557.
30. Lu, Z., Y. Zhang, H. Liu, *et al.* 2007. Transport of a cancer chemopreventive polyphenol, resveratrol: interaction with serum albumin and hemoglobin. *J. Fluoresc.* **17:** 580–587.
31. Bertelli, A.A.E., L. Giovanni, R. Stradi, *et al.* 1996. Kinetics of *trans*- and *cis*-resveratrol (3,4′,5-trihydroxystilbene) after red wine oral administration in rats. *Int. J. Clin. Pharm. Res.* **16:** 77–81.
32. Abd El-Mohsen, M., H. Bayele, G. Kuhnle, *et al.* 2006. Distribution of [^{3}H]*trans*-resveratrol in rat tissues following oral administration. *Br. J. Nutr.* **96:** 62–70.
33. Juan, M.E., M. Maijo & J.M. Planas. 2010. Quantitation of *trans*-resveratrol and its metabolites in rat plasma and tissues by HPLC. *J. Pharmaceut. Biomed. Anal.* **51:** 391–398.
34. Vitrac, X., A. Desmouliere, B. Brouillaud, *et al.* 2003. Distribution of [^{14}C]-*trans*-resveratrol, a cancer chemopreventive polyphenol, in mouse tissues after oral administration. *Life Sci.* **72:** 2219–2233.
35. Sale, S., R.D. Verschoyle, D. Boocock, *et al.* 2004. Pharmacokinetics in mice and growth-inhibitory properties of the putative cancer chemopreventive agent resveratrol and the synthetic analogue*trans* 3,4,5,4′-tetramethoxystilbene. *Br. J. Cancer*. **90:** 736–744.
36. Walle, T., U.K. Walle, D. Sedmera, *et al.* 2006. Benzo[*a*]pyrene-induced oral carcinogenesis and chemoprevention—studies in bioengineered human tissue. *Drug Metab. Dispos.* **34:** 346–350.
37. Suh, N., S. Paul, X. Hao, *et al.* 2007. Pterostilbene, an active constituent of blueberries, suppresses aberrant crypt foci formation in the azomethane-induced colon carcinogenesis model in rats. *Clin. Cancer Res.* **13:** 350–355.
38. Lin, H.-S., B.-D. Yue & P.C. Ho. 2009. Determination of pterostilbene in rat plasma by a simple HPLC-UV method and its application in pre-clinical pharmacokinetic study. *Biomed. Chromatogr.* **23:** 1308–1315.
39. Lin, H.-S. & P.C. Ho. 2009. A rapid HPLC method for the quantification of 3,5,4′-trimethoxy-*trans*-stilbene (TMS) in rat plasma and its application in pharmacokinetic study. *J. Pharm. Biomed. Anal.* **49:** 387–392.
40. Cai, H., S. Sale, R.G. Britton, *et al.* 2010. Pharmacokinetics in mice and metabolism in murine and human liver fractions of the putative cancer chemopreventive agents 3′,4′,5′,5,7-pentamethoxyflavone and tricin (4′,5,7-trihydroxy-3′,5′-dimethoxyflavone). *Cancer Chemother. Pharm.* 2010 April 04 [Epub ahead of print] doi: 10.1007/s00280-010-1313-1.
41. Walle, T., X. Wen & U.K. Walle. 2007. Improving metabolic stability of cancer chemopreventive polyphenols. *Expert Op. Drug Metab. Toxicol.* **3:** 379–388.
42. Aumont, V., S. Krisa, E. Battaglia, *et al.* 2001. Regioselective and stereospecific glucuronidation of *trans*- and *cis*-resveratrol in human. *Arch. Biochem. Biophys.* **393:** 281–289.

Ann. N.Y. Acad. Sci. ISSN 0077-8923

ANNALS OF THE NEW YORK ACADEMY OF SCIENCES
Issue: *Resveratrol and Health*

Resveratrol: a cardioprotective substance

Joseph M. Wu and Tze-chen Hsieh
Department of Biochemistry and Molecular Biology, New York Medical College, Valhalla, New York

Address for correspondence: Joseph M. Wu, Room 147, Department of Biochemistry and Molecular Biology, New York Medical College, Valhalla, NY 10595. Joseph_Wu@nymc.edu

Coronary heart disease (CHD) is a major and preventable cause of morbidity and death in the United States. Recently, significant research efforts have been directed at an epidemiological phenomenon known as the "French paradox." This observation refers to the coexistence of high risk factors with unanticipated low incidence of CHD, and is postulated to be associated with low-to-moderate consumption of red wine. *In vivo* studies have shown that red wine intake is more CHD-preventative in comparison to other alcoholic drinks; enhanced cardioprotection may be attributed to grape-derived polyphenols, e.g., resveratrol, in red wine. This review summarizes results of *in vitro* and animal studies showing that resveratrol exerts multifaceted cardioprotective activities, as well as evidence demonstrating the presence of proteins specifically targeted by resveratrol, as exemplified by *N*-ribosyldihydronicotinamide:quinone oxidoreductase, NQO2. A mechanism encompassing nongenomic and genomic effects and a research roadmap is proposed as a framework for uncovering further insights on cardioprotection by resveratrol.

Keywords: resveratrol; cardioprotection; quinone reductase 2

Introduction: red wine confers protection against CHD

Coronary heart disease (CHD), a primary cause of morbidity and death in developed countries, is regarded to be preventable through changes in lifestyle and diet. A documented decline in age-adjusted CHD-related mortality over the years has been observed in the United States, in concert with advances in strategies designed to control CHD, e.g., cessation in cigarette use, increase in exercise frequency, reduction in fat intake and increase in consumption of fruits and vegetables in the diet. Epidemiological studies of a phenomenon commonly referred to as the French paradox show an inverse relationship between low-to-moderate consumption of red wine and the risk for CHD (Fig. 1).[1] The unanticipated low incidence of CHD co-existing with high risk factors has been associated with the cultural habit in France of drinking red wine in moderation with meals. *In vivo* studies have demonstrated that red wine is more effective in preventing CHD compared to other alcoholic beverages; enhanced protection may attribute to polyphenols in red wine, including resveratrol.[2,3] Thus, a 30–45% reduction in CHD risk is found in population-based studies among low/moderate drinkers compared to those who drink excessively or individuals who abstain from use of alcohol-containing beverages. Animal and *in vitro* studies show that ethanol/red wine intake is beneficial to the cardiovascular system; rabbits fed ethanol or red wine and challenged with balloon angioplasty showed inhibition of neointimal hyperplasia in coronary arteries. Moreover, balloon-injured coronary arteries exposed to alcohol are accompanied by suppression of intimal hyperplasia. Exposure to ethanol or resveratrol *in vitro* also results in inhibition of vascular smooth muscle cells (VSMC) proliferation. In a recently published study, "Alcohol and the cardiovascular system: research challenges and opportunities," CHD protection by moderate intake of red wine is identified as a research need of considerable urgency.[35]

Studies supporting cardioprotection by resveratrol

Concerted effort by studies in our and other laboratories on the cardioprotective effects of resveratrol, and red wine with and without alcohol, have

doi: 10.1111/j.1749-6632.2010.05854.x

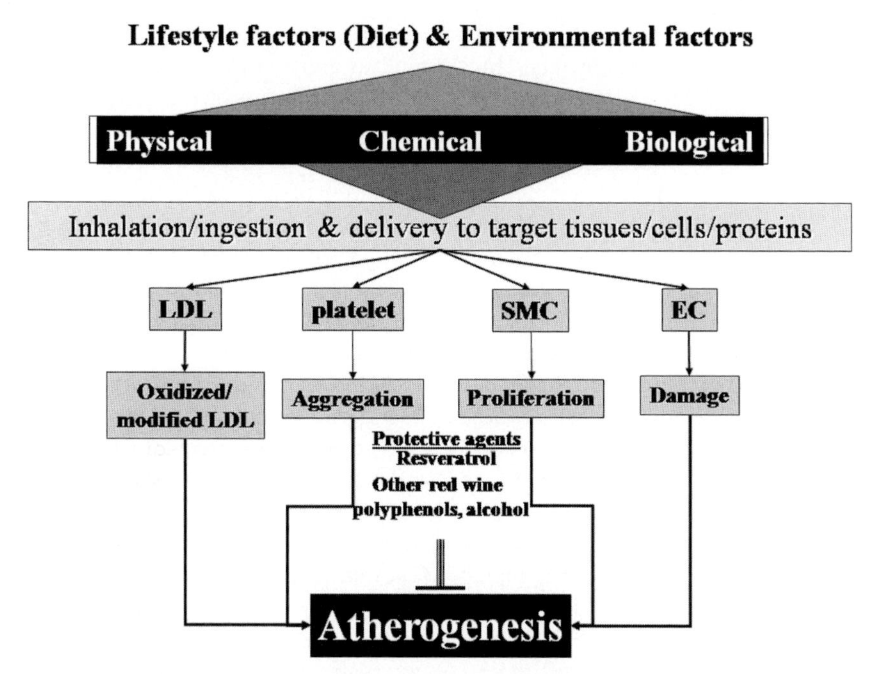

Figure 1. Scheme depicting atherogenesis as the clinical manifestation of multiple cellular dysfunctions resulting from exposure to lifestyle and environmental events in the form of physical, chemical, and biological challenges. Cardioprotection by resveratrol may be attributed to its ability to act on the same multiple cellular targets adversely affected by extrinsic and intrinsic risk factors. These include the inhibition of LDL oxidation, suppression of platelet aggregation, and inhibition of smooth muscle and endothelial cell proliferation and function, by resveratrol.

shown the following. First, resveratrol inhibits low-density lipoprotein (LDL) oxidation. Oxidation of LDL is considered a key primary event in the initiation of atherosclerosis. Because resveratrol has antioxidant properties, we tested whether it affected LDL oxidation. LDL isolated from normolipidemic adult males was oxidatively modified using Cu^{2+}, with and without the addition of resveratrol. LDL oxidation was monitored by reactivity to thiobarbituric acid, agarose gel electrophoresis, and uptake into macrophages. Resveratrol significantly inhibited LDL oxidation.[4–6] Secondly, resveratrol is shown to inhibit platelet aggregation—platelets are actively involved in the process of hemostasis, by which injury in the vascular endothelium is rapidly repaired in order not to compromise the fluidity of the blood. In normal endothelium injury, platelets adhere to the subendothelial matrix of a damaged vessel, spread over the surface and recruit additional platelets to form a thrombus. Improper regulation or over reactivity of this repair system can lead to pathological thrombosis. Studies show that platelet aggregation by a number of agonists is suppressed by resveratrol, both *in vitro* and *in vivo*.[7–9] Administration of resveratrol (4 mg/kg/day) to rabbits fed a high cholesterol diet caused a 35% reduction in the average ADP-induced platelet aggregation rate (PAR), which was indistinguishable from animals fed a normal diet.[7,10] Third, resveratrol suppresses proliferation of smooth muscle cells and pulmonary aortic endothelial cells. Migration and proliferation of smooth muscle cells in the intima of susceptible vessels is a requisite for atherogenesis. We found that exposure to resveratrol reduces proliferation of smooth muscle cells accompanied by $G_1 \rightarrow S$ block.[11–13] Similarly, resveratrol also inhibits proliferation of cultured bovine pulmonary aortic endothelial cells (BPAEC) concomitant with induction of nitric oxide synthase in a dose-dependent manner.[14–16] Fourth, in rabbits intragastrically fed resveratrol (4 mg/kg/day) for a duration of five weeks beginning one week prior to induced endothelial injury by denudation in the iliac artery, hyperplasia in the damaged endothelial vessel wall was inhibited, as evidenced by reduction of intimal proliferation index [scored as the ratio of intimal to (intimal + medial) area] from 0.41 ± 0.13 in control animals to 0.28 ± 0.07 in resveratrol-fed animals ($P < 0.01$). In addition, the relative content of smooth muscle cells, characterized by their "hills and valley" shape and positive reactivity for α-actin, in the intima of resveratrol-fed animals was also proportionately suppressed.[17,18] Fifth, the mean area of atherosclerotic plaques was reduced from 56.4 ± 13.5 in rabbits fed a hypercholesterolemic diet to 33.6 ± 19.6 (arbitrary units) in animals fed resveratrol.[19] Lastly, resveratrol pronouncedly inhibits proliferation of cultured human aortic smooth muscle cells,[20,21] concomitant with dose-dependent increase in the expression of tumor suppressor gene p53, heat shock protein HSP27, quinone reductase 1 and 2, and altered subcellular distribution of nitric oxide synthase and apoptosis inducing factor.[20]

Taken together, these results provide support for the conclusion that resveratrol acts as a cardioprotective agent by a plethora of activities impinging on events key to the prevention of atherosclerosis and CHD, thereby reinforcing the notion that age-adjusted CHD deaths may be independently modulated by diet-based strategies including use of grape polyphenol resveratrol.

Discovering target proteins of resveratrol

How does resveratrol exert its multi-cell type CHD-protective activities? One might envisage possibilities that include transcriptional, posttranscriptional control of gene expression, differential cellular compartmentalization and trafficking of proteins, and co- or post-translational processing mechanisms, which can be studied by approaches that focus on DNA/gene aspects (by genomics), RNA transcription alterations (using microarrays), or protein level changes (by proteomics). Since proteins (as enzymes and structural/regulatory elements) play a direct role on maintaining and regulating the homeostatic and dynamic overture of biological systems, we have focused on the identification of protein targets that might underpin or contribute to cardioprotection by resveratrol using a *ligand (resveratrol)-directed proteomic* strategy. This strategy is based on the basic tenet that cardioprotective activities stemming from exposure to resveratrol relate to its ability to interact with distinct resveratrol target proteins, (RTPs).[20,22–24] In this approach, resveratrol is appended to a solid matrix to facilitate retention, selection, and purification of distinct RTPs. Of note, affinity isolation of RTPs using resveratrol-directed proteomics from cells and tissues has several novel features: the discovery of RTPs with possible significant and integral links to observed cellular responses to resveratrol; the concentration and identification of scarce, low-abundance proteins not easily amenable to detection using crude extracts; and the facilitation of the generation of RTP profiles relevant to the biological effects of resveratrol.

To experimentally test and validate the presence of RTPs, resveratrol was immobilized on epoxy-activated agarose, generating a biospecific affinity platform for the rapid and specific capture of RTPs.[20,22,24,25] Extracts prepared from various sources were fractionated on resveratrol affinity agarose columns by sequential elution with 0.35 M and 1 M NaCl, followed by 1 mM ATP and, finally, 2 mM resveratrol. The variously eluted fractions were concentrated and resolved by 10% SDS-PAGE, followed by silver staining. A specific protein migrating with a molecular weight of 22 kDa, denoted RTP-22 was found in the resveratrol-eluted fraction, based on our studies using extracts prepared from human prostate[22,25] and melanoma cells,[23] and from cultured human aortic smooth muscle cells.[20] In all cell extracts tested, the retention of RTP-22 on the resveratrol affinity column was effectively and almost completely competed when extracts were first incubated with excess resveratrol prior to fractionation.[20,22,23,25] The RTP-22 was subsequently identified as quinone reductase NQO2 by combining resveratrol-directed affinity chromatography purification strategy with MALDI-TOF mass spectrometry and cloning.[26] Notably, purified recombinant NQO2 showed a high affinity for resveratrol, with a dissociation constant $K_D \leq 50$ nM.[26] Moreover, X-ray crystallographic analysis of NQO2 in complex with resveratrol shows that resveratrol binds to a hydrophobic interaction cavity located on the interface between dimeric NQO2.[26] The boundary of this interaction cavity is defined by the isoalloxazine ring of cofactor FAD in NQO2, as well as by side chains of Y132′, F178′, F126′, M164′, and C121′ from the first protomer, and Y155 and F106 from the second protomer.[26] It was found that the deep and narrow nature of the interaction cleft perfectly accommodates the physical dimension of resveratrol, in support of the notion that a stilbene-binding pocket exists within NQO2.[26] In addition to hydrophobic and van der Waals interactions, all three hydroxyl groups of resveratrol form hydrogen bonds with key amino acid residues in the protein,

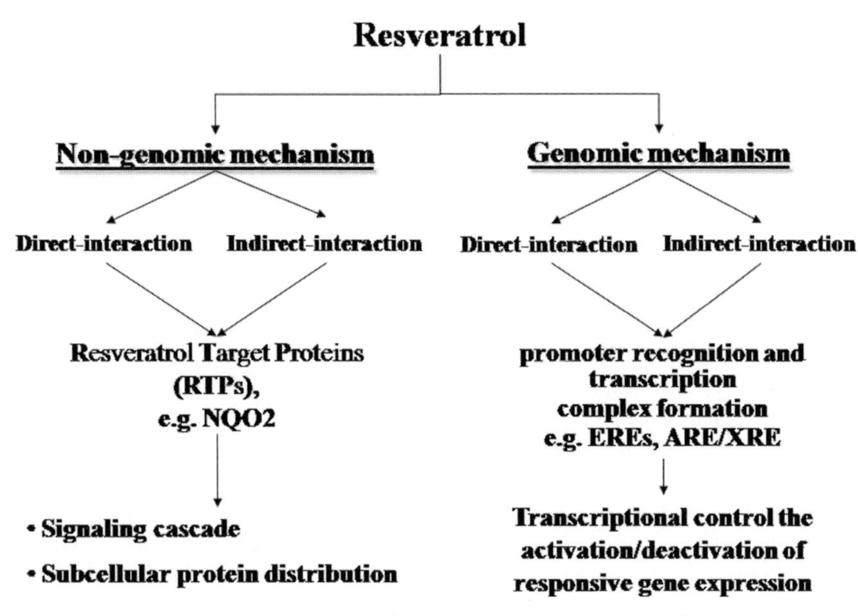

Figure 2. Proposed genomic and nongenomic activities of resveratrol contributing to its cardioprotective activities, both directly and indirectly. Cardioprotection by resveratrol is hypothesized to occur by multi-active genomic mechanisms. For instance, resveratrol may modulate transcription by binding to estrogen response elements (EREs) as a phytoestrogen. In addition, resveratrol may target antioxidant-response-element (ARE) and modulate transcription by promoting the cytosol-to-nucleus translocation of transcription factor Nrf2. The nongenomic activity of resveratrol may involve its direct and indirect interaction with target protein NQO2; NQO2 may modulate the cardioprotective effects of resveratrol by acting as a direct and indirect cardio-active sensor and mediator of resveratrol.

thus providing not only enthalpy for the high affinity binding to NQO2, but also maintaining the planar conformation of resveratrol.[26]

Proposed mechanisms of cardioprotection by resveratrol

Our current hypothesis regarding cardioprotection by grape polyphenol resveratrol is that it acts by both nongenomic and genomic mechanisms, directly and indirectly. The nongenomic cardioprotective activities of resveratrol may encompass its direct binding and interaction with cellular RTPs, as well as occur indirectly via its ability to function as effective modulator of enzyme activities (Fig. 2). In the direct mechanistic theme, for instance, resveratrol has been shown to bind to integrins on cell membrane,[27,28] leading to integrin-coupled signaling cascades and subcellular protein distribution. A key RTP we have chosen to focus on is NQO2. Our working hypothesis is that binding of resveratrol to NQO2 induces a conformational change that directly alters its molecular fate in cellular trafficking and gene expression; conformationally altered NQO2.resveratrol complex may facilitate additional interaction with NQO2-binding proteins to effect subsequent changes in gene expression. In contrast, the indirect nongenomic mechanism of resveratrol action may be by its ability to exert potent inhibition of enzymes possibly linked to the development of atherogenesis.

Cardioprotection by resveratrol may also involve genomic mechanisms (Fig. 2), postulated as involving multiple transcription regulatory features that include (1) its phytoestrogenic attributes, where it affects transcription via the classical mechanism of binding to estrogen response elements (EREs);[29,30] (2) its ability to regulate antioxidant-response-element (ARE)-dependent transcription by promoting translocation of transcription factor Nrf2 from the cytosol-located Nrf2/Keap1 complex to the nucleus to facilitate transcription of antioxidant response element, ARE-responsive genes;[31–34] and

(3) ERE/ARE-independent, RTP-dependent or mediated transcriptional events, details of which remain to be elucidated in future studies.

Proposed future research maps

We propose herewith a limited research roadmap that might facilitate further discovery and elucidation of the mechanism of cardioprotection by resveratrol: (1) structure/functional, uptake and biotransformation experiments combined with *in silico* analysis, using resveratrol analogues in appropriate tissue culture and animal models systems; (2) structural studies of freshly isolated and preserved specimens from animals challenged with cardiovascular damaging pharmacological agents and diets, with and without dietary supplementation by resveratrol $\pm$ other red wine polyphenols; (3) use of laser capture microdissection (LCM) microsystems to isolate stable and unstable plaques from archived diseased human tissues; (4) resveratrol affinity chromatography directed proteomic analysis to identify qualitative and quantitative RTP profile differences between normal and diseased tissues, for further identification by mass spectrometry, interactome, and microgenomic analysis; and (5) small inhibitory RNA and pharmacological agent intervention experiments. It is hoped that these proposed strategies and approaches will provide a better understanding of the role NQO2 plays in cardioprotection as well as illuminating insights on the mechanism of cardioprotection by resveratrol in red wine.

Acknowledgments

We thank Dr. Ole Vang in the Department of Life Sciences and Chemistry, Roskilde University, Denmark, for the invitation to write this review. A portion of this review was presented by JMW as an invited speaker at the 1st International Conference of Resveratrol and Health, held in Elsinore, Denmark, September 13–15, 2010. Studies on cardioprotection by resveratrol in our laboratories over the years were supported in part by the Intramural Sponsored Research Program of New York Medical College, California Grape Trade Commission, Phillip Morris USA Inc., and Phillip Morris International, to JMW.

Conflicts of interest

The authors declare no conflicts of interest.

References

1. Wu, J.M. *et al.* 2001. Mechanism of cardioprotection by resveratrol, a phenolic antioxidant present in red wine (Review). *Int. J. Mol. Med.* **8:** 3–17.
2. Das, D.K. & N. Maulik. 2006. Resveratrol in cardioprotection: a therapeutic promise of alternative medicine. *Mol. Interv.* **6:** 36–47.
3. Das, D.K. *et al.* 1999. Cardioprotection of red wine: role of polyphenolic antioxidants. *Drugs Exp. Clin. Res.* **25:** 115–120.
4. Zou, J.G. *et al.* 1999. Resveratrol inhibits copper ion-induced and azo compound-initiated oxidative modification of human low density lipoprotein. *Biochem. Mol. Biol. Int.* **47:** 1089–1096.
5. Zou, J. *et al.* 2000. Effects of resveratrol on oxidative modification of human low density lipoprotein. *Chin. Med. J. (Engl).* **113:** 99–102.
6. Belguendouz, L., L. Fremont & A. Linard. 1997. Resveratrol inhibits metal ion-dependent and independent peroxidation of porcine low-density lipoproteins. *Biochem. Pharmacol.* **53:** 1347–1355.
7. Wang, Z. *et al.* 2002. Effect of resveratrol on platelet aggregation in vivo and in vitro. *Chin. Med. J. (Engl).* **115:** 378–380.
8. Kirk, R.I. *et al.* 2000. Resveratrol decreases early signaling events in washed platelets but has little effect on platalet in whole food. *Blood Cells Mol. Dis.* **26:** 144–150.
9. Lin, K.H. *et al.* 2009. Mechanisms of resveratrol-induced platelet apoptosis. *Cardiovasc. Res.* **83:** 575–585.
10. Wang, Z. *et al.* 2002. Effects of red wine and wine polyphenol resveratrol on platelet aggregation in vivo and in vitro. *Int. J. Mol. Med.* **9:** 77–79.
11. Zou, J. *et al.* 1999. Suppression of mitogenesis and regulation of cell cycle traverse by resveratrol in cultured smooth muscle cells. *Int. J. Oncol.* **15:** 647–651.
12. Poussier, B. *et al.* 2005. Resveratrol inhibits vascular smooth muscle cell proliferation and induces apoptosis. *J. Vasc. Surg.* **42:** 1190–1197.
13. Brito, P.M. *et al.* 2009. Resveratrol inhibits the mTOR mitogenic signaling evoked by oxidized LDL in smooth muscle cells. *Atherosclerosis* **205:** 126–134.
14. Hsieh, T.C. *et al.* 1999. Resveratrol increases nitric oxide synthase, induces accumulation of p53 and p21(WAF1/CIP1), and suppresses cultured bovine pulmonary artery endothelial cell proliferation by perturbing progression through S and G2. *Cancer Res.* **59:** 2596–2601.
15. Bruder, J.L. *et al.* 2001. Induced cytoskeletal changes in bovine pulmonary artery endothelial cells by resveratrol and the accompanying modified responses to arterial shear stress. *BMC Cell Biol.* **2:** 1.
16. Klinge, C.M. *et al.* 2008. Resveratrol stimulates nitric oxide production by increasing estrogen receptor alpha-Src-caveolin-1 interaction and phosphorylation in human umbilical vein endothelial cells. *FASEB J.* **22:** 2185–2197.
17. Zou, J. *et al.* 2000. Effect of resveratrol on intimal hyperplasia after endothelial denudation in an experimental rabbit model. *Life Sci.* **68:** 153–163.
18. Zou, J.G. *et al.* 2003. Effect of red wine and wine polyphenol resveratrol on endothelial function in hypercholesterolemic rabbits. *Int. J. Mol. Med.* **11:** 317–320.

19. Wang, Z. *et al.* 2005. Dealcoholized red wine containing known amounts of resveratrol suppresses atherosclerosis in hypercholesterolemic rabbits without affecting plasma lipid levels. *Int. J. Mol. Med.* **16:** 533–540.
20. Wang, Z. *et al.* 2006. Regulation of proliferation and gene expression in cultured human aortic smooth muscle cells by resveratrol and standardized grape extracts. *Biochem. Biophys. Res. Commun.* **346:** 367–376.
21. Juan, S.H. *et al.* 2005. Mechanism of concentration-dependent induction of heme oxygenase-1 by resveratrol in human aortic smooth muscle cells. *Biochem. Pharmacol.* **69:** 41–48.
22. Wang, Z. *et al.* 2004. Identification and purification of resveratrol targeting proteins using immobilized resveratrol affinity chromatography. *Biochem. Biophys. Res. Commun.* **323:** 743–749.
23. Hsieh, T.C. *et al.* 2005. Inhibition of melanoma cell proliferation by resveratrol is correlated with upregulation of quinone reductase 2 and p53. *Biochem. Biophys. Res. Commun.* **334:** 223–230.
24. Hsieh, T.C. *et al.* 2008. Identification of glutathione sulfotransferase-pi (GSTP1) as a new resveratrol targeting protein (RTP) and studies of resveratrol-responsive protein changes by resveratrol affinity chromatography. *Anticancer Res.* **28:** 29–36.
25. Hsieh, T.C. 2009. Uptake of resveratrol and role of resveratrol-targeting protein, quinone reductase 2, in normally cultured human prostate cells. *Asian J. Androl.* **11:** 653–661.
26. Buryanovskyy, L. *et al.* 2004. Crystal structure of quinone reductase 2 in complex with resveratrol. *Biochemistry* **43:** 11417–11426.
27. Lin, H.Y. *et al.* 2006. Integrin alphaVbeta3 contains a receptor site for resveratrol. *FASEB J.* **20:** 1742–1744.
28. Lin, H.Y. *et al.* 2008. Resveratrol is pro-apoptotic and thyroid hormone is anti-apoptotic in glioma cells: both actions are integrin and ERK mediated. *Carcinogenesis* **29:** 62–69.
29. Yoon, K. *et al.* 2001. Differential activation of wild-type and variant forms of estrogen receptor alpha by synthetic and natural estrogenic compounds using a promoter containing three estrogen-responsive elements. *J. Steroid Biochem. Mol. Biol.* **78:** 25–32.
30. Wu, F. & S. Safe. 2007. Differential activation of wild-type estrogen receptor alpha and C-terminal deletion mutants by estrogens, antiestrogens and xenoestrogens in breast cancer cells. *J. Steroid Biochem. Mol. Biol.* **103:** 1–9.
31. Hsieh, T.C. *et al.* 2006. Induction of quinone reductase NQO1 by resveratrol in human K562 cells involves the antioxidant response element ARE and is accompanied by nuclear translocation of transcription factor Nrf2. *Med Chem.* **2:** 275–285.
32. Chen, C.Y. *et al.* 2005. Resveratrol upregulates heme oxygenase-1 expression via activation of NF-E2-related factor 2 in PC12 cells. *Biochem. Biophys. Res. Commun.* **331:** 993–1000.
33. Ungvari, Z. *et al.* 2010. Resveratrol confers endothelial protection via activation of the antioxidant transcription factor Nrf2. *Am. J. Physiol. Heart Circ. Physiol.* **299:** H18–H24.
34. Hasko, G. & P. Pacher. 2010. Endothelial Nrf2 activation: a new target for resveratrol? *Am. J. Physiol. Heart Circ. Physiol.* **299:** H10–H12.
35. Lucas, D.L. *et al.* 2005. Alcohol and the cardiovascular system: research challenges and opportunities. *J. Am. Coll. Cardiol.* **45:** 1916–1924.

Ann. N.Y. Acad. Sci. ISSN 0077-8923

ANNALS OF THE NEW YORK ACADEMY OF SCIENCES
Issue: *Resveratrol and Health*

Resveratrol in cardiovascular health and disease

Goran Petrovski,[1] Narasimman Gurusamy,[2] and Dipak K. Das[2]

[1]Department of Biochemistry and Molecular Biology, The Apoptosis and Genomics Research Group of the Hungarian Academy of Sciences, University of Debrecen, Hungary. [2]Cardiovascular Research Center, University of Connecticut School of Medicine, Farmington, Connecticut

Address for correspondence: Dipak K. Das, Ph.D., Sc.D., M.D. (hon), FAHA, Cardiovascular Research Center, University of Connecticut, School of Medicine, Farmington, CT 06030-1110. ddas@neuron.uchc.edu

Resveratrol, initially used for cancer therapy, has shown beneficial effects against most degenerative and cardiovascular diseases from atherosclerosis, hypertension, ischemia/reperfusion, and heart failure to diabetes, obesity, and aging. The cardioprotective effects of resveratrol are associated with its preconditioning-like action potentiated by its adaptive response. During preconditioning, small doses of resveratrol can exert an adaptive stress response, forcing the expression of cardioprotective genes and proteins such as heat shock and antioxidant proteins. Similarly, resveratrol can induce autophagy, another form of stress adaptation for degrading damaged or long-lived proteins, as a first line of protection against oxidative stress. Resveratrol's interaction with multiple molecular targets of diverse intracellular pathways (e.g., action on sirtuins and FoxOs through multiple transcription factors and protein targets) intertwines with those of the autophagic pathway to give support in the modified redox environment after stem cell therapy, which leads to prolonged survival of cells. The successful application of resveratrol in therapy is based upon its hormetic action similar to any toxin: exerting beneficial effects at lower doses and cytotoxic effects at higher doses.

Keywords: resveratrol; autophagy; cardiovascular disease; atherosclerosis; angiogenesis; myocardial regeneration; hypertension; stem cells; longevity genes; hormesis

Introduction

Cardiovascular diseases (CVDs) are the major causes of morbidity and mortality in the developed nations, including the United States.[1] About 100 million Americans suffer from some kind of CVD including high blood pressure, atherosclerosis, coronary heart disease, or stroke. Thirty-nine percent of all deaths—or one in every three deaths—is claimed to be caused by a heart disease. Although proper medicines are abundant in developing countries to combat CVDs, preventive measures appear to be necessary to lower the incidence of heart problems. Maintaining a healthy heart is particularly important for those with a family history of coronary heart disease including hypertension, heart attack, and atherosclerosis. Fortunately, a number of natural cures are available that appear to maintain cardiovascular health. For example, maintaining a healthy lifestyle, daily exercise, and choosing proper diets certainly help maintain a healthy heart.[2,3] Certain mineral-rich foods including those high in potassium, calcium, and magnesium can control blood pressure. Low-fat milks are rich sources of calcium, while magnesium and potassium can be found in green vegetables and fish oil. Among the fatty foods, limiting saturated, polyunsaturated, monounsaturated, and *trans-* fats would help reduce cholesterol and associated forms of coronary heart diseases. Major sources of saturated fat include beef, butter, cheese, whole milk, and coconut and palm oils. In contrast, polyunsaturated fats like omega-3-fatty acids containing foods including fish oil, flax seed oil, and canola oil may reduce the risk of heart attack. Some of the examples for maintaining a healthy heart include garlic, olive oil, broccoli, capsicum, coenzyme Q10, *Terminalia arjuna*, Ashwagandha (*Withania somnifera*), cocoa, and fish oil, most of which are included in the Mediterranean diet. Numerous reports are available in the literature about heart healthy diet from Mediterranean foods.[4,5]

First published in 1992 as the French Paradox, the health benefits of daily drinking of wine in moderation has been proven by most of the scientists and

doi: 10.1111/j.1749-6632.2010.05843.x

clinicians all over the world.[6] Wine contains 11–14% alcohol and many polyphenolic compounds, most of which possess antioxidant properties. Although alcohol possesses cardioprotective properties, it is universally believed that it is not the alcohol but some of the polyphenols present in wine that are responsible for the cardioprotective properties of wine.[7] Among the polyphenols present in wine, especially red wine, resveratrol has drawn major attention. Resveratrol is present in the skins of grapes and thus in red wine. In addition to grapes, resveratrol is present in a large variety of fruits including cranberry, mulberry, lingonberry, bilberry, partridgeberry, sparkleberry, deerberry, blueberry, jackfruit, peanut, as well as in a wide variety of flowers and leaves including gnetum, butterfly orchid tree, white hellebore, scots pine, corn lily, eucalyptus, and spruce. Commercially, resveratrol is extracted from the dried roots of *Polygonum cuspidatum*, mainly found in Japan and China. *Polygonum* extract has been used in Japanese and Chinese traditional medicine to treat fungal infections, various skin inflammations, liver disease, and cardiovascular problems.[8] Resveratrol is also an antifungal compound that is synthesized in response to environmental stressors including water deprivation, UV radiation, and especially fungal infection.[9]

A growing body of evidence now supports the notion that resveratrol is the major factor responsible for the cardioprotective effects of wine. In the heart, resveratrol blocks low-density lipoprotein (LDL) peroxidation, increases high-density lipoprotein (HDL) levels, induces vasorelaxation presumably through the induction of nitric oxide (NO) synthesis, inhibits endothelin (ET), modifies angiogenic response, reduces ventricular arrhythmias, possesses antithrombin activity and prevents platelet aggregation, inhibits formation of soluble adhesion molecules, reduces reactive oxygen species (ROS), reduces blood pressure, and ameliorates ischemic reperfusion injury (Fig. 1).[8,10–12] It is believed that resveratrol-mediated cardioprotection is not due to its direct drug-like effect on the diseased heart, rather, it potentiates a preconditioning (PC)-like effect, a state-of-the art for cardioprotection. The PC effect results from an adaptive response, which receives further support from a recent discovery that resveratrol promotes autophagy (Fig. 2).[13,14] More recently, resveratrol has been found to regenerate the infracted myocardium.[11]

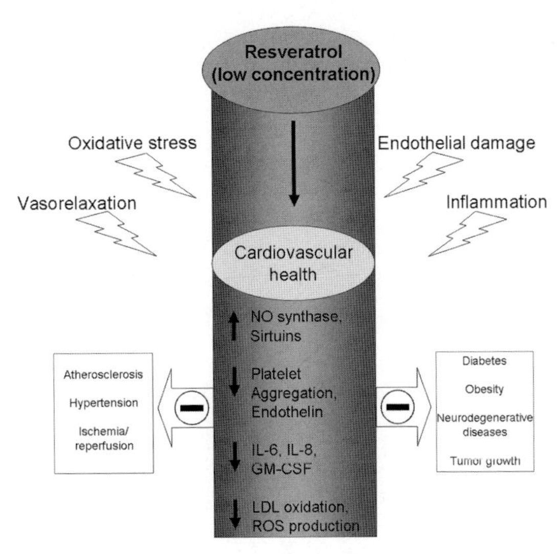

Figure 1. Effects of resveratrol on cardiovascular health and other diseases. Granulocyte-macrophage colony-stimulating factor (GM-CSF), interleukin (IL), low-density lipoprotein (LDL), nitric oxide (NO), and reactive oxygen species (ROS).

Resveratrol appears to maintain a healthy heart in numerous ways, and can provide numerous health benefits, which will be the subject of discussion in this review.

Resveratrol and cardiovascular health

As mentioned earlier, CVDs—being the leading cause of death and illness worldwide—are greatly dependent not only on nonmodifiable predisposing factors (age, sex, and genetic composition), but also on a modifiable factor—the lifestyle. Indeed, risk factors other than increased plasma cholesterol[15] and obesity,[16] such as oxidative stress, inflammation, and endothelial damage, all have great relevance in the development of CVD. One of the most well-known benefits of resveratrol is improvement of cardiovascular health (Fig. 1).[17] Resveratrol, in physiologic concentrations after red wine consumption, increases the expression of NO synthase, an enzyme responsible for synthesizing the potent vasodilator NO in human vascular endothelial cells.[18] Resveratrol also decreases the expression of the potent vasoconstrictor ET.[18] Cerebral ischemic damage in rat brains can be prevented via pretreatment with resveratrol[19] by activating sirtuin (Sirt) 1, coupled with a decrease in mitochondrial uncoupling protein 2 and an increase in mitochondrial ATP synthesizing efficiency.[19] Resveratrol can also exert cardioprotection

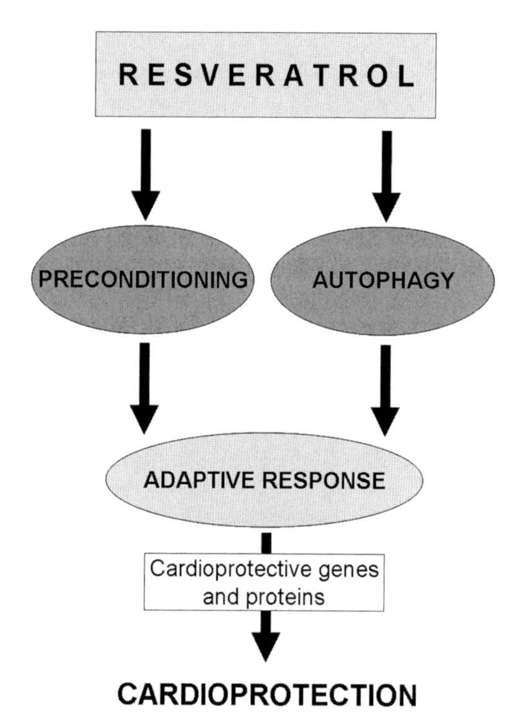

Figure 2. Cardioprotective pathways of resveratrol.

by inhibiting platelet aggregation, similar to how aspirin works.[17] In the following sections, the beneficial effects of resveratrol, as summarized in Figure 1, will be briefly discussed in relation to some major heart problems.

Atherosclerosis

Atherosclerosis is an inflammatory disease of the arterial wall caused by endothelial wall damage during continuous hemodynamic and redox stress conditions.[20] Under such conditions, modified LDLs, lipids, and fibrous elements can accumulate in the wall. Its progression into plaque disruption can lead to local thrombosis and acute coronary syndrome.[21,22] As previously mentioned, resveratrol can regulate the production of NO, a potent vasodilator, by counterbalancing the effect of the vasoconstrictor ET-1,[18,23–25] thus providing thromboresistance and preventing atherogenesis.[24] Resveratrol can potentially inhibit stress-induced ET-1 gene expression by interfering with its promoting extracellular-signal-regulated kinases (ERK) 1/2 pathway[26] and therefore improve endothelial function.[27]

Inflammation mediates all stages of atherosclerosis from initiation to progression and, eventually, plaque rupture. Prostaglandin E_2 (PGE_2) plays a key role in inflammation—its synthesis is catalyzed by cyclooxygenase (COX) 2. Resveratrol can inhibit the atherosclerosis-associated inflammation via regulating the COX-2 activity at a transcriptional level, thereby inhibiting the production of PGE_2.[28] In cultured murine macrophages, resveratrol can suppress the proinflammatory interleukin (IL)-6 gene expression, protein synthesis, and secretion,[29,30] and inhibit the release of both IL-8 and granulocyte-macrophage colony-stimulating factor.[31] Oxidation of LDL and uptake into the vascular wall as a main cause of endothelial injury and inflammation can be prevented by resveratrol.[32] Endothelial cell injury, platelet aggregation, and adherence to the injured cell surface can be blocked by this stilbene, inhibiting further the adhesion of collagen to platelets[33] and ADP-induced platelet aggregation[34] in a concentration-dependent manner.[35] Low concentrations of resveratrol can significantly inhibit intracellular and extracellular ROS production[36] by enhancing the intracellular free radical scavenger glutathione.[37]

Oxidized lipids, metabolic stress, and inflammation in atherosclerotic plaques can also induce autophagy as a safeguard against cellular distress by degrading damaged intracellular material. Basal autophagy can be enhanced by certain drugs for cardioprotection, which can reach damaging and autophagic cell death levels under excessive stimulation.[38] Loss of smooth muscle cells may lead to plaque destabilization and rupture under excessive cell death and autophagy conditions.[39] However, therapeutic induction of autophagy in atherosclerotic plaque resolution may have a counterbalancing effect on other survival or cell death pathways such as apoptosis and necrosis.

Hypertension

Higher systemic blood pressure (SBP) associates with increased heart weight, serum ET-1, angiotensin II (AngII) concentrations, and decreased serum NO. Resveratrol decreases SBP, heart weight, ET-1, and AngII concentrations while increases the vasodilator NO concentration, which protects against increased SBP and subsequent cardiac hypertrophy in mice.[40,41] Mean arterial pressure after resveratrol treatment dropped significantly in spontaneously hypertensive rats.[42] In another study on rats treated with resveratrol, echocardiographic analysis of cardiac structure and function after hypertensive stress showed improvement in the interventricular septal wall thickness and left ventricular posterior wall thickness, as well as isovolumetric relaxation time at systole and diastole. Endothelial NO synthase (eNOS), inducible nitric oxide synthase (iNOS), and redox factor-1 (ref-1) were significantly decreased under these stress conditions. Resveratrol may therefore be beneficial against certain types of cardiac hypertrophy found in clinical settings of hypertension and aortic valve stenosis.

For the last 30 years it has been know that lysosomal pathways are involved in the different models of heart disease.[43–47] More recently, advances in our understanding of autophagy have revealed that hemodynamic load-induced aggregation of intracellular proteins may have a protective role.[48] However, excessive pressure overload may elicit a robust autophagic response in cardiomycytes that is maladaptive and contributes to disease progression.

Ischemia/reperfusion injury and PC

PC is a protective and adaptive phenomenon whereby brief episodes of ischemia and reperfusion (I/R) render the heart resistant to subsequent ischemic injury or stress. A large number of stimuli such as short cyclic episodes of I/R, a number of pharmacological agents (agonists of adenosine, bradykinin, adrenergic and muscarinic receptor, NO donors, and phosphodiesterase inhibitors), and various noxious stimuli (endotoxins, cytokines, and ROS) have been found to generate PC-like phenotype.[49] Resveratrol can similarly precondition the heart in a NO-dependent manner (Fig. 3), inducing the expression of iNOS, eNOS, and vascular endothelial growth factor (VEGF) in rats.[50] Resveratrol provided cardioprotection as evidenced by superior postischemic ventricular recovery, reduced myocardial infarct size, and decreased number of apoptotic cardiomyocytes. Resveratrol likely activates both adenosine A_1 and A_3 receptors that phosphorylate phosphatidylinositol-3 kinase (PI3K), which then phosphorylates protein kinase B (Akt) and thus preconditions the heart by producing NO, as well as by the activation of antioxidant Bcl-2.[50,51] Activation of adenosine A_3 receptors could also precondition the heart by a survival signal through the cAMP response element-binding protein (CREB) phosphorylation via PI3K/Akt and via MERK (mitogen-activated extracellular signal-regulated protein kinase)/CREB pathways.[52] Recent studies have also demonstrated that NO can induce the expression of heme oxygenase-1. Tin protoporphyrin (SnPP), a heme oxygenase-1 (HO-1) inhibitor, abolished increased cardiac function parameters, reduced myocardial infarct size, and decreased cardiomyocyte apoptosis that characterize the cardioprotection with resveratrol.[53] The HO-1–mediated mechanisms were related to the p38 MAP kinase and Akt survival signaling but independent of nuclear factor-kappaB (NF-κB) activation. A polyphenol, resveratrol, protects the heart by its antioxidative properties through various redox signaling mechanisms. The ability of resveratrol to modulate redox signaling has been extensively reviewed,[54,55] but the role of Sirt activation in the heart is yet to be determined. White wine lacking polyphenols but containing antioxidant compounds such as caffeic acid and tyrosol mediated cardioprotection against I/R injury in rat hearts work via similar survival pathway involving Akt/FOXO3a/NF-κB.[56] Our recent results also indicate that at lower doses, resveratrol-mediated survival of cardiac myoblasts is, in part, mediated through the induction of autophagy,[57] which, along other enhanced survival signals, helps to recover the cells from injury.[58]

Angiogenesis

Resveratrol in general has angiosuppressive effects—rat gliomas significantly decrease in size upon resveratrol treatment.[59] Proliferation and migration of vascular endothelial cells under resveratrol treatment activates eukaryotic elongation factor-2 kinase, which in turn inactivates elongation factor-2, an important factor in protein translation.[60] In mice that develop abnormal angiogenesis in the retina after laser treatment, when given resveratrol, the abnormal blood vessels began to disappear. Similarly, resveratrol suppressed the growth of new blood vessels in animals by directly inhibiting capillary endothelial cell growth[61]—it blocked the VEGF and fibroblast growth factor (FGF) receptor-mediated angiogenic responses and significantly delayed angiogenesis-dependent wound healing. Exposure of human umbilical vein endothelial cells (HUVECs) to low doses of resveratrol significantly blocks VEGF-mediated migration and tube formation but not cell proliferation.[26]

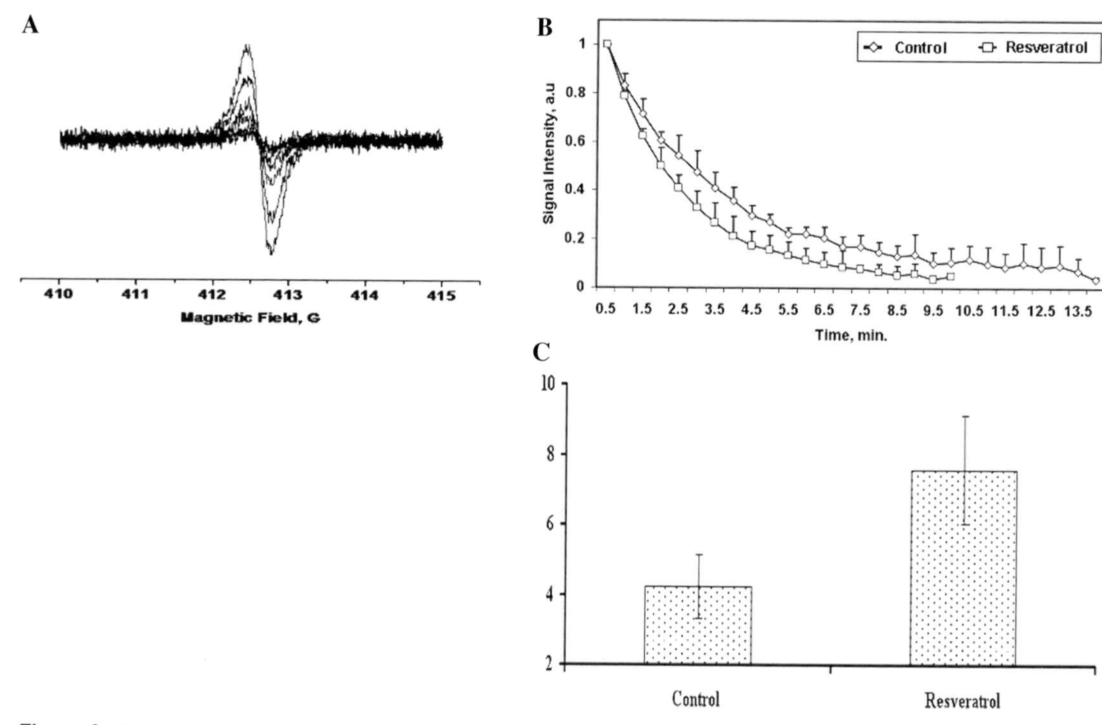

Figure 3. (A) Typical electron spin resonance (ESR) spectra recorded from the reduction of nitroxide during ischemia of the isolated heart. The reduction in nitroxide resulted in a decrease in the ESR signal intensity over time during the ischemic phase. (B) The decrease in signal intensity over time was fit to determine the rate of decay of the ischemic heart in control and resveratrol-treated hearts. (C) The rate of decay of the TEMPO nitroxide (2, 2, 6, 6-tetramethyl-piperidine-1-oxyl) during ischemic challenge in control and resveratrol-pretreated isolated hearts. Results are shown as mean ± SEM of 4 hearts per group.

Under the same concentrations, resveratrol fails to affect VEGF-stimulated activation of VEGF receptor, extracellular signal-regulated protein kinase 1/2, p38 mitogen-activated protein kinase, and Akt. Interestingly, at low micromolar doses, resveratrol effectively abrogates VEGF-mediated tyrosine phosphorylation of vascular endothelial cadherin and its complex partner, β-catenin.

Insufficient angiogenesis can result in nutritional and metabolic deficiency, igniting prosurvival pathways of autophagy in cells and tissues. In this case as well, the double-edged sword of autophagy is evident, since low-level induction of this pathway may be prosurviving for the deficiently supplied tissue, but excessive autophagy in cells can be destructive and possibly proinflammatory.[62,63]

Metabolic syndrome: diabetes and obesity

Although most of the research done on the positive effect of resveratrol and its antidiabetic potential has been conducted on animals, the results from our own laboratory show that injecting streptozotosin diabetic rats with this compound results in a significant decrease in the levels of blood glucose.[64] Similarly, direct injection of resveratrol into the brains of diabetic rats resulted in reduction of insulin levels like it does in hyperinsulinemic rats. A recent study reported that long-term intracerebroventricular infusion of resveratrol-normalized hyperglycemia and greatly improved hyperinsulinemia in diet-induced obese and diabetic mice[65] independent of changes in body weight, food intake, and circulating leptin levels. In addition, central nervous system resveratrol delivery improves hypothalamic NF-κB inflammatory signaling by reducing acetylated-RelA/p65 and total RelA/p65 protein contents and inhibitor of NF-κB alpha and beta mRNA levels. Furthermore, this treatment leads to reduced hepatic phosphoenolpyruvate carboxykinase 1 mRNA and protein levels and ameliorates pyruvate-induced hyperglycemia in this mouse model of type 2 diabetes. The mechanism of resveratrol's action is by far complex and still evolving. Obesity studies in our laboratory demonstrate that resveratrol renders the hearts of *ob/ob* mice resistant to ischemic reperfusion injury.[66] The antidiabetic potential of resveratrol appears to warrant further promise of imposing the same effect on humans.

Insulin signaling, which enhances protein synthesis, also inhibits autophagy in insulin target tissues.[67] Thus, there is the potential for alterations in this control in insulin-resistant states such as diabetes, and defects in autophagy may adversely affect cell function in diabetes. Such defects may lead to inappropriate accumulation of dysfunctional mitochondria, insulin granules, or activation of apoptotic pathways. Furthermore, various cellular stresses can induce dysfunctional protein folding, which can cause the accumulation of misfolded and aggregated proteins in cells.[68,69] Inability of the autophagy pathway to clear protein aggregates may be intimately related to cell death in diabetes.

Longevity

The first revolutionary discovery that resveratrol can activate the antiaging gene SIRT1[70] has been followed by recent studies supporting this antiaging function. However, animal studies of the dose effect of resveratrol would mean that one needs to consume several bottles of red wine per day to reach such a dose.[71] More recently, low dose of resveratrol has been found to activate not only SIRT1, but also SIRT 3 and SIRT 4 as well as FoxOs and pre-B cell colony-enhancing factor (PBEF)—all being proteins linked or associated with longevity.[72] In addition, white wine and its components tyrosol and hydroxytyrosol have been found to activate same longevity genes, suggesting that activation of SIRT1 could be a nonspecific phenomenon. Resveratrol prevented age-related and obesity-related cardiovascular functional decline in mice, but did not affect the overall survival or maximum life span for mice on a standard diet, compared to mice on the same diet with resveratrol.[73] Different species appear to increase their life span upon resveratrol, from yeast and nematodes to flies (*Drosophila melanogaster*), as well as mice kept on a high fat diet. In another related study, SIRT1 overexpression enhanced the autophagic flux in cancer cells, and this effect was blocked by inhibitors of its catalytic activity. Resveratrol induced autophagy in *Caenorhabditis elegans* also via activation of SIRT2; this effect was abolished by siRNA mediated depletion of SIRT2.[74] Autophagy, in turn, increases life expectancy of the worm.

A substantial amount of evidence suggests that resveratrol mimics calorie restriction and increases the life span. Ongoing research to identify an ideal antiaging compound has not been successful. To date, calorie restriction remains the only effective mean to increase the life span,[75] although the mechanisms whereby calorie restriction increases life span remain speculative. Several reports can be found in the literature supporting the role of exercise and certain chemicals such as rapamycin and resveratrol in expanding life span.[76] A study using microarray revealed that out of 6,347 modified genes by calorie restriction, about 58 genes displayed at least twofold alterations in gene expression.[77] There is a striking similarity between resveratrol and calorie restriction on the alteration of metabolic pathways. For example, both resveratrol and calorie restriction improve insulin sensitivity thereby reducing the insulin and glucose levels in the body,[78] which in turn reduce the life-threatening cardiovascular risk factors. Both resveratrol and calorie restriction can potentiate the expression of Glucose transporter type 4 (GLUT4).[66] As mentioned earlier, the most important antiaging gene that is upregulated by calorie restriction and resveratrol is SIRT1, although it works in coordination with the FoxOs. Indeed, resveratrol and red wine containing significant amount of resveratrol that can inactivate the FoxOs by phosphorylation.[72]

In a recent study, life extension of fish by resveratrol was associated with the change in the slope of mortality trajectory,[79] indicating lowering of the time-dependent increase in death risk with resveratrol. In the same study, the authors noticed an initial increase in death rate after resveratrol treatment, suggesting that the weak toxic effect of resveratrol was associated with a stress response that ultimately increased the expression of longevity genes, thereby increasing the life span. A similar effect was observed in *Drosophila*, when hypothermia was associated with a reduction in the accumulation of age-dependent irreversible injury.[80] This effect was quite different from the life extension induced by calorie restriction, which is reflected in time-dependent reduction in acute risk of death without changing the slope of mortality rate.[80]

Resveratrol and autophagy

On the basis of our previous observation that cardioprotection induced by ischemic PC induces autophagy and that resveratrol induces PC-like effects, we sought to determine if resveratrol could induce autophagy. Macroautophagy, commonly known as autophagy, involves a bulk degradation process clearing organelles, long-lived proteins, and protein complexes. Cytosolic constituents destined to be degraded become enclosed by double membrane structures, known as autophagosomes or autophagic vacuoles, which are fused with lysosomes followed by the degradation of its contents. In fact, the name explains the term autophagy, a housekeeping process through self-cannibalism or eating inside-out.

Three types of autophagy are known: (i) microautophagy involving "dumping" of cytosolic constituents into the lysosome by direct invagination of the lysosomal membrane followed by budding of vesicles into the lysosomal lumen; (ii) macroautophagy, or autophagy involving the formation of a double membrane structure known as an autophagosome, which sequesters cytosolic constituents to be delivered into the lysosomes for digestion; and (iii) chaperone-mediated autophagy characterized by selectivity of cytosolic proteins to be degraded. Any autophagic process undergoes four stages: (i) induction by external stress, such as environmental stress (e.g., oxidative stress), nutritional stress (e.g., nutritional deprivation), and physical stress (e.g., ischemia or hypoxia)—the gatekeeper for induction is mTOR, which regulates transcriptional activation of downstream target genes; (ii) autophagosome formation, as described earlier, where a number of autophagy (Atg) genes participate and recruit beclin-1 and microtubule-associated protein light-chain 3 (LC3); (iii) autophagosomes undergo docking and fusion with the lysosome; and finally (iv) autophagic vesicles are broken down by lysosomal proteases, where lysosomal-associated membrane protein 2 (LAMP-2) plays a crucial role in the degradation process.

In a recent publication, we demonstrated that resveratrol at lower doses (0.1 and 1 μM in H9c2 cardiac myoblast cells and 2.5 mg/kg/day in rats) induced cardiac autophagy shown by enhanced formation of autophagosomes and its component LC3-II after hypoxia-reoxygenation or ischaemia-reperfusion.[11] Autophagy was attenuated with the higher dose of resveratrol, suggesting hormetic action of resveratrol, and the induction of autophagy was correlated with enhanced cell survival and decreased apoptosis. Treatment with rapamycin (100 nM), a known inducer of autophagy, did not further increase autophagy compared with resveratrol alone.[11] Autophagic inhibitors, wortmannin (2 μM), and 3-methyladenine (10 mM) significantly attenuated resveratrol-induced autophagy and induced cell death.[11] The activation of mammalian target of rapamycin (mTOR) was differentially regulated by low-dose resveratrol, i.e., the phosphorylation of mTOR at serine 2448 was inhibited, whereas the phosphorylation of mTOR at serine 2481 was increased, which was attenuated with a higher dose of resveratrol.[11] Although resveratrol attenuated the activation of mTOR complex 1, low-dose resveratrol significantly induced the expression of rictor, a component of mTOR complex 2, and activated its downstream survival kinase Akt (Ser 473).[11] Resveratrol-induced rictor was found to bind with mTOR. Furthermore, treatment with rictor siRNA attenuated the resveratrol-induced autophagy. These results indicate that at lower doses, resveratrol-mediated cell survival is, in part, mediated through the induction of autophagy involving the mTOR-rictor survival pathway.[11]

It appears that resveratrol can generate a survival signal through autophagy. The important question is, "can we use such autophagy clinically for the health benefits?" The simple answer is yes, if the amount of stress can be controlled. There is no doubt that a small amount of stress induced by resveratrol can induce autophagy generating a survival signal, while the same autophagy (induced by higher doses of resveratrol) will potentiate a death signal if the amount of such stress is large and becomes cytotoxic to the cells. Thus, a "therapeutic amount of stress" (5–10 mM resveratrol) must be defined for the induction of autophagy, the same amount that has been known to induce an "adaptive response" for the cells subjected to hostile environment.

Resveratrol in regeneration of infracted myocardium

Another recent study from our laboratory using cardiac stem cells demonstrated that resveratrol can potentiate the regeneration of infracted myocardium.[57] A major problem in the effectiveness of stem cell therapy is the death of stem cells due to the oxidative environment present in the normal tissue. Reduction of oxidative stress or maintaining a reduced environment in the target tissue can enhance stem cell survival and cardiac regeneration after stem cell therapy. In the study mentioned,[57] we pretreated rats with resveratrol (2.5 mg/kg/day gavaged for 2 weeks) after which a left anterior descending coronary artery (LAD) occlusion was carried, followed by direct injection of adult cardiac stem cells stably expressing enhanced green-fluorescent protein (EGFP) on the border zone of the myocardium through survival surgery. The prevalence of cardiac-reduced environment was seen in resveratrol-treated rat hearts via significantly enhanced redox signaling observed through the nuclear factor-E2-related factor-2 (Nrf2), stromal cell-derived factor-1 (SDF1), and NF-κB, as well as ref-1, seven days after LAD occlusion.[57] Significantly improved cardiac functional parameters (left ventricular ejection fraction and fractional shortening), enhanced stem cell survival and proliferation (expression of cell proliferation marker Ki67), and differentiation of stem cells toward the regeneration of the myocardium (expression of EGFP) was evident 28 days after LAD occlusion in rats treated with resveratrol, compared to control rats.[57] Our study clearly demonstrated that resveratrol can modify the physiological redox environment within the myocardium. Maintaining a reduced tissue environment by treatment with resveratrol in rats enhanced the cardiac regeneration by adult cardiac stem cells via improved cell survival, proliferation, and differentiation, leading to improved cardiac function. This study is the first demonstrating that nutritional modification of the redox environment with resveratrol can prolong the regeneration of the infracted myocardium.[57]

In addition cardiac stem cells were modified with resveratrol by preincubating the cells stably expressing EGFP with 10 mM resveratrol for 60 min followed by washing the cells with buffer to get rid of any free resveratrol.[57] Rats were anesthetized, their hearts were opened, and LAD was occluded to induce a heart attack. The animals were divided into two groups: one group was treated with resveratrol-modified stem cells, while the other group was treated with stem cells alone. One week after the LAD occlusion, the cardiac-reduced environment was confirmed in resveratrol-treated rat hearts by the enhanced expression of Nrf2 and Ref-1. M-mode echocardiography was performed to determine cardiac function up to three months after the stem cell therapy. Initially (after 72 h), both groups revealed improvement in cardiac function, but only the resveratrol-modified stem cell group revealed improvement in cardiac function (left ventricular ejection fraction, fractional shortening, and cardiac output) at the end of the 1, 2, and 4 months period. The improvement in cardiac function was accompanied by the enhanced stem cell survival and proliferation as evidenced by the expression of cell proliferation marker Ki67 and differentiation of stem cells toward the regeneration of the myocardium as evidenced by the expression of EGFP up to six months after LAD occlusion in the resveratrol-treated stem cell group of hearts.[57] Again, our results demonstrate that resveratrol maintained a reduced tissue environment by overexpressing Nrf2 and Ref-1 in rat hearts up to four months, resulting in an enhancement of the regeneration of the adult cardiac stem cells as evidenced by increased cell survival and differentiation leading to improved cardiac function.

Resveratrol and hormones

Several recent studies implicate that resveratrol displays hormetic action, protecting the cells at lower doses while killing them at relatively higher doses. Since such a hormetic behavior might have significant impact on epidemiological and clinical studies, we sought to determine the dose–response curve for resveratrol action. We fed by gavaging up to 30 days a group of rats three different doses of resveratrol 2.5, and 100 mg/kg while the control group was given vehicles only. Our results indeed showed hormesis for resveratrol, being cardioprotective at lower doses only and detrimental for higher doses (D. Das, *J. Clin. Exp. Cardiol.*, in press). At 100 mg/kg dose, 100% of the hearts died in case of resveratrol.[a] The results clearly demonstrate that resveratrol is beneficial to the heart only at low doses and is detrimental at higher doses. Also, the action of resveratrol is quickly realized, in most cases within 14 days; up to 30 days, resveratrol does not add any additional benefit. Such hormesis has been known for more

[a]Dudley *et al.* 2009, *J. Nutr. Biochem.*, **20**(6), 2009, 443–452.

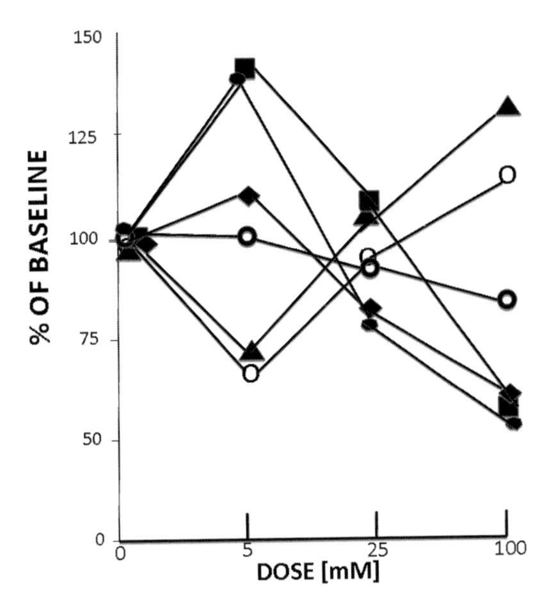

Figure 4. Hormetic action of resveratrol. Dose of resveratrol [*X*-axis] is plotted against the values of cardiac function, infarct size, and apoptosis. Resveratrol at a dose of 2.5 mg/kg provided maximum protection [peak], which then progressively and steadily declined.

than 100 years and frequently observed among the toxins. As mentioned earlier, resveratrol is a phytoalexin, whose growth is stimulated by environmental stress such as fungal infection, ultraviolet (UV) radiation, and water deprivation.[81] Cardioprotective effects of resveratrol are exerted through its ability to precondition a heart through adaptive response, which causes the development of intracellular stress leading to the upregulation of intracellular defense system such as antioxidants and heat shock protein.[82] Preconditioning is another example of hormesis, which is potentiated by subjecting an organ like heart to cyclic episodes of short durations of ischemia, each followed by another short durations of reperfusion.[83] Such small but therapeutic amount of stress renders the heart resistant to subsequent lethal ischemic injury. Such an adaptive response is commonly observed with aging. Autophagy also constitutes another form of adaptive response. Consistent with this idea, resveratrol has been found to stimulate longevity genes, and at least in prokaryotic species extend the life span. In this respect, resveratrol may fulfill the definition of a hormetin.[84] There is no doubt that alcohol, wine, and wine-derived resveratrol all display hormesis. It has been known for quite some time that cardioprotective effects of alcohol or wine intake follow a J-shaped curve.[85] This study (D. Das, *J. Clin. Exp. Cardiol.* in press) echoed this concept that cardioprotective effects of resveratrol follow a J-shaped curve and display hormesis (Fig. 4). At lower doses, resveratrol acts as an antiapoptotic agent, providing cardioprotection as evidenced by increased expression in cell survival proteins, improved postischemic ventricular recovery, and reduction of myocardial infarct size and cardiomyocyte apoptosis by maintaining a stable redox environment compared to control. At higher doses, however, resveratrol depresses cardiac function, elevates levels of apoptotic protein expressions, results in an unstable redox environment, and increases myocardial infarct size and number of apoptotic cells. A significant number of reports are available in the literature to show that at a high dose, resveratrol not only hinders tumor growth, but also inhibits the synthesis of RNA, DNA, and protein; causes structural chromosome abberrations, chromatin breaks, chromatin exchanges, weak aneuploidy, and higher S-phase arrest; blocks cell proliferation; and decreases wound healing, endothelial cell growth by VEGF and FGF-2, and angiogenesis in healthy tissue cells leading to cell death.[86–89]

Summary and conclusion

Resveratrol, a grape- and wine-derived phytoalexin polyphenol, provides diverse health benefits, the most prominent being the best natural medicine to cure diverse CVDs including atherosclerosis, hypertension, ischemia/reperfusion, heart failure, diabetes, obesity, and aging. Resveratrol-mediated cardioprotection is not due to its drug-like action neutralizing the toxins introduced by the disease; rather, the protection is realized from its PC-like effects potentiated by its adaptive response. During the PC, a small amount of resveratrol (defined as therapeutic dose) exerts adaptive stress response thereby enabling the cardiac cells to make cardioprotective genes and proteins such as heat shock and antioxidant proteins. The ability of resveratrol to exert adaptive response is further supported by its ability to induce autophagy, another form of adaptation to stress. In that context, autophagy may be viewed as "cry for survival," and such survival is a result of adaptive response to fight against stress. If the stress—such as oxidative stress—is mild, it generates an adaptive response to survive against it. Autophagy results in a survival signal by inducing a number of genes and transcription factors that alter the stress-induced death signal into a survival signal, leading to production of antiapoptotic or antideath proteins. On the other hand, if the stress is overwhelming, the adaptive response fails and the cells die due to induction of apoptotic signals. It is important to remember that to survive, cells must get rid of damaged, detrimental, and unwanted components, and they do so through autophagy. Thus, resveratrol may be used clinically to induce autophagy leading to cardioprotection. Finally, resveratrol can modify the redox environment within the cells through its ability to induce redox protein thioredoxin, which can pave the way for stem cell survival. Recent studies documented its ability to regenerate the infracted myocardium up to four months after the cell therapy. An important thing to remember is that for successful application of resveratrol in therapy, one must recognize the fact that resveratrol possesses hormetic action similar to any toxins, exerts beneficial action at lower doses, and becomes cytotoxic at higher doses, generating a J-shaped or inverted U-shape curve like wine and alcohol.

Further research and more clinical studies are necessary in order to ensure the safety of resveratrol and also for ascertaining the optimum doses and ways of inducing low-level autophagy for prevention and treatment. The findings that resveratrol can potentiate diverse health benefits from chemoprevention to cardioprotection—through its ability to induce autophagy and cardiac regeneration through prolonging the stem cell survival—certainly warrants an opening of a new era of pharmaceutical intervention in cardiovascular medicine.

Conflicts of interest

The authors declare no conflicts of interest.

References

1. Murray, C.J. & A.D. Lopez. 1997. Global mortality, disability, and the contribution of risk factors: global burden of disease study. *Lancet* **349:** 1436–1442.
2. Fontana, L. *et al.* 2007. Calorie restriction or exercise: effects on coronary heart disease risk factors. A randomized, controlled trial. *Am. J. Physiol. Endocrinol. Metab.* **293:** E197–E202.
3. Twisk, J.W. *et al.* 1997. Which lifestyle parameters discriminate high- from low-risk participants for coronary heart disease risk factors. Longitudinal analysis covering adolescence and young adulthood. *J. Cardiovasc. Risk* **4:** 393–400.
4. Bellisle, F. 2009. Infrequently asked questions about the Mediterranean diet. *Public Health Nutr.* **12:** 1644–1647.
5. Willett, W.C. *et al.* 1995. Mediterranean diet pyramid: a cultural model for healthy eating. *Am. J. Clin. Nutr.* **61:** 1402S–1406S.
6. Renaud, S. & M. de Lorgeril. 1992. Wine, alcohol, platelets, and the French paradox for coronary heart disease. *Lancet* **339:** 1523–1526.
7. Sato, M., N. Maulik & D.K. Das. 2002. Cardioprotection with alcohol: role of both alcohol and polyphenolic antioxidants. *Ann. N.Y. Acad. Sci.* **957:** 122–135.
8. Ghanim, H. *et al.* 2010. An antiinflammatory and reactive oxygen species suppressive effects of an extract of polygonum cuspidatum containing resveratrol. *J. Clin. Endocrinol. Metab.* **91:** E1–E8.
9. Roupe, K.A. *et al.* 2006. Pharmacometrics of stilbenes: seguing towards the clinic. *Curr. Clin. Pharmacol.* **1:** 81–101.
10. Berrougui, H. *et al.* 2009. A new insight into resveratrol as an atheroprotective compound: inhibition of lipid peroxidation and enhancement of cholesterol efflux. *Atherosclerosis* **207:** 420–427.
11. Gurusamy, N. *et al.* Cardioprotection by resveratrol: a novel mechanism via autophagy involving the mTORC2 pathway. *Cardiovasc. Res.* **86:** 103–112.
12. Holthoff, J.H. *et al.* 2010. Resveratrol, a dietary polyphenolic phytoalexin, is a functional scavenger of peroxynitrite. *Biochem. Pharmacol.* **80:** 1260–1265.
13. Gurusamy, N. *et al.* 2009. Cardioprotection by adaptation to ischaemia augments autophagy in association with BAG-1 protein. *J. Cell Mol. Med.* **13:** 373–387.
14. Morselli, E. *et al.* 2009. Autophagy mediates pharmacological lifespan extension by spermidine and resveratrol. *Aging (Albany NY).* **1:** 961–970.
15. Abeywardena, M.Y. 2003. Dietary fats, carbohydrates and vascular disease: Sri Lankan perspectives. *Atherosclerosis.* **171:** 157–161.
16. Wildman, R.P. *et al.* Cardiovascular disease risk of abdominal obesity vs. metabolic abnormalities. *Obesity* (Silver Spring). 2010 Aug 19 [Epub ahead of print] doi:10.1038/oby.2010.168.
17. Markus, M.A. & B.J. Morris. 2008. Resveratrol in prevention and treatment of common clinical conditions of aging. *Clin. Interv. Aging* **3:** 331–339.
18. Nicholson, S.K., G.A. Tucker & J.M. Brameld. 2008. Effects of dietary polyphenols on gene expression in human vascular endothelial cells. *Proc. Nutr. Soc.* **67:** 42–47.
19. Della-Morte, D. *et al.* 2009. Resveratrol pretreatment protects rat brain from cerebral ischemic damage via a sirtuin 1-uncoupling protein 2 pathway. *Neuroscience* **159:** 993–1002.
20. Altman, R. 2003. Risk factors in coronary atherosclerosis athero-inflammation: the meeting point. *Thromb. J.* **1:** 4.
21. Ross, R. 1999. Atherosclerosis—an inflammatory disease. *N. Engl. J. Med.* **340:** 115–126.
22. Vorchheimer, D.A. & V. Fuster. 2001. Inflammatory markers in coronary artery disease: let prevention douse the flames. *JAMA* **286:** 2154–2156.
23. Kinlay, S. *et al.* 2001. Role of endothelin-1 in the active constriction of human atherosclerotic coronary arteries. *Circulation* **104:** 1114–1118.

24. Davignon, J. & P. Ganz. 2004. Role of endothelial dysfunction in atherosclerosis. *Circulation* **109:** III27–III32.
25. Rubbo, H. *et al.* 2002. Antioxidant and diffusion properties of nitric oxide in low-density lipoprotein. *Methods Enzymol.* **359:** 200–209.
26. Lin, M.T. *et al.* 2003. Inhibition of vascular endothelial growth factor-induced angiogenesis by resveratrol through interruption of Src-dependent vascular endothelial cadherin tyrosine phosphorylation. *Mol. Pharmacol.* **64:** 1029–1036.
27. Zou, J.G. *et al.* 2003. Effect of red wine and wine polyphenol resveratrol on endothelial function in hypercholesterolemic rabbits. *Int. J. Mol. Med.* **11:** 317–320.
28. O'Leary, K.A. *et al.* 2004. Effect of flavonoids and vitamin E on cyclooxygenase-2 (COX-2) transcription. *Mutat. Res.* **551:** 245–254.
29. Wang, M.J. *et al.* 2001. Resveratrol inhibits interleukin-6 production in cortical mixed glial cells under hypoxia/hypoglycemia followed by reoxygenation. *J. Neuroimmunol.* **112:** 28–34.
30. Zhong, M. *et al.* 1999. Inhibitory effect of resveratrol on interleukin 6 release by stimulated peritoneal macrophages of mice. *Phytomedicine* **6:** 79–84.
31. Culpitt, S.V. *et al.* 2003. Inhibition by red wine extract, resveratrol, of cytokine release by alveolar macrophages in COPD. *Thorax* **58:** 942–946.
32. Fremont, L., L. Belguendouz & S. Delpal. 1999. Antioxidant activity of resveratrol and alcohol-free wine polyphenols related to LDL oxidation and polyunsaturated fatty acids. *Life Sci.* **64:** 2511–2521.
33. Godichaud, S. *et al.* 2000. Deactivation of cultured human liver myofibroblasts by trans-resveratrol, a grapevine-derived polyphenol. *Hepatology* **31:** 922–931.
34. Olas, B. *et al.* 2002. Effect of resveratrol, a natural polyphenolic compound, on platelet activation induced by endotoxin or thrombin. *Thromb. Res.* **107:** 141–145.
35. Wang, Z. *et al.* 2002. Effects of red wine and wine polyphenol resveratrol on platelet aggregation *in vivo* and *in vitro*. *Int. J. Mol. Med.* **9:** 77–79.
36. Jang, J.H. & Y.J. Surh. 2001. Protective effects of resveratrol on hydrogen peroxide-induced apoptosis in rat pheochromocytoma (PC12) cells. *Mutat. Res.* **496:** 181–190.
37. Jang, D.S. *et al.* 1999. Inhibitory effects of resveratrol analogs on unopsonized zymosan-induced oxygen radical production. *Biochem. Pharmacol.* **57:** 705–712.
38. Martinet, W. & G.R. De Meyer. 2009. Autophagy in atherosclerosis: a cell survival and death phenomenon with therapeutic potential. *Circ. Res.* **104:** 304–317.
39. Clarke, M.C. *et al.* 2006. Apoptosis of vascular smooth muscle cells induces features of plaque vulnerability in atherosclerosis. *Nat. Med.* **12:** 1075–1080.
40. Liu, Z. *et al.* 2005. Effects of trans-resveratrol on hypertension-induced cardiac hypertrophy using the partially nephrectomized rat model. *Clin. Exp. Pharmacol. Physiol.* **32:** 1049–1054.
41. Miatello, R., M. Cruzado & N. Risler. 2004. Mechanisms of cardiovascular changes in an experimental model of syndrome X and pharmacological intervention on the renin-angiotensin-system. *Curr. Vasc. Pharmacol.* **2:** 371–377.
42. Miatello, R. *et al.* 2005. Chronic administration of resveratrol prevents biochemical cardiovascular changes in fructose-fed rats. *Am. J. Hypertens* **18:** 864–870.
43. Dammrich, J. & U. Pfeifer. 1983. Cardiac hypertrophy in rats after supravalvular aortic constriction. II. Inhibition of cellular autophagy in hypertrophying cardiomyocytes. *Virchows Arch. B Cell Pathol. Incl. Mol. Pathol.* **43:** 287–307.
44. Decker, R.S. *et al.* 1980. Lysosomal vacuolar apparatus of cardiac myocytes in heart of starved and refed rabbits. *J. Mol. Cell. Cardiol.* **12:** 1175–1189.
45. Pfeifer, U. *et al.* 1987. Short-term inhibition of cardiac cellular autophagy by isoproterenol. *J. Mol. Cell Cardiol.* **19:** 1179–1184.
46. Sybers, H.D., J. Ingwall & M. DeLuca. 1976. Autophagy in cardiac myocytes. *Recent Adv. Stud. Cardiac. Struct. Metab.* **12:** 453–463.
47. Yamamoto, S. *et al.* 2000. On the nature of cell death during remodeling of hypertrophied human myocardium. *J. Mol. Cell Cardiol.* **32:** 161–175.
48. Rothermel, B.A. & J.A. Hill. 2008. Autophagy in load-induced heart disease. *Circ. Res.* **103:** 1363–1369.
49. Bolli, R. 2000. The late phase of preconditioning. *Circ. Res.* **87:** 972–983.
50. Das, S. *et al.* 2005. Coordinated induction of iNOS-VEGF-KDR-eNOS after resveratrol consumption: a potential mechanism for resveratrol preconditioning of the heart. *Vascul Pharmacol.* **42:** 281–289.
51. Das, S. *et al.* 2005. Resveratrol-mediated activation of cAMP response element-binding protein through adenosine A3 receptor by Akt-dependent and -independent pathways. *J. Pharmacol. Exp. Ther.* **314:** 762–769.
52. Das, S. *et al.* 2005. Pharmacological preconditioning with resveratrol: role of CREB-dependent Bcl-2 signaling via adenosine A3 receptor activation. *Am. J. Physiol. Heart Circ. Physiol.* **288:** H328–H335.
53. Das, S., C.G. Fraga & D.K. Das. 2006. Cardioprotective effect of resveratrol via HO-1 expression involves p38 map kinase and PI-3-kinase signaling, but does not involve NFkappaB. *Free Radic. Res.* **40:** 1066–1075.
54. Das, D.K. & N. Maulik. 2006. Resveratrol in cardioprotection: a therapeutic promise of alternative medicine. *Mol. Interv.* **6:** 36–47.
55. Vidavalur, R. *et al.* 2006. Significance of wine and resveratrol in cardiovascular disease: French paradox revisited. *Exp. Clin. Cardiol.* **11:** 217–225.
56. Thirunavukkarasu, M. *et al.* 2008. White wine induced cardioprotection against ischemia-reperfusion injury is mediated by life extending Akt/FOXO3a/NFkappaB survival pathway. *J. Agric. Food Chem.* **56:** 6733–6739.
57. Gurusamy N. *et al.* 2010. Red wine antioxidant resveratrol-modified cardiac stem cells regenerate infarcted myocardium. *J. Cell Mol. Med.***14:** 2235–2239.
58. Lekli, I. *et al.* 2009. Co-ordinated autophagy with resveratrol and gamma-tocotrienol confers synergetic cardioprotection. *J. Cell Mol. Med.*
59. Chen, J.C. *et al.* 2006. Resveratrol suppresses angiogenesis in gliomas: evaluation by color Doppler ultrasound. *Anticancer Res.* **26:** 1237–1245.

60. Khan, A.A. *et al.* (Resveratrol regulates pathologic angiogenesis by a eukaryotic elongation factor-2 kinase-regulated pathway. *Am. J. Pathol.* **177:** 481–492.
61. Brakenhielm, E., R. Cao & Y. Cao. 2001. Suppression of angiogenesis, tumor growth, and wound healing by resveratrol, a natural compound in red wine and grapes. *FASEB J.* **15:** 1798–1800.
62. Fesus, L., M.A. Demeny & G. Petrovski. Autophagy Shapes Inflammation. *Antioxid Redox Signal.* 2010 Sep 2 [Epub ahead of print] doi:10.1089/ars.2010.3485.
63. Petrovski, G. *et al.* 2007. Phagocytosis of cells dying through autophagy evokes a pro-inflammatory response in macrophages. *Autophagy* **3:** 509–511.
64. Lekli, I. *et al.* Functional recovery of diabetic mouse hearts by glutaredoxin-1 gene therapy: role of Akt-FoxO-signaling network. *Gene Ther.* **17:** 478–485.
65. Ramadori, G. *et al.* 2009. Central administration of resveratrol improves diet-induced diabetes. *Endocrinology* **150:** 5326–5333.
66. Lekli, I. *et al.* 2008. Protective mechanisms of resveratrol against ischemia-reperfusion-induced damage in hearts obtained from Zucker obese rats: the role of GLUT-4 and endothelin. *Am. J. Physiol. Heart Circ. Physiol.* **294:** H859–H866.
67. Codogno, P. & A.J. Meijer. 2005. Autophagy and signaling: their role in cell survival and cell death. *Cell Death Differ.* **12**(Suppl 2): 1509–1518.
68. Hayden, M.R. *et al.* 2005. Type 2 diabetes mellitus as a conformational disease. *Jop.* **6:** 287–302.
69. Wiseman, R.L. & W.E. Balch. 2005. A new pharmacology–drugging stressed folding pathways. *Trends Mol. Med.* **11:** 347–350.
70. Howitz, K.T. *et al.* 2003. Small molecule activators of sirtuins extend Saccharomyces cerevisiae lifespan. *Nature* **425:** 191–196.
71. Baur, J.A. & D.A. Sinclair. 2006. Therapeutic potential of resveratrol: the *in vivo* evidence. *Nat. Rev. Drug Discov.* **5:** 493–506.
72. Mukherjee, S. *et al.* 2009. Expression of the longevity proteins by both red and white wines and their cardioprotective components, resveratrol, tyrosol, and hydroxytyrosol. *Free Radic. Biol. Med.* **46:** 573–578.
73. Pearson, K.J. *et al.* 2008. Resveratrol delays age-related deterioration and mimics transcriptional aspects of dietary restriction without extending life span. *Cell Metab.* **8:** 157–168.
74. Dillin, A. & J.W. Kelly. 2007. Medicine. The yin-yang of sirtuins. *Science* **317:** 461–462.
75. Minor, R.K. *et al.* Dietary interventions to extend life span and health span based on calorie restriction. *J. Gerontol. A Biol. Sci. Med. Sci.* **65:** 695–703.
76. Kaeberlein M. Resveratrol and rapamycin: are they anti-aging drugs? *Bioessays* **32:** 96–99.
77. Weindruch, R. *et al.* 2001. Microarray profiling of gene expression in aging and its alteration by caloric restriction in mice. *J. Nutr.* **131:** 918S–923S.
78. Baur, J.A. *et al.* 2006. Resveratrol improves health and survival of mice on a high-calorie diet. *Nature* **444:** 337–342.
79. Valenzano, D.R. *et al.* 2006. Resveratrol prolongs lifespan and retards the onset of age-related markers in a short-lived vertebrate. *Curr. Biol.* **16:** 296–300.
80. Partridge, L., S.D. Pletcher & W. Mair. 2005. Dietary restriction, mortality trajectories, risk and damage. *Mech. Ageing Dev.* **126:** 35–41.
81. Soleas, G.J., E.P. Diamandis & D.M. Goldberg. 1997. Resveratrol: a molecule whose time has come? And gone? *Clin. Biochem.* **30:** 91–113.
82. Collins, M.A. *et al.* 2009. Alcohol in moderation, cardioprotection, and neuroprotection: epidemiological considerations and mechanistic studies. *Alcohol Clin. Exp. Res.* **33:** 206–219.
83. Chiueh, C.C., T. Andoh & P.B. Chock. 2005. Induction of thioredoxin and mitochondrial survival proteins mediates preconditioning-induced cardioprotection and neuroprotection. *Ann. N.Y. Acad. Sci.* **1042:** 403–418.
84. Mattson, M.P. 2008. Hormesis defined. *Ageing Res. Rev.* **7:** 1–7.
85. Prickett, C.D. *et al.* 2004. Alcohol: Friend or Foe? Alcoholic Beverage Hormesis for Cataract and Atherosclerosis is Related to Plasma Antioxidant Activity. *Nonlinearity Biol. Toxicol. Med.* **2:** 353–370.
86. Athar, M. *et al.* 2009. Multiple molecular targets of resveratrol: anti-carcinogenic mechanisms. *Arch. Biochem. Biophys.* **486:** 95–102.
87. Bishayee, A. 2009. Cancer prevention and treatment with resveratrol: from rodent studies to clinical trials. *Cancer Prev. Res. (Phila Pa)* **2:** 409–418.
88. Li, Y. *et al.* Regulation of microRNAs by natural agents: an emerging field in chemoprevention and chemotherapy research. *Pharm. Res.* **27:** 1027–1041.
89. Sen, C.K. *et al.* 2002. Oxygen, oxidants, and antioxidants in wound healing: an emerging paradigm. *Ann. N.Y. Acad. Sci.* **957:** 239–249.

Ann. N.Y. Acad. Sci. ISSN 0077-8923

ANNALS OF THE NEW YORK ACADEMY OF SCIENCES
Issue: *Resveratrol and Health*

Anti-diabetic effects of resveratrol

Tomasz Szkudelski and Katarzyna Szkudelska

Department of Animal Physiology and Biochemistry, Poznan University of Life Sciences, Wolynska, Poland

Address for correspondence: Tomasz Szkudelski, Department of Animal Physiology and Biochemistry, Poznan University of Life Sciences, Wolynska 35, 60-637 Poznan, Poland. tszkudel@jay.up.poznan.pl

Diabetes mellitus is a complex metabolic disease affecting about 5% of people all over the world. Data from the literature indicate that resveratrol is a compound exerting numerous beneficial effects in organisms. Rodent studies, for example, have demonstrated that resveratrol decreases blood glucose in animals with hyperglycemia. This effect seems to predominantly result from increased intracellular transport of glucose. Resveratrol was also demonstrated to induce effects that may contribute to the protection of β cells in diabetes. In experiments on pancreatic islets, the ability of resveratrol to reduce insulin secretion was demonstrated; this effect was confirmed in animals with hyperinsulinemia, in which resveratrol decreased blood insulin levels. Moreover, inhibition of cytokine action and attenuation of the oxidative damage of the pancreatic tissue by resveratrol were recently shown. Studies of animals with insulin resistance indicate that resveratrol may also improve insulin action. The mechanism through which resveratrol improves insulin action is complex and involves reduced adiposity, changes in gene expression, and changes in the activities of some enzymes. These data indicate that resveratrol may be useful in preventing and treating diabetes.

Keywords: resveratrol; diabetes; glycemia; β cells; insulin resistance

Introduction

Diabetes mellitus is a complex metabolic disease and, according to the classification of the American Diabetes Association,[1] is divided into different types: type 1 and type 2 diabetes are the most frequent. Type 1 diabetes accounts for less than 10% of all diabetic cases and results from the autoimmune destruction of β cells. Patients with type 1 diabetes are dependent on exogenous insulin. Type 2 diabetes affects about 90% of all people with diabetes and is characterized by defects in insulin secretion and action.

In the last few years, rodent studies and experiments *in vitro* provided evidence that resveratrol (3,5,4′-trihydroxystilbene)—a naturally occurring phytoalexin present in numerous plant species—exerts beneficial effects in the organism and may be helpful in preventing and treating some metabolic diseases, including diabetes.[2,3] In general, the management of diabetes involves three main aspects: reduction of blood glucose, preservation of β cells, and, in the case of type 2 diabetes, improvement in insulin action. Data from the literature indicate that the beneficial effects of resveratrol in relation to diabetes comprise all these aspects (Fig. 1).

Reduction of blood glucose

The maintenance of blood glucose in the physiological range is pivotal in diabetes, since increased glycemia causes numerous diabetic complications.[4] Therefore, different strategies, including treatment with hypoglycemic agents, are introduced in clinical practice to decrease blood glucose to physiological level. However, the prolonged administration of some agents reducing blood glucose (e.g., metformin or α-glucosidase inhibitors) induces unfavorable effects.[5,6] In this context, a compound reducing blood glucose without any side effects, even if administered for a long period of time, would be very useful in both type 1 and type 2 diabetes.

Numerous studies on diabetic rats revealed the anti-hyperglycemic action of resveratrol. Among different beneficial effects of resveratrol found in diabetes, the ability of this compound to reduce hyperglycemia seems to be the best documented. The

doi: 10.1111/j.1749-6632.2010.05844.x

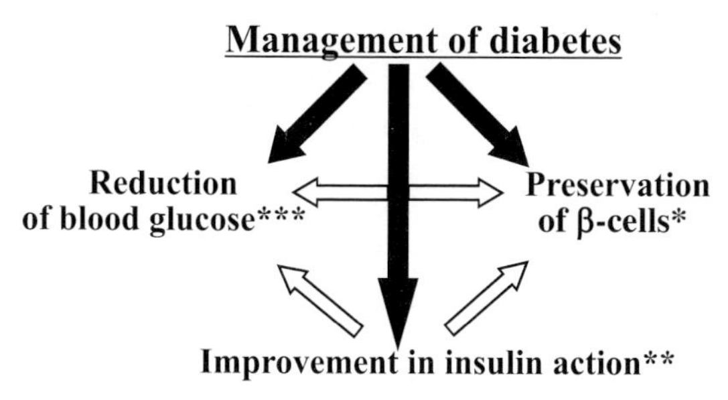

Figure 1. The general management of diabetes and the beneficial effects of resveratrol documented in the literature: ***-well documented, **-poorly documented, *-very poorly documented.

anti-hyperglycemic action of resveratrol was demonstrated in obese rodents[7,8] and in two animal models of diabetes: in rats with streptozotocin-induced diabetes or with streptozotocin-nicotinamide-induced diabetes.[9–15] Some studies also revealed that administration of resveratrol to diabetic rats resulted in diminished levels of glycosylated hemoglobin (HbA_{1C}), which reflects the prolonged reduction of glycemia.[10,15]

The anti-hyperglycemic effect of resveratrol observed in diabetic animals is thought to result from its stimulatory action on intracellular glucose transport. Increased glucose uptake by different cells isolated from diabetic rats was found in the presence of resveratrol. Interestingly, in experiments on isolated cells, resveratrol was able to stimulate glucose uptake in the absence of insulin.[9] The stimulation of glucose uptake induced by resveratrol seems to be due to increased action of glucose transporter in the plasma membrane. Studies on rats with experimentally induced diabetes demonstrated increased expression of the insulin-dependent glucose transporter, GLUT4, as a result of resveratrol ingestion, compared with diabetic animals not given resveratrol.[14,16] It should be mentioned, however, that in some experiments on rats with streptozotocin-induced diabetes, resveratrol appeared to be ineffective and failed to decrease blood glucose.[17,18]

Preservation of β cells

Type 2 diabetes develops slowly, may be undetected for many years, and is usually accompanied by insulin resistance. Initially, blood glucose is maintained in the physiological range because of the compensatory increase in insulin secretion. This compensatory mechanism impedes and delays the diagnosis of diabetes. Moreover, chronic overstimulation of β cells causes their exhaustion and degradation, leading over time to insufficient secretion of insulin.[19] According to this scenario, results of many studies demonstrate that in individuals with the chronic stimulation of β cells, temporary resting of these cells ameliorates their ability to secrete insulin and may delay the onset of the overt diabetes. These beneficial effects were observed in rodents and humans with type 1 diabetes and in humans with type 2 diabetes. A temporary inhibition of insulin secretion was also reported to delay the progress of type 2 diabetes.[20] However, in clinical practice, only a few inhibitors of insulin secretion are used, and side effects appear during their prolonged administration. Therefore, numerous natural compounds are intensively studied in the context of their potential influence on insulin secretion.[21,22]

Rodent studies revealed that resveratrol may affect blood insulin concentrations. In animals with hyperinsulinemia, resveratrol was found to effectively reduce blood insulin. This effect was noticed in mice on a high-fat diet,[23–26] in rats on a high-cholesterol-fructose diet,[27] and in obese Zucker rats.[28]

According to some animal studies, experiments *in vitro* demonstrated the ability of resveratrol to reduce insulin secretion by freshly isolated rat pancreatic islets.[29–31] The inhibition of insulin secretion caused by resveratrol was found to result from metabolic changes in β cells. Under physiological conditions, glucose-induced insulin secretion is preceded by a sequence of events involving intracellular transport of glucose and its oxidative metabolism, hyperpolarisation of the inner mitochondrial membrane, increased formation of ATP and an increase in the ATP/ADP ratio, closure of the ATP-sensitive potassium channels, depolarization of the plasma membrane, opening of voltage-sensitive calcium channels, and the rise in cytosolic Ca^{2+}. The increase in cytosolic Ca^{2+} triggers secretion of insulin. Moreover, other signals in β cells are generated to maintain the sustained insulin-secretory response.[32,33] In this sequence of events leading to increased secretion of insulin, resveratrol was demonstrated to act at the level of ATP formation. Pancreatic islets exposed to resveratrol released more lactate, and glucose oxidation was deeply decreased compared with control islets.[30] Resveratrol was also reported to attenuate hyperpolarization of the inner mitochondrial membrane, indicating reduced activity of the mitochondrial respiratory chain. The metabolic changes induced by resveratrol resulted in decreased ATP levels in islet cells. Since the increase in ATP/ADP ratio is pivotal for insulin secretion, diminution of ATP formation in the presence of resveratrol resulted in attenuated secretion of the pancreatic hormone.[30] Importantly, the inhibition of insulin secretion caused by resveratrol appeared to be reversible and was not due to permanent disturbances in β cells.[29,30]

Since chronic overstimulation of β cells is known to induce their degradation, inhibition of insulin secretion by resveratrol may attenuate these unfavorable effects. This assumption is, however, not proven in animal studies and requires further investigations.

The protective action of resveratrol on the endocrine pancreas may also involve other mech anisms. One of them is the inhibitory influence of resveratrol on cytokine action. Lee *et al.*[34] recently reported that exposure of isolated rat pancreatic islets to cytokines resulted in numerous unfavorable effects, such as increased DNA binding of NF-κB, increased production of NO, and expression of iNOS. All these deleterious effects appeared to be suppressed by resveratrol. The protective action of resveratrol against cytokine-induced toxicity was additionally confirmed in experiments demonstrating increased viability of islets exposed to cytokines and resveratrol, compared with islets incubated with cytokines but without resveratrol. Importantly, resveratrol was also demonstrated to restore secretory function of β cells disrupted by cytokine action; the decrease in glucose-stimulated insulin secretion resulting from exposure to cytokines appeared to be fully restored when pancreatic islets were pretreated with resveratrol. This protective action of resveratrol against cytokine-induced dysfunction of β cells is thought to result from the ability of resveratrol to activate NAD^+-dependent protein deacetylase Sirt1.[34]

Studies *in vivo* on animals with experimentally induced diabetes confirmed the important role of the inhibition of cytokine action in the mechanism thereby resveratrol protects pancreatic β cells. In streptozotocin-nicotinamide-induced diabetic rats, oral administration of resveratrol significantly reduced blood TNF-α, IL-1β, and IL-6 compared with diabetic rats non-treated with resveratrol.[15]

The other mechanism through which resveratrol exerts its protective action in diabetes is related to the anti-oxidant defense. It is known that the anti-oxidant defense in β cells is significantly lower compared with other kinds of cells, making them more susceptible to the oxidative damage.[35] Palsamy and Subramanian[15] recently reported that in streptozotocin-nicotinamide-induced diabetic rats, the levels of lipid peroxides, hydroperoxides, and protein carbonyls in the pancreatic tissue were significantly increased compared with non-diabetic animals, indicating oxidative damage. Simultaneously, the activities of enzymes participating in the anti-oxidant defense—such as superoxide dismutase, catalase, glutathione peroxidase, and glutathione-S-transferase—in the pancreatic tissue of diabetic animals were deeply reduced. However, resveratrol administered to diabetic rats substantially ameliorated parameters of the oxidative damage and increased the activities of the above-mentioned enzymes.[15] This effect of resveratrol seems to be very important, since oxidative stress is one of the factors leading to β cell failure in type 2 diabetes.[36]

These data indicate that resveratrol is able to attenuate cytokine-induced toxicity and effectively reduce oxidative damage of the pancreas, and thereby may ameliorate the endocrine function of this gland. This was confirmed by results demonstrating that in diabetic rats with hypoinsulinemia and with preserved ability of β cells to secrete insulin, administration of resveratrol resulted in the significant increase in blood insulin.[11,15,16]

Moreover, in streptozotocin-nicotinamide-induced diabetic rats ingesting resveratrol, degenerative changes of islets were less marked, and degranulation of β cells was lower. In addition, an increase in secretory granules and no vacuolarization were shown in islets of diabetic rats receiving resveratrol compared with cells of diabetic rats not treated with this compound.[15] Although some studies demonstrated protective action of resveratrol on pancreas, experimental data on the beneficial influence of this compound on β cells in diabetes are still lacking, and thus further animal studies are required.

Improvement in insulin action

Type 2 diabetes is usually accompanied by insulin resistance, defined as the impaired action of insulin on target cells, mainly adipocytes, hepatocytes, and muscle cells. The insulin resistance develops mainly in overweight or obese individuals with type 2 diabetes.[37] It is well documented that decreased adiposity improves insulin action.[38] Moreover, a low-calorie diet and increased physical activity are known to improve metabolic control and insulin sensitivity in type 2 diabetes.

Animal studies provided evidence that resveratrol may be useful as a compound improving insulin action in type 2 diabetes. The ability of resveratrol to increase insulin sensitivity was studied in mice fed a high-fat diet and manifesting insulin resistance. The ingestion of resveratrol by these animals substantially improved action of insulin.[23–25,39] Interestingly, a similar effect of resveratrol was also observed in obese Zucker rats.[28] The improvement in insulin action in animals with genetic obesity-induced type 2 diabetes indicates that resveratrol is effective not only when insulin resistance is induced by a high-calorie diet.

The improvement in insulin action caused by resveratrol seems to result from different effects. One of them is reduced adiposity. Resveratrol-induced reduction in body fat content was demonstrated in mice and rats on a hypercaloric diet.[24,39–43] Moreover, resveratrol ingestion was found to cause effects that are similar to those induced by calorie restriction.[44,45]

Consistent with animal studies, experiments *in vitro* revealed reduced ATP content[46] and decreased accumulation of triglycerides in isolated rat adipocytes exposed to resveratrol.[47] In these cells, resveratrol increased lipolytic response to epinephrine and decreased lipogenesis.[47] Reduced accumulation of triglycerides in fat cells observed in the presence of resveratrol may contribute to decreased adiposity in the whole organism.

In another study performed on isolated human adipocytes, resveratrol effectively prevented insulin resistance induced by cell exposure to conjugated linoleic acid. In these experiments, insulin-stimulated glucose transport was significantly increased in adipocytes preincubated with conjugated linoleic acid and resveratrol, compared with cells exposed to conjugated linoleic acid alone. Moreover, resveratrol prevented inflammation induced by conjugated linoleic acid. The mechanism of this action is proposed to involve, among others, attenuation of cellular stress, prevention of activation of extracellular signal-related kinase, inhibition of inflammatory gene expression, and increase in peroxisome proliferator-activated receptor γ (PPAR-γ) activity.[48]

In resveratrol-induced improvement in insulin action a pivotal role is ascribed to the activation of Sirt1 and 5′-AMP-activated protein kinase (AMPK). Activation of these enzymes by resveratrol was demonstrated in numerous animal studies.[23,24,34,40,41] The importance of AMPK in the mechanism of resveratrol action was additionally confirmed in experiments on AMPK-deficient mice. In AMPK-deficient mice fed a high-fat diet, resveratrol was ineffective and neither reduced body fat nor improved insulin action.[39] However, in light of the most recent studies of Pacholec *et al.*,[49] the role of Sirt1 in the mechanism of resveratrol action should be reconsidered, since these authors demonstrated that resveratrol is not a direct activator of Sirt1. Although the exact mechanism of resveratrol action is still poorly elucidated, there is no doubt that this compound is able to improve insulin action in different animal models of insulin resistance.

Conclusions

Data from the literature clearly demonstrate that resveratrol exerts pleiotropic action in organisms. The preventive and therapeutic action of this compound in relation to diabetes is complex and involves different effects. Elucidation of these beneficial properties of resveratrol is necessary to enable clinical human studies. Is seems quite possible that resveratrol, alone or in combination with current anti-diabetic therapies, will be used in preventing and treating diabetes.

Conflicts of interest

The authors declare no conflicts of interest.

References

1. American Diabetes Association. 2006. Diagnosis and classification of diabetes mellitus. *Diabetes Care* **29:** S43–S48.
2. Fröjdö, S., C. Durand & L. Pirola. 2008. Metabolic effects of resveratrol in mammals – a link between improved insulin action and aging. *Curr. Aging Sci.* **1:** 145–151.
3. Szkudelska, K. & T. Szkudelski. 2010. Resveratrol, obesity and diabetes. *Eur. J. Pharmacol.* **635:** 1–8.
4. Bloomgarden, Z.T. 2004. Diabetes complications. *Diabetes Care* **27:** 1506–1514.

5. Palumbo, P.J. 2001. Glycemic control, mealtime glucose excursions, and diabetic complications in type 2 diabetes mellitus. *Mayo Clin Proc.* **76:** 609–618.
6. Cheng, A.Y. & I.G. Fantus. 2005. Oral antihyperglycemic therapy for type 2 diabetes mellitus. *CMAJ* **172:** 213–226.
7. Lekli, I., G. Szabo, B. Juhasz, *et al.* 2008. Protective mechanisms of resveratrol against ischemia-reperfusion-induced damage in hearts obtained from Zucker obese rats: the role of GLUT-4 and endothelin. *Am. J. Physiol. Heart Circ. Physiol.* **294:** H859–H866.
8. Sharma, S., C.S. Misra, S. Arumugam, *et al.* 2010. Antidiabetic activity of resveratrol, a known SIRT1 activator in a genetic model for type-2 diabetes. *Phytother. Res.* In press.
9. Su, H.C., L.M. Hung & J.K. Cheng. 2006. Resveratrol, a red wine antioxidant, possesses an insulin-like effect in streptozotocin-induced diabetic rats. *Am. J. Physiol. Endocrinol. Metab.* **290:** 1339–1346.
10. Palsamy, P. & S. Subramanian. 2008. Resveratrol, a natural phytoalexin, normalizes hyperglycemia in streptozotocin-nicotinamide induced experimental diabetic rats. *Biomed. Pharmacother.* **62:** 598–605.
11. Palsamy, P. & S. Subramanian. 2009. Modulatory effects of resveratrol on attenuating the key enzymes activities of carbohydrate metabolism in streptozotocin-nicotinamide-induced diabetic rats. *Chem. Biol. Interact.* **179:** 356–362.
12. Silan, C. 2008. The effects of chronic resveratrol treatment on vascular responsiveness of streptozotocin-induced diabetic rats. *Biol. Pharm. Bull.* **31:** 897–902.
13. Thirunavukkarasu, M., S.V. Penumathsa, S. Koneru, *et al.* 2007. Resveratrol alleviates cardiac dysfunction in streptozotocin-induced diabetes: role of nitric oxide, thioredoxin, and heme oxygenase. *Free Radic. Biol. Med.* **43:** 720–729.
14. Penumathsa, S.V., M. Thirunavukkarasu, L. Zhan, *et al.* 2008. Resveratrol enhances GLUT-4 translocation to the caveolar lipid raft fractions through AMPK/Akt/eNOS signalling pathway in diabetic myocardium. *J. Cell. Mol. Med.* **12:** 2350–2361.
15. Palsamy, P. & S. Subramanian. 2010. Ameliorative potential of resveratrol on proinflammatory cytokines, hyperglycemia mediated oxidative stress, and pancreatic beta-cell dysfunction in streptozotocin-nicotinamide-induced diabetic rats. *J. Cell. Physiol.* **224:** 423–432.
16. Chi, T.C., W.P. Chen, T.L. Chi, *et al.* 2007. Phosphatidylinositol-3-kinase is involved in the antihyperglycemic effect induced by resveratrol in streptozotocin-induced diabetic rats. *Life Sci.* **80:** 1713–1720.
17. Schmatz, R., M.R. Schetinger, R.M. Spanevello, *et al.* 2009. Effects of resveratrol on nucleotide degrading enzymes in streptozotocin-induced diabetic rats. *Life Sci.* **84:** 345–350.
18. Schmatz, R., C.M. Mazzanti, R. Spanevello, *et al.* 2010. Ectonucleotidase and acetylcholinesterase activities in synaptosomes from the cerebral cortex of streptozotocin-induced diabetic rats and treated with resveratrol. *Brain Res. Bull.* **80:** 371–376.
19. Hansen, J.B., P.O. Arkhammar, T.B. Bodvardottir & P. Wahl. 2004. Inhibition of insulinsecretion as a new drug target in the treatment of metabolic disorders. *Curr. Med. Chem.* **11:** 1595–1615.
20. Brown, R.J. & K.I. Rother. 2008. Effects of beta-cell rest on beta-cell function: a review of clinical and preclinical data. *Pediatr. Diabetes* **9:** 14–22.
21. Pinent, M, A Castell, I Baiges, *et al.* 2008. Bioactivity of flavonoids on insulin-secreting cells. *Compr. Rev. Food Sci. F.* **7:** 299–308.
22. Nadal, A., P. Alonso-Magdalena, S. Soriano, *et al.* 2009. The pancreatic β-cell as a target of estrogens and xenoestrogens: implications for blood glucose homeostasis and diabetes. *Mol. Cell. Endocrinol.* **304:** 63–68.
23. Baur, J.A., K.J. Pearson, N.L. Price, *et al.* 2006. Resveratrol improves health and survival of mice on high-calorie diet. *Nature* **444:** 337–342.
24. Lagouge, M., C. Argmann, Z Grhart-Hines, *et al.* 2006. Resveratrol improves mitochondrial function and protects against metabolic disease by activating SIRT1 and PGC-1α. *Cell* **721:** 1–14.
25. Sun, C., F. Zhang, X. Ge, *et al.* 2007. SIRT1 improves insulin sensitivity under insulin-resistant conditions by repressing PTP1B. *Cell Metab.* **6:** 307–319.
26. Ramadori, G., L. Gautron, T. Fujikawa, *et al.* 2009. Central administration of resveratrol improves diet-induced diabetes. *Endocrinology* **150:** 5326–5333.
27. Deng, J.Y., P.S. Hsieh, J.P. Huang, *et al.* 2008. Activation of estrogen receptor is crucial for resveratrol-stimulating muscular glucose uptake via both insulin-dependent and -independent pathways. *Diabetes* **57:** 1814–1823.
28. Rivera, L., R. Morón, A. Zarzuelo & M. Galisteo. 2009. Long-term resveratrol administration reduces metabolic disturbances and lowers blood pressure in obese Zucker rats. *Biochem. Pharmacol.* **77:** 1053–1063.
29. Szkudelski, T. 2006. Resveratrol inhibits insulin secretion from rat pancreatic islets. *Eur. J. Pharmacol.* **55:** 176–181.
30. Szkudelski, T. 2007. Resveratrol-induced inhibition of insulin secretion from rat pancreatic islets: evidence for pivotal role of metabolic disturbances. *Am. J. Physiol. Endocrinol. Metab.* **293:** 901–907.
31. Szkudelski, T. 2008. The insulin-suppressive effect of resveratrol – an *in vitro* and *in vivo* phenomenon. *Life Sci.* **82:** 430–435.
32. Henquin, J.C. 2000. Triggering and amplifying pathways of regulation of insulin secretion by glucose. *Diabetes* **49:** 1751–1760.
33. Maechler, P. 2002. Mitochondria as the conductor of metabolic signals for insulin exocytosis in pancreatic beta-cells. *Cell. Mol. Life Sci.* **59:** 1803–1818.
34. Lee, J.H., M.Y. Song, E.K. Song, *et al.* 2009. Overexpression of SIRT1 protects pancreatic beta-cells against cytokine toxicity by suppressing the nuclear factor-kappaB signaling pathway. *Diabetes* **58:** 344–351.
35. Lenzen, S. 2008. Oxidative stress: the vulnerable β-cell. *Biochem. Soc. Trans.* **36:** 343–347.
36. Robertson, R.P. 2006. Oxidative stress and impaired insulin secretion in type 2 diabetes. *Curr. Opin. Pharmacol.* **6:** 615–619.
37. Kahn, B.B. & J.S. Flier. 2000. Obesity and insulin resistance. *J. Clin. Invest.* **106:** 473–481.

38. Goodpaster, B.H., D.E. Kelley, R.R. Wing, *et al.* 1999. Effects of weight loss on regional fat distribution and insulin sensitivity in obesity. *Diabetes* **48:** 839–847.
39. Um, J.H., S.J. Park, H. Kang, *et al.* 2010. AMP-activated protein kinase-deficient mice are resistant to the metabolic effects of resveratrol. *Diabetes* **59:** 554–563.
40. Shang, J., L.L. Chen & F.X. Xiao. 2008. Resveratrol improves high-fat induced nonalcoholic fatty liver in rats. *Zhonghua Gan. Zang Bing Za. Zhi.* **16:** 616–619.
41. Shang, J., L.L. Chen, F.X. Xiao, *et al.* 2008. Resveratrol improves non-alcoholic fatty liver disease by activating AMP-activated protein kinase. *Acta Pharmacol. Sin.* **29:** 698–706.
42. Rocha, K.K., G.A. Souza, G.X. Ebaid, *et al.* 2009. Resveratrol toxicity: effects on risk factors for atherosclerosis and hepatic oxidative stress in standard and high-fat diets. *Food Chem. Toxicol.* **47:** 1362–1367.
43. Macarulla, M.T., G. Alberdi, S. Gómez, *et al.* 2009. Effects of different doses of resveratrol on body fat and serum parameters in rats fed a hypercaloric diet. *J. Physiol. Biochem.* **65:** 369–376.
44. Barger, J.L., T. Kayo, J.M. Vann, *et al.* 2008. A low dose of dietary resveratrol partially mimics caloric restriction and retards aging parameters in mice. *PloS ONE* **3:** e2264.
45. Pearson, K.J., J.A. Baur, K.N. Lewis, *et al.* 2008. Resveratrol delays age-related deterioration and mimics transcriptional aspects of dietary restriction without extending life span. *Cell Metab.* **8:** 157–168.
46. Szkudelska, K., L. Nogowski & T. Szkudelski. 2010. Resveratrol and genistein as adenosine triphosphate-depleting agents in fat cells. *Metabolism* (In press).
47. Szkudelska, K., L. Nogowski & T. Szkudelski. 2009. Resveratrol, a naturally occurring diphenolic compound affects lipogenesis, lipolysis and antilipolytic action of insulin in isolated rat adipocytes. *J. Steroid Biochem. Mol. Biol.* **113:** 17–24.
48. Kennedy, A., A. Overman, K. Lapoint, *et al.* 2009. Conjugated linoleic acid-mediated inflammation and insulin resistance in human adipocytes are attenuated by resveratrol. *J. Lipid Res.* **50:** 225–232.
49. Pacholec, M., J.E. Bleasdale, B. Chrunyk, *et al.* 2010. SRT1720, SRT2183, SRT1460, and resveratrol are not direct activators of SIRT1. *J. Biol. Chem.* **285:** 8340–8351.

Ann. N.Y. Acad. Sci. ISSN 0077-8923

ANNALS OF THE NEW YORK ACADEMY OF SCIENCES
Issue: *Resveratrol and Health*

Effect of resveratrol on fat mobilization

Clifton A. Baile,[1] Jeong-Yeh Yang,[1] Srujana Rayalam,[1] Diane L. Hartzell,[1] Ching-Yi Lai,[1] Charlotte Andersen,[2] and Mary Anne Della-Fera[1]

[1]University of Georgia, Athens, Georgia. [2]Department of Veterinary Disease Biology, Faculty of Life Sciences, University of Copenhagen, Frederiksberg C, Denmark

Address for correspondence: Clifton A. Baile, 425 River Rd., University of Georgia, Athens, GA 30602. cbaile@uga.edu

Higher levels of body fat are associated with increased risk for development of numerous adverse health conditions. Phytochemicals are potential agents to inhibit differentiation of preadipocytes, stimulate lipolysis, and induce apoptosis of existing adipocytes, thereby reducing adipose tissue mass. Resveratrol decreased adipogenesis and viability in maturing preadipocytes; these effects were mediated not only through down-regulating adipocyte specific transcription factors and enzymes but also by genes that modulate mitochondrial function. Additionally, resveratrol increased lipolysis and reduced lipogenesis in mature adipocytes. In addition, combining resveratrol with other natural products produced synergistic activities from actions on multiple molecular targets in the adipocyte life cycle. Treatment of mice with resveratrol alone was shown to improve resistance to weight gain caused by a high-fat diet. Moreover, dietary supplementation of aged ovariectomized rats with a combination of resveratrol and vitamin D, quercetin, and genistein not only decreased weight gain but also inhibited bone loss. Combining several phytochemicals, including resveratrol, or using them as templates for synthesizing new drugs, provides a large potential for using phytochemicals to target adipocyte adipogenesis, apoptosis, and lipolysis.

Keywords: adipocytes; natural products; rodents; phytochemicals; obesity; osteoporosis

The obesity epidemic

Higher levels of body fat are associated with increased risk for development of numerous adverse health conditions. Currently, more than one billion adults are overweight and at least 300 million are clinically obese.[1] Greater degrees of obesity are associated with increased risks for noninsulin-dependent diabetes mellitus, cancer, cardiovascular disease, and of numerous other adverse health conditions.[2] Furthermore, aging results in relative increases in body fat content accompanied by accumulation of adipocytes in bone marrow, which is directly implicated in age-related bone loss. Bone loss disorders are also on the increase worldwide. In the U.S. today an estimated 55% of people 50 years of age and older are at risk for developing osteoporosis, and osteoporotic fractures alone cost the United States around $17.9 billion per annum.[3]

There are only two anti-obesity medications approved presently in the U.S. for the long-term treatment of obesity, and both drugs have side effects, such as increasing blood pressure and causing dry mouth, constipation, headache, and insomnia.[4] While a number of medications like bisphosphonates, teriparatide, and selective estrogen receptor modulators are available for osteoporosis prevention and therapy in the U.S., there is evidence for side effects with these drugs;[5] moreover, it is difficult to establish the effects of these drugs on disease progression owing to slowly changing primary bone biomarkers and bone mineral density.[6] Due to potentially hazardous side effects of pharmaceuticals and dissatisfaction with high costs, a larger percentage of people in the U.S. are purchasing and exploring the applications of medicinal plants than before.[7]

Various natural molecules are potential agents, and identifying the active chemicals and their molecular targets is one way to find new active compounds to use for modern drug development. Plants have always been a source of drugs, and

doi: 10.1111/j.1749-6632.2010.05845.x

many currently available drugs have been derived directly or indirectly from them. In both 2001 and 2002, approximately one quarter of the best-selling drugs worldwide were natural products or derived thereof.[5] Natural products have been used worldwide as traditional medicines for thousands of years to treat various forms of diseases including obesity-related metabolic disorders. Several studies have revealed that natural products exhibit an extensive spectrum of biological activities such as stimulation of the immune system, antibacterial, antiviral, anti-hepatotoxic, anti-inflammatory, antioxidant, anti-mutagenic, and anti-cancer effects.[6,7] Another motivation for finding beneficial natural agents is epidemiological studies showing benefits of phytochemicals, for example, red wine, soy, and fruit.

Herbal preparations for the treatment of obesity-related disorders and osteoporosis have been used for centuries; however, there is little rigorous testing of natural product formulations for these disorders. Regardless, so-called nutraceuticals are considered a good alternative strategy for the development of effective and safe anti-obesity drugs.[8] Many natural products have anti-obesity effects, natural products such as polyphenols, a major group of plant metabolites widely found in fruits, vegetables, cereals, legumes, and wine. Scientific investigation of compounds from natural products has a history of successful identification of biochemical actions that have been exploited for the treatment of many diseases.[9]

The adipocyte life cycle

The biologic events leading to obesity are characterized by changes in cell properties of adipocytes and may include an increase in the number or size, or both.[10] Adipocytes are derived from mesenchymal stem cells, which have the potential to differentiate into myoblasts, chondroblasts, osteoblasts, or adipocytes. The adipocyte life cycle includes alteration of cell shape and growth arrest, clonal expansion, and a complex sequence of changes in gene expression leading to storage of lipid and, finally, cell death.[11]

During the growth phase, preadipocytes resemble fibroblasts morphologically. PREF-1, a preadipocyte secreted factor, serves as a marker for preadipocytes and is extinguished during adipocyte differentiation.[12] At confluence, preadipocytes enter a resting phase, called "growth arrest," before undergoing the differentiation process. Two transcription factors, CCAAT/enhancer binding protein-alpha (C/EBP-α) and peroxisome proliferator-activated receptor-gamma (PPAR-γ), were shown to be involved in preadipocyte growth arrest required for adipocyte differentiation.[13] Following growth arrest, preadipocytes must receive an appropriate combination of mitogenic and adipogenic signals to continue through the subsequent differentiation steps. During the process of differentiation, preadipocytes undergo one round of DNA replication leading to clonal amplification of committed cells.[14] The induction of differentiation also results in drastic change in cell shape, as the cells convert from fibroblastic to spherical shape.

Resveratrol and adipocyte proliferation, adipogenesis, lipolysis, and apoptosis

Resveratrol (3,5,4′-trihydroxystilbene) has been shown to reduce the synthesis of lipids in rat liver[15] and 3T3-L1 adipocytes.[16] Resveratrol decreased lipid accumulation in maturing 3T3-L1 preadipocytes and decreased cell viability.[17] In mature 3T3-L1 adipocytes, resveratrol increased lipolysis and reduced lipogenesis, contributing to reduced lipid accumulation *in vitro*; in addition, resveratrol decreased cell viability dose dependently and induced apoptosis (Fig. 1).[17,18] Resveratrol has also been shown to inhibit proliferation and adipogenic differentiation in a SIRT1-dependent manner in human preadipocytes. In mature human adipocytes, resveratrol increased basal and insulin-stimulated glucose uptake, but inhibited lipogenesis in parallel with a down-regulation of lipogenic gene expression.[19] Interestingly, Mader *et al.* recently showed that resveratrol synergistically enhanced tumor necrosis factor-alpha (TNF)-related apoptosis-inducing ligand (TRAIL)-induced apoptosis of human adipocytes, and that this effect was independent of SIRT1.[20] Resveratrol has also been shown to decrease proliferation, induce apoptosis, and cell cycle arrest in various other cell lines.[21,22]

Differentiation of preadipocytes and the induction of metabolic pathways related to lipid metabolism includes expression of several adipocyte specific genes like PPARγ, C/EBPα,[23] sterol regulatory element binding proteins-1c (SREBP-1c),[24] fatty acid synthase (FAS),[24] lipoprotein lipase

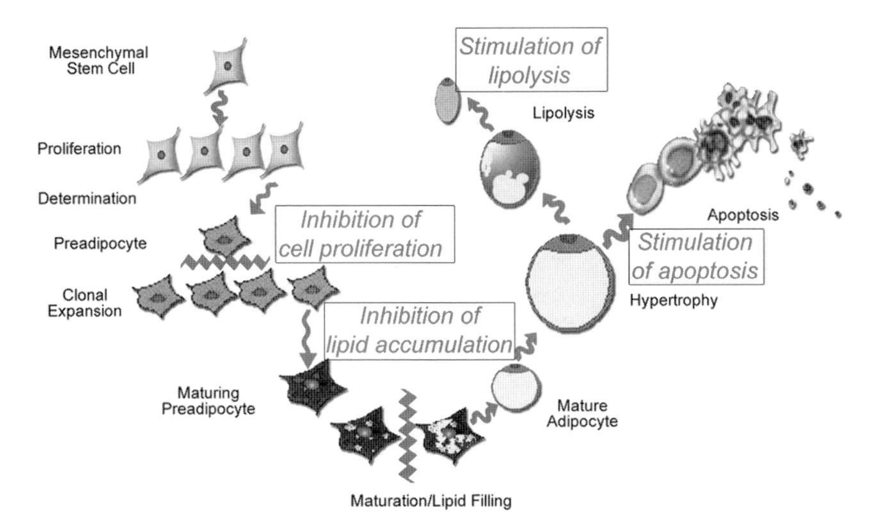

Figure 1. Effect of resveratrol on the adipocyte life cycle. Resveratrol has effects on multiple stages of the adipocyte life cycle, including inhibition of cell proliferation during early preadipocyte development, inhibition of differentiation, and lipid accumulation during later preadipocyte development, promotion of lipolysis in mature adipocytes, and promotion of apoptosis of mature adipocytes.[17–20]

(LPL),[25] and hormone-sensitive lipase (HSL).[26] Resveratrol has been shown to down-regulate the expression of PPARγ, C/EBPα, SREBP-1c, FAS, HSL, and LPL,[17] indicating that resveratrol may alter fat mass by directly affecting biochemical pathways involved in adipogenesis in maturing preadipocytes.

Resveratrol decreases adipogenesis in maturing preadipocytes not only through down-regulating adipocyte-specific transcription factors and enzymes but also by altering expression of genes that modulate mitochondrial function. Resveratrol has been shown to increase the activity of sirtuins, a family of key enzymes in calorie restriction, e.g., the SIR2-family histone deacetylases.[27] Activated SIRT1 further deacetylates the transcriptional coactivator peroxisome proliferator-activated receptor gamma coactivator 1-alpha (PGC-1α) at promoter regions to induce expression of genes involved in mitochondrial biogenesis and fatty acid oxidation.[28]

The mitochondrial sirtuin deacetylase SIRT3 influences mitochondrial function by reducing membrane potential.[29] The expression of SIRT3 was shown to be correlated with uncoupling protein 1 (UCP1), which resides on the mitochondrial inner membrane and mediates adaptive thermogenesis.[29] Mitofusin 2 (MFN2), a mitochondrial membrane protein that participates in mitochondrial fusion in mammalian cells, has been shown to play an important role in glucose oxidation[30] and was also reported to have a role in the pathophysiology of obesity.[31] Resveratrol increased expression of SIRT3, UCP1, and MFN2 *in vitro* in maturing 3T3-L1 preadipocytes.[17]

Resveratrol has also been shown to have specific effects on mature adipocytes that likely contribute to its overall anti-inflammatory effects. For example, resveratrol reduced expression of TNF-α, interleukin-6 (IL-6), and COX-2, key mediators of the inflammatory response in 3T3-L1 adipocytes, and inhibited TNF-α–activated NF-kappaB signaling, thereby reducing cytokine expression.[32] Resveratrol also reversed the TNF-α–induced secretion and mRNA expression of plasminogen activator inhibitor-1 (PAI-1), IL-6, and adiponectin in 3T3-L1 adipocytes.[33] Using conditioned medium from lipopolysaccharide-stimulated macrophages, Kang *et al.*[34] showed that resveratrol pretreatment decreased secretion of TNF-α and IL-6 into the medium, and in 3T3-L1 adipocytes incubated with this conditioned medium, extracellular receptor-activated kinase (ERK) and NF-kappaB (NF-κB) activation was suppressed compared to conditioned medium from macrophages not treated with resveratrol. Furthermore, conditioned medium from resveratrol-treated macrophages modulated insulin signaling transduction by

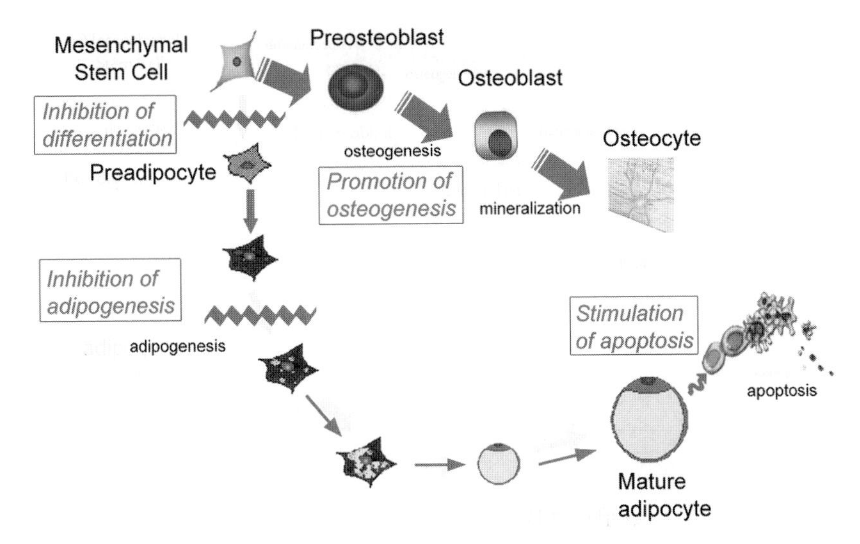

Figure 2. Effect of resveratrol on osteogenic and adipogenic development of mesenchymal stem cells. Resveratrol both inhibits adipogenic differentiation and promotes osteogenic differentiation of mesenchymal stem cells.[17,49–52]

modification of Ser/Thr phosphorylation of insulin receptor substrate-1 and downstream Akt, thus improving insulin sensitivity.[34] In human adipocytes, treatment with resveratrol was also shown to reverse IL-1β–stimulated secretion and gene expression of the proinflammatory adipokines IL-6, IL-8, monocyte chemoattractant protein 1 (MCP-1) 3, and PAI-1.[35]

Anti-obesity effects of resveratrol in rodents

Lagouge *et al.*[36] showed that dietary treatment of mice with either 200 or 400 mg/kg/day resveratrol delivered in either chow or high-fat diets significantly increased their resistance to high-fat diet induced-obesity and their aerobic capacity, substantiated by increased running time and consumption of oxygen in muscle fibers. Resveratrol's effects were associated with induction of genes for oxidative phosphorylation and mitochondrial biogenesis, which protected mice against diet-induced-obesity and insulin resistance. The body weight reduction was the result of less adipose tissue in the resveratrol high-fat diet treated mice, even though food intake was similar to that of the control group. Resveratrol treatment induced marked mitochondrial morphological changes and also increased UCP-1 expression levels in brown adipose tissue.[36] These mitochondria, with activated futile heat-producing pathways, likely contributed to increased energy expenditure and resistance to weight gain. The resveratrol-treated mice had increased oxygen consumption but maintained the same RQ. These mice were also much more resistant to an environmental cold test. This resistance was clearly due to the enhanced mitochondrial activity of the brown adipose tissue. It is interesting to note that energy balance regulatory systems of the mice did not drive the mice to eat more of the high-fat diet to compensate for the increased energy expenditure induced by the resveratrol treatment. Thus, in mice, resveratrol has many of the ideal anti-obesity qualities of long sought after obesity prophylactics and therapies.

It is important to understand how resveratrol regulates metabolism, given its potential as a therapeutic molecule for metabolic disorders. A number of studies have been conducted to identify a central target of resveratrol. While resveratrol has been shown to be a highly potent activator of SIRT1, and to a lesser extent, of yeast SIR2,[37] the role of resveratrol's effects on sirtuins in mediating its metabolic effects has recently been questioned.[38] A recent study suggests that activation of 5' AMP-activated protein kinase (AMPK) may have a primary role in the metabolic changes induced by resveratrol.[39] In this study resveratrol failed to diminish the high-fat diet induced obesity and increase insulin sensitivity,

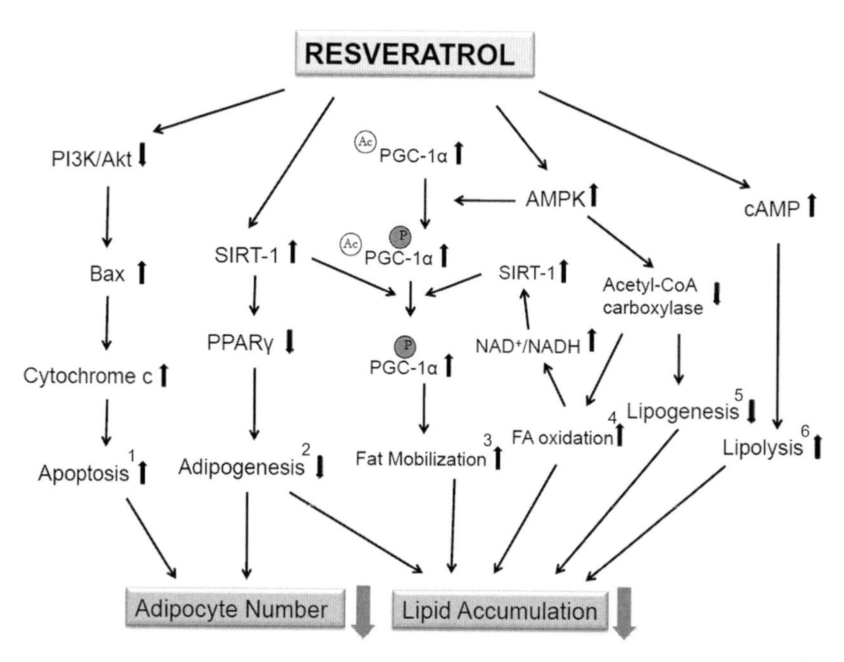

Figure 3. Signaling pathways modulated by resveratrol leading to decreased adipocyte number and diminished lipid accumulation. 1) Resveratrol induced apoptosis in several cell types including adipocytes and cancer cells via modulation of Akt pathway and Bcl-2.[17,58] 2) An increase in SIRT1 activation by resveratrol leads to a decrease in adipogenesis by repressing PPAR-γ.[16] 3) PGC-1alpha plays a role in the processes of differentiation and fat mobilization in cultured adipocytes. AMPK and SIRT1 directly affect PGC-1α activity through phosphorylation and deacetylation, respectively leading to its activation, which in turn promotes fat mobilization.[59,60] 4–5) Resveratrol phosphorylates AMPK and, once activated, AMPK inhibits acetyl-CoA carboxylase enhancing oxidation of fatty acids and decreasing lipid synthesis.[61] 6) Resveratrol increased the lipolytic rate in adipocytes via an increase in cAMP levels.[61]

glucose tolerance, and mitochondrial biogenesis in AMP-activated protein kinase-deficient mice.

Anti-obesity effects of resveratrol in combinations with other phytochemicals

Combining multiple natural products that have synergistic activity resulting from actions on multiple molecular targets in the adipocyte life cycle may offer advantages over treatments with single compounds. Although targeting several signaling pathways simultaneously using phytochemicals to achieve synergistic effects has been investigated for cancer, a similar possibility for obesity or osteoporosis has not been investigated. Several phytochemicals alone or in combination have been shown to enhance effects on promoting adipocyte apoptosis, inhibiting lipid accumulation, and promoting osteogenesis.[17,40–48]

Phytochemicals may act by blocking one or more targets in signal transduction pathways, by potentiating effects of other phytochemicals, or by increasing the bioavailability or half-life of other chemicals. For example, flavonoids such as genistein and resveratrol in combination exert an enhanced effect in 3T3-L1 adipocytes on inhibiting adipogenesis, inducing lipolysis, and triggering apoptosis.[44] Similarly, genistein, quercetin, and resveratrol each produced anti-adipogenic activities in adipocytes; however, in combination, they caused enhanced inhibition of lipid accumulation in both murine 3T3-L1 adipocytes and primary human adipocytes, greater than the responses to individual compounds and to the calculated additive response. In addition, combined treatment with genistein, quercetin, and resveratrol decreased cell viability and induced apoptosis in early- and mid-phase maturing and lipid-filled mature primary human adipocytes. In contrast, no compound alone at the concentrations tested induced apoptosis.[42]

A combination treatment including resveratrol, genistein, quercetin, and vitamin D was tested as a dietary treatment in ovariectomized rats to determine effects on weight gain, adiposity, and bone density. The combination treatment was found to reduce weight gain and adiposity without affecting food intake, compared to both control, and vitamin D alone. Furthermore, trabecular bone density was markedly increased along with a decrease in bone marrow adipocyte density and a decrease in marrow osteoclast number compared to both control and vitamin D alone.[41] Resveratrol alone has been shown to stimulate osteogenesis *in vitro*,[49–52] and to reduce bone loss in several rodent models (Fig. 2).[53–55] Although the combination study of Lai *et al.* did not include treatment with resveratrol alone, *in vitro* studies indicate that resveratrol can act synergistically with other phytochemicals to inhibit adipogenesis.[42] Inhibition of adipogenesis in bone marrow can be particularly beneficial because of the resulting increase in osteogenesis that occurs,[51] as well as the decrease in production of adipocyte-derived factors that stimulate osteoclastogenesis.[56,57] Thus, resveratrol alone or in combination with other phytochemicals with activity in the life cycle of bone forming cells may be beneficial in the treatment of bone loss disorders. Since multiple pathways are dysfunctional in obesity and osteoporosis, an ideal approach for prevention of these disorders would be using combinations of phytochemicals, including resveratrol, to address several targets simultaneously.

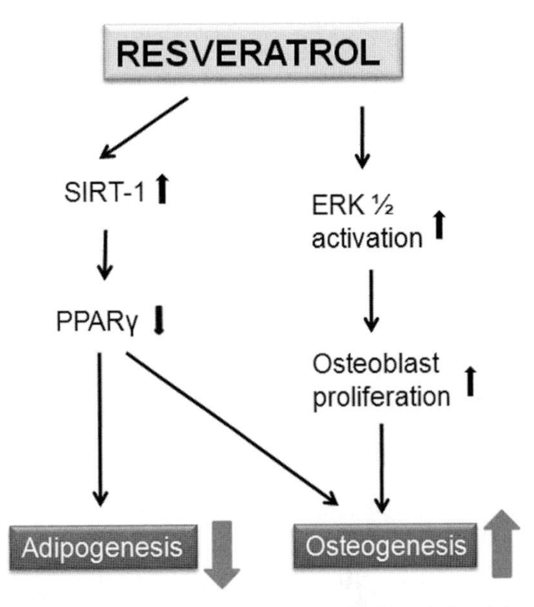

Figure 4. Schematic representation of effects induced by resveratrol leading to a decrease in adipogenesis and an increase in osteogenesis. Activation of SIRT1 by resveratrol inhibits and decreases PPAR-γ, blocks adipocyte development, and increases the expression of osteoblast markers, leading to osteoblast proliferation.[62] Rapid activation of ERK1/2 and p38-MAPK signaling by resveratrol stimulates osteoblastic maturation and osteogenesis in human bone marrow stromal cell cultures.[50]

Summary

Resveratrol has many desired anti-obesity effects on adipocytes in body storage tissues and on bone marrow (Figs. 3 and 4). Resveratrol decreased adipogenesis and viability in maturing preadipocytes, mediated not only through down-regulating adipocyte specific transcription factors and enzymes but also by genes that modulate mitochondrial function. Resveratrol also increased lipolysis and reduced lipogenesis in mature adipocytes. Furthermore, these responses were shown to be greatly enhanced when resveratrol was combined with other phytochemicals. Lagouge, *et al.*,[36] showed that dietary treatment of mice with resveratrol provided the mice protection from obesity induced by the consumption of high-fat diets. Resveratrol-treated mice did not develop insulin resistance and were more tolerant to cold as a result of increased UCP-1 expression levels in brown adipose tissue. Thus, energy expenditure was increased but without an accompanying increased food intake. Dietary supplementation with combinations of resveratrol with vitamin D, quercetin, and genistein inhibited bone loss and decreased adiposity in aged ovariectomized female rats. Combining several phytochemicals, including resveratrol, or using them as templates for synthesizing new drugs, provides an extraordinary potential for using phytochemicals to target adipocyte adipogenesis, apoptosis, and lipolysis for the prevention and treatment of obesity and many associated diseases, including osteoporosis. Recent resveratrol developments are prime examples of these opportunities.

Conflicts of interest

The authors declare no conflicts of interest.

References

1. World Health Organization. World Health Statistics. 2009. Available at: http://www.who.int/whosis/whostat/2009/en/index.html Accessed July 1, 2010.
2. Visscher, T.L. & J.C. Seidell. 2001. The public health impact of obesity. *Annu. Rev. Public Health* **22:** 355–375.
3. Cooper, C. 1999. Epidemiology of osteoporosis. *Osteoporos Int.* **9**(Suppl 2): S2–S8.

4. de Simone, G. & G. D'Addeo. 2008. Sibutramine: balancing weight loss benefit and possible cardiovascular risk. *Nutr. Metab. Cardiovasc. Dis.* **18:** 337–341.
5. Gennari, L., D. Merlotti, F. Valleggi & R. Nuti. 2009. Ospemifene use in postmenopausal women. *Expert Opin. Investig. Drugs* **18:** 839–849.
6. Post, T.M., S.C. Cremers, T. Kerbusch & M. Danhof. 2010. Bone physiology, disease and treatment: towards disease system analysis in osteoporosis. *Clin. Pharmacokinet* **49:** 89–118.
7. Kessler, R.C., R.B. Davis, D.F. Foster, *et al.* 2001. Long-term trends in the use of complementary and alternative medical therapies in the United States. *Ann. Intern. Med.* **135:** 262–268.
8. Mayer, M.A., C. Hocht, A. Puyo & C.A. Taira. 2009. Recent advances in obesity pharmacotherapy. *Curr. Clin. Pharmacol.* **4:** 53–61.
9. Calixto, J.B., M.M. Campos, M.F. Otuki & A.R. Santos. 2004. Anti-inflammatory compounds of plant origin. Part II. modulation of pro-inflammatory cytokines, chemokines and adhesion molecules. *Planta Med.* **70:** 93–103.
10. Flier, J.S. 1995. The adipocyte: storage depot or node on the energy information superhighway? *Cell* **80:** 15–18.
11. Gregoire, F.M. 2001. Adipocyte differentiation: from fibroblast to endocrine cell. *Exp. Biol. Med. (Maywood)* **226:** 997–1002.
12. Wang, Y., K.A. Kim, J.H. Kim & H.S. Sul. 2006. Pref-1, a preadipocyte secreted factor that inhibits adipogenesis. *J. Nutr.* **136:** 2953–2956.
13. Umek, R.M., A.D. Friedman & S.L. McKnight. 1991. CCAAT-enhancer binding protein: a component of a differentiation switch. *Science* **251:** 288–292.
14. Pairault, J. & H. Green. 1979. A study of the adipose conversion of suspended 3T3 cells by using glycerophosphate dehydrogenase as differentiation marker. *Proc. Natl. Acad. Sci. USA* **76:** 5138–5142.
15. Arichi, H., Y. Kimura, H. Okuda, *et al.* 1982. Effects of stilbene components of the roots of Polygonum cuspidatum Sieb. et Zucc. on lipid metabolism. *Chem. Pharm. Bull. (Tokyo).* **30:** 1766–1770.
16. Picard, F., M. Kurtev, N. Chung, *et al.* 2004. Sirt1 promotes fat mobilization in white adipocytes by repressing PPAR-gamma. *Nature* **429:** 771–776.
17. Rayalam, S., J.Y. Yang, S. Ambati, *et al.* 2008. Resveratrol induces apoptosis and inhibits adipogenesis in 3T3-L1 adipocytes. *Phytotherapy Res.* **22:** 1367–1371.
18. Szkudelska, K., L. Nogowski & T. Szkudelski. 2009. Resveratrol, a naturally occurring diphenolic compound, affects lipogenesis, lipolysis and the antilipolytic action of insulin in isolated rat adipocytes. *J. Steroid Biochem. Mol. Biol.* **113:** 17–24.
19. Fischer-Posovszky, P., V. Kukulus, D. Tews, *et al.* 2010. Resveratrol regulates human adipocyte number and function in a Sirt1-dependent manner. *Am. J. Clin. Nutr.* **92:** 5–15.
20. Mader, I., M. Wabitsch, K.M. Debatin, *et al.* 2010. Identification of a novel proapoptotic function of resveratrol in fat cells: SIRT1-independent sensitization to TRAIL-induced apoptosis. *Faseb J.* **24:** 1997–2009.
21. Ferry-Dumazet, H., O. Garnier, M. Mamani-Matsuda, *et al.* 2002. Resveratrol inhibits the growth and induces the apoptosis of both normal and leukemic hematopoietic cells. *Carcinogenesis* **23:** 1327–1333.
22. Liang, Y.C., S.H. Tsai, L. Chen, *et al.* 2003. Resveratrol-induced G2 arrest through the inhibition of CDK7 and p34CDC2 kinases in colon carcinoma HT29 cells. *Biochem. Pharmacol.* **65:** 1053–1060.
23. Lazar, M.A. 2002. Becoming fat. *Genes Dev.* **16:** 1–5.
24. Kim, J.B., H.M. Wright, M. Wright & B.M. Spiegelman. 1998. ADD1/SREBP1 activates PPARgamma through the production of endogenous ligand. *Proc. Natl. Acad. Sci. USA* **95:** 4333–4337.
25. Auwerx, J., P. Leroy & K. Schoonjans. 1992. Lipoprotein lipase: recent contributions from molecular biology. *Crit. Rev. Clin. Lab. Sci.* **29:** 243–268.
26. Sztalryd, C., M.C. Komaromy & F.B. Kraemer. 1995. Overexpression of hormone-sensitive lipase prevents triglyceride accumulation in adipocytes. *J. Clin. Invest.* **95:** 2652–2661.
27. Zhang, J. 2006. Resveratrol inhibits insulin responses in a SirT1-independent pathway. *Biochem. J.* **397:** 519–527.
28. Gerhart-Hines, Z., J.T. Rodgers, O. Bare, *et al.* 2007. Metabolic control of muscle mitochondrial function and fatty acid oxidation through SIRT1/PGC-1alpha. *Embo J.* **26:** 1913–1923.
29. Shi, T., F. Wang, E. Stieren & Q. Tong. 2005. SIRT3, a mitochondrial sirtuin deacetylase, regulates mitochondrial function and thermogenesis in brown adipocytes. *J. Biol. Chem.* **280:** 13560–13567.
30. Mingrone, G., M. Manco, M. Calvani, *et al.* 2005. Could the low level of expression of the gene encoding skeletal muscle mitofusin-2 account for the metabolic inflexibility of obesity? *Diabetologia* **48:** 2108–2114.
31. Zorzano, A., D. Bach, S. Pich & M. Palacin. 2004. [Role of novel mitochondrial proteins in energy balance]. *Rev. Med. Univ. Navarra.* **48:** 30–35.
32. Gonzales, A.M. & R.A. Orlando. 2008. Curcumin and resveratrol inhibit nuclear factor-kappaB-mediated cytokine expression in adipocytes. *Nutr. Metab. (Lond).* **5:** 17–30.
33. Ahn, J., H. Lee, S. Kim & T. Ha. 2007. Resveratrol inhibits TNF-alpha-induced changes of adipokines in 3T3-L1 adipocytes. *Biochem. Biophys. Res. Commun.* **364:** 972–977.
34. Kang, L., W. Heng, A. Yuan, *et al.* 2010. Resveratrol modulates adipokine expression and improves insulin sensitivity in adipocytes: Relative to inhibition of inflammatory responses. *Biochimie* **92:** 789–796.
35. Olholm, J., S.K. Paulsen, K.B. Cullberg, *et al.* 2010. Anti-inflammatory effect of resveratrol on adipokine expression and secretion in human adipose tissue explants. *Int. J. Obes (Lond).* **34:** 1546–1553.
36. Lagouge, M., C. Argmann, Z. Gerhart-Hines, *et al.* 2006. Resveratrol improves mitochondrial function and protects against metabolic disease by activating SIRT1 and PGC-1alpha. *Cell* **127:** 1109–1122.
37. Howitz, K.T., K.J. Bitterman, H.Y. Cohen, *et al.* 2003. Small molecule activators of sirtuins extend Saccharomyces cerevisiae lifespan. *Nature* **425:** 191–196.

38. Kaeberlein, M., T. McDonagh, B. Heltweg, *et al.* 2005. Substrate-specific activation of sirtuins by resveratrol. *J. Biol. Chem.* **280:** 17038–17045.
39. Um, J.H., S.J. Park, H. Kang, *et al.* 2010. AMP-activated protein kinase-deficient mice are resistant to the metabolic effects of resveratrol. *Diabetes* **59:** 554–563.
40. Ambati, S., J.Y. Yang, S. Rayalam, *et al.* 2009. Ajoene exerts potent effects in 3T3-L1 adipocytes by inhibiting adipogenesis and inducing apoptosis. *Phytother. Res.* **23:** 513–518.
41. Lai, C.Y., J.Y. Yang, S. Rayalam, *et al.* 2010. Preventing Bone Loss and Weight Gain with Combinations of Vitamin D and Phytochemicals. *J. Med. Food* In press.
42. Park, H.J., J.Y. Yang, S. Ambati, *et al.* 2008. Combined Effects of Genistein, Quercetin and Resveratrol in Human and 3T3-L1 Adipocytes. *J. Med. Food* **11:** 773–783.
43. Rayalam, S., M.A. Della-Fera, S. Ambati, *et al.* 2007. Enhanced effects of guggulsterone plus 1,25(OH)2D3 on 3T3-L1 adipocytes. *Biochem. Biophys. Res. Commun.* **364:** 450–456.
44. Rayalam, S., M.A. Della-Fera, J.Y. Yang, *et al.* 2007. Resveratrol potentiates genistein's anti-adipogenic and pro-apoptotic effects in 3T3-L1 adipocytes. *J. Nutr.* **137:** 2668–2673.
45. Rayalam, S., J.Y. Yang, M.A. Della-Fera, *et al.* 2009. Anti-obesity effects of xanthohumol plus guggulsterone in 3T3-L1 adipocytes. *J. Med. Food* **12:** 846–853.
46. Yang, J.Y., M.A. Della-Fera, S. Rayalam, *et al.* 2008. Enhanced pro-apoptotic and anti-adipogenic effects of genistein plus guggulsterone in 3T3-L1 adipocytes. *Biofactors* **30:** 159–169.
47. Yang, J.Y., M.A. Della-Fera, S. Rayalam, *et al.* 2008. Enhanced inhibition of adipogenesis and induction of apoptosis in 3T3-L1 adipocytes with combinations of resveratrol and quercetin. *Life Sci.* **82:** 1032–1039.
48. Yang, J.Y., M.A. Della-Fera, S. Rayalam & C.A. Baile. 2008. Enhanced effects of xanthohumol plus honokiol on apoptosis and lipolysis in 3T3-L1 adipocytes. *Obesity (Silver Spring)* **16:** 1232–1238.
49. Mizutani, K., K. Ikeda, Y. Kawai & Y. Yamori. 1998. Resveratrol stimulates the proliferation and differentiation of osteoblastic MC3T3-E1 cells. *Biochem. Biophys. Res. Commun.* **253:** 859–863.
50. Dai, Z., Y. Li, L.D. Quarles, *et al.* 2007. Resveratrol enhances proliferation and osteoblastic differentiation in human mesenchymal stem cells via ER-dependent ERK1/2 activation. *Phytomedicine* **14:** 806–814.
51. Backesjo, C.M., Y. Li, U. Lindgren & L.A. Haldosen. 2009. Activation of Sirt1 decreases adipocyte formation during osteoblast differentiation of mesenchymal stem cells. *Cells Tissues Organs* **189:** 93–97.
52. Kao, C.L., L.K. Tai, S.H. Chiou, *et al.* 2009. Resveratrol promotes osteogenic differentiation and protects against dexamethasone damage in murine induced pluripotent stem cells. *Stem. Cells Dev.* **19:** 247–258.
53. Liu, Z.P., W.X. Li, B. Yu, *et al.* 2005. Effects of trans-resveratrol from Polygonum cuspidatum on bone loss using the ovariectomized rat model. *J. Med. Food* **8:** 14–19.
54. Habold, C., I. Momken, A. Ouadi, *et al.* 2010. Effect of prior treatment with resveratrol on density and structure of rat long bones under tail-suspension. *J. Bone Miner. Metab.* In press.
55. Mizutani, K., K. Ikeda, Y. Kawai & Y. Yamori. 2000. Resveratrol attenuates ovariectomy-induced hypertension and bone loss in stroke-prone spontaneously hypertensive rats. *J. Nutr. Sci. Vitaminol (Tokyo)* **46:** 78–83.
56. Sakaguchi, K., I. Morita & S. Murota. 2000. Relationship between the ability to support differentiation of osteoclast-like cells and adipogenesis in murine stromal cells derived from bone marrow. *Prostaglandins Leukot Essent Fatty Acids* **62:** 319–327.
57. Rosen, C.J. & M.L. Bouxsein. 2006. Mechanisms of disease: is osteoporosis the obesity of bone? *Nat. Clin. Pract. Rheumatol.* **2:** 35–43.
58. Aziz, M.H., M. Nihal, V.X. Fu, *et al.* 2006. Resveratrol-caused apoptosis of human prostate carcinoma LNCaP cells is mediated via modulation of phosphatidylinositol 3'-kinase/Akt pathway and Bcl-2 family proteins. *Mol. Cancer Ther.* **5:** 1335–1341.
59. Canto, C. & J. Auwerx. 2009. PGC-1alpha, SIRT1 and AMPK, an energy sensing network that controls energy expenditure. *Curr. Opin. Lipidol.* **20:** 98–105.
60. Fullerton, M.D. & G.R. Steinberg. 2010. SIRT1 takes a backseat to AMPK in the regulation of insulin sensitivity by resveratrol. *Diabetes* **59:** 551–553.
61. Szkudelska, K. & T. Szkudelski. 2010. Resveratrol, obesity and diabetes. *Eur. J. Pharmacol.* **635:** 1–8.
62. Backesjo, C.M., Y. Li, U. Lindgren & L.A. Haldosen. 2006. Activation of Sirt1 decreases adipocyte formation during osteoblast differentiation of mesenchymal stem cells. *J. Bone Miner. Res.* **21:** 993–1002.

Ann. N.Y. Acad. Sci. ISSN 0077-8923

ANNALS OF THE NEW YORK ACADEMY OF SCIENCES
Issue: *Resveratrol and Health*

Transport, stability, and biological activity of resveratrol

Dominique Delmas,[1,2] Virginie Aires,[1,2] Emeric Limagne,[1,2] Patrick Dutartre,[1,2,3] Frédéric Mazué,[1,2] François Ghiringhelli,[1,4] and Norbert Latruffe[1,2]

[1]Inserm U866, Dijon, France. [2]Université de Bourgogne, Faculté des Sciences Gabriel, Centre de Recherche-Biochimie Métabolique et Nutritionnelle, Dijon, France. [3]COHIRO, Faculté de Médecine, Dijon, France. [4]Université de Bourgogne, Faculté de Médecine, Centre de Recherche-Immunothérapie et Chimiothérapie des Cancers, Dijon, France

Address for correspondence: Dominique Delmas, Inserm UMR866 "Lipids, Nutrition, Cancers," Faculté des Sciences de la Vie, 6, boulevard Gabriel, 21000 Dijon, France. ddelmas@u-bourgogne.fr

Numerous studies have reported interesting properties of *trans*-resveratrol, a phytoalexin, as a preventive agent of several important pathologies: vascular diseases, cancers, viral infections, and neurodegenerative processes. These beneficial effects of resveratrol have been supported by observations at the cellular and molecular levels in both cellular and *in vivo* models, but the cellular fate of resveratrol remains unclear. We suggest here that resveratrol uptake, metabolism, and stability of the parent molecule could influence the biological effects of resveratrol. It appears that resveratrol stability involves redox reactions and biotransformation that influence its antioxidant properties. Resveratrol's pharmacokinetics and metabolism represent other important issues, notably, the putative effects of its metabolites on pathology models. For example, some metabolites, mainly sulfate-conjugated resveratrol, show biological effects in cellular models. The modifications of resveratrol stability, chemical structure, and metabolism could change its cellular and molecular targets and could be crucial for improving or decreasing its chemopreventive properties.

Keywords: resveratrol; transport; stability; resveratrol analogues; metabolites; biotransformation

Introduction

Resveratrol, or 3,4′,5-trihydroxystilbene (Fig. 1), is a secondary metabolite produced in limited plant species and is found in many natural foods (e.g., grapes, red wine, purple grape juice, and some berries).[1] In plants, a major form of resveratrol is *trans*-resveratrol-3-*O*-β-D-glucoside, often referred to as piceid.[2] Like many other plant polyphenols (i.e., flavonoids, epicatechins), resveratrol is considered to be a preventive food component.[3] Indeed, numerous studies have reported that *trans*-resveratrol is a preventive agent of several important pathologies: neurodegenerative processes, viral infections, vascular diseases, and cancers.[4,5] These beneficial effects of resveratrol have been supported by observations at the cellular and molecular levels in cellular and in *in vivo* models; however, resveratrol presents a low bioavailability and a rapid clearance from the circulation.[6] Concerning resveratrol metabolism, studies in human and animals reveal that resveratrol is mainly metabolized into glucuronide and sulfate conjugates, which might retain some biological activity. Recently, five sulfate metabolites have been synthesized and their biological effects tested in a set of assays that are associated with cancer chemopreventive activity. Resveratrol uptake, cellular fate, metabolism, stability of the parent molecule and that of its metabolites need further investigation to gain insight into resveratrol's biological activity and to provide the scientific basis for nutritional or clinical applications. In this review, we will detail (1) the stability of resveratrol, with a special focus on reveratrol's solubility, *cis*-isomerization, and anti- and pro-oxidant properties resulting from its biotransformation; (2) the action of resveratrol metabolites; (3) the current and challenging strategies to increase resveratrol bioavailability, which is determined by its lipophilicity, chemical structure, and site of delivery; and (4) the patents available in databases for resveratrol structure modifications and formulation to

doi: 10.1111/j.1749-6632.2010.05871.x

(A)

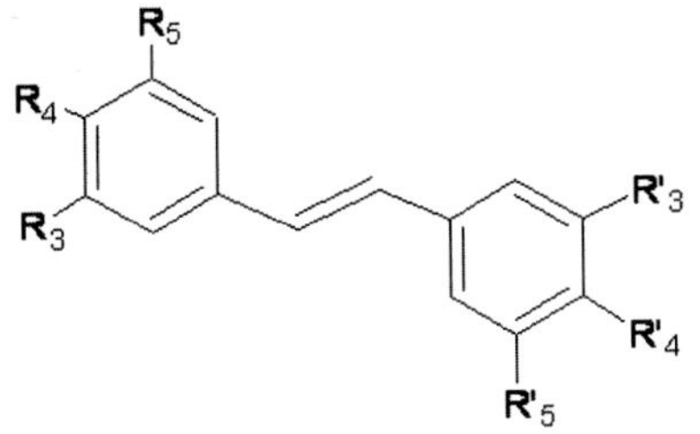

3,5,4'-trihydroxy *trans*-stilbene (*trans*-Resveratrol) (E) $R_3 = R_5 = R'_4 = OH$; $R_4 = R'_3 = R'_5 = H$

3,3',4,5'-tetrahydroxy *trans*-stilbene (Piceatannol) (E) $R_3 = R_4 = R'_3 = R'_5 = OH$; $R'_4 = R_5 = H$

*Trans*Resveratrol-3-O-β-D-glucoside (*trans*-Piceid) (E) $R_5 = R'_3 = R'_4 = OH$; R_3 = O-β-D-glucoside; $R_4 = R'_3 = R'_5 = H$

Trans-Resveratrol-3-O-D-sulfate (E) $R_5 = R'_4 = OH$; $R_3 = SO_3$; $R_4 = R'_3 = R'_5 = H$

Trans-Resveratrol-4'-O-D-sulfate (E)) $R_3 = R_5 = OH$; $R'_4 = SO_3$; $R_4 = R'_3 = R'_5 = H$

Trans-Resveratrol-3-O-D-glucuronide (E) $R_5 = R'_4 = OH$; R_3 = glucuronide; $R_4 = R'_3 = R'_5 = H$

Trans-Resveratrol-4'-O-D-glucuronide (E) $R_3 = R_5 = OH$; R'_4 = glucuronide; $R_4 = R'_3 = R'_5 = H$

Trans-Resveratrol-4'-O-hexanoic acid (E) $R_3 = R_5 = OH$; $R'_4 = O\text{-}(CH_2)_5COOH$; $R_4 = R'_3 = R'_5 = H$

(B)

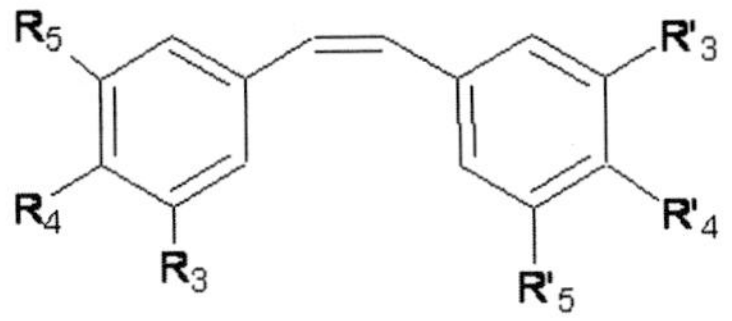

Cis-3,5,4'-trihydroxy stilbene (*Cis*-Resveratrol) (Z) $R_3 = R_5 = R'_3 = R'_4 = OH$; $R_4 = R'_3 = R'_5 = H$

Cis-3,3',4,5'-tetrahydroxystilbene (*Cis*-Piceatannol) (Z) $R_3 = R_4 = R'_3 = R'_5 = OH$; $R'_4 = R_5 = H$

Cis-Resveratrol-3-O-β-D-glucoside (*Cis*-Piceid) (Z) $R_5 = R'_3 = R'_4 = OH$; R_3 = O-β-D-glucoside; $R_4 = R'_3 = R'_5 = H$

Figure 1. Chemical structures of resveratrol, resveratrol metabolites, and resveratrol derivatives in their *trans*- (A) or *cis*- (B) form.

exhibit better activity. The modifications of resveratrol stability, chemical structure, and metabolism could change its cellular and molecular targets and could be crucial for improving or decreasing the efficiency of its chemopreventive properties.

Resveratrol stability and transport, crucial events for its biological action

Role of biophysical parameters in resveratrol stability

Several biophysical parameters of resveratrol could be involved in its activities and uptake, in particular (1) solubility, (2) *cis*-isomerization, and (3) redox reactions.

Resveratrol solubility and aggregation. The solubility and molar absorption spectrum of *trans*- and *cis*-resveratrol isomers in aqueous solvents are poorly described; nevertheless, it appears that the polyphenol has a low water solubility (less than 0.05 mg/mL),[7] which subsequently may affect its distribution in humans. Ethanol is mainly used to increase resveratrol's solubility (50 mg/mL), though DMSO is also used (at least 16 mg/mL). We have shown that ethanol as solvent lowers the threshold of resveratrol antiproliferative effect, compared to DMSO.[8] However, the variation in ethanol percentage, ranging from 0.025% to 0.1% (in the culture medium), did not change the inhibitory effect on cancer cell proliferation.[8] So far, the action mechanism of ethanol has not been elucidated.

Another way to improve resveratrol water solubility and stability can involve different vectors to form a complex with resveratrol. The use of cyclodextrins is promising in this respect. The most important functional property of cyclodextrins is their ability to form inclusion complexes with a wide range of organic molecules.[9] Complex formation of *trans*-resveratrol with β-cyclodextrin ($\Delta G^\circ = -17.01$ kJ/mol) is largely driven by enthalpy ($\Delta H^\circ = -30.62$ kJ/mol) and slight entropy changes ($\Delta S^\circ = -45.68$ kJ/mol). This complexation depends of several factors such as the concentration and type of cyclodextrin or the *trans*-resveratrol pKa values (pKa1 = 8.8, pKa2 = 9.8, pKa3 = 11.4).[10] Knowledge of the protonation state of *trans*-resveratrol is fundamental for understanding its biological activities because when the pH of the medium is higher than the first pKa the biological activity of *trans*-resveratrol will be strongly reduced.

The existence of lipophilic derivatives of polyphenols and flavonoids via esterification of the hydroxyl functions with aliphatic molecules can also be used as a tool to increase their lipohilicity and, therefore, to improve their intestinal absorption and cell permeability.[11] For example, the acetylation of resveratrol, which could increase its absorption,[11] enhances its cellular uptake without loss of activities.[12,13] Similarly, to improve the solubility in water and the binding affinity to human serum albumin (HSA), resveratrol aliphatic acid (resveratrol hexanoic acid) (Fig. 1) has been synthesized and presents a higher binding affinity to HSA.[14,15] For these reasons, resveratrol aliphatic derivatives and solid lipid nanoparticles such as zinc/calcium-pectinate–optimized beads are being developed to improve its stability, its bioavailability, and site of delivery for human applications.

***Cis*-isomerization.** In natural foods, plants, or wine, resveratrol exists naturally as both *cis*- and *trans*-isomers, the *trans* isomer being the major and more stable natural form. *Cis*-isomerization can also occur when the *trans* isoform is exposed to sunlight[16] or to artificial or natural UV[17] radiation at wavelengths of 254 nm[18] or 366 nm.[19] On the basis of their distinct λ_{max} and retention time, HPLC assays can distinguish the two isomers.[17] *Cis*-metabolites have been identified in human urine samples, mainly *cis*-resveratrol-4′-sulfate, *cis*-resveratrol-3-*O*-glucuronide, and *cis*-resveratrol-4′-*O*-glucuronide.[20,21] Most studies have used *trans*-resveratrol for administration due to instability of the *cis* isomer;[16] however, data indicates that both isomers may have different biological effects.[19,22–25]

Redox reactions. As with many other polyphenols, resveratrol may undergo an auto-oxidation process that occurs with many polyphenols, which leads to the production of $O_2^{\cdot -}$, H_2O_2, and a complex mixture of semiquinones and quinines that may be cytotoxic[26]. This has been particularly shown in culture media and may indicate a pitfall in the interpretation of *in vitro* results. The evaluation of resveratrol oxidation *in vitro* shows that 96% of 200 μM resveratrol was degraded, which produced 90 μM of H_2O_2 after 24 h at 37°C.[27] This degradation is markedly and dose-dependently enhanced by bicarbonate ions.[27] In contrast, Long *et al.* show that resveratrol degradation can occur without any H_2O_2 production.[28] Concerning crystalline resveratrol and its glucoside, a recent study shows that both molecules are stable up to three months with negligible degradation in various conditions (e.g., high temperature, ambient fluorescent light, UV/fluorescent light, and air).[29] These auto-oxidation or degradation events could be very important, because oxidized resveratrol could generate complexes with others molecules, such as copper ions. Indeed, the oxidative product of resveratrol is a dimer, and the initial electron transfer generates the reduction of Cu(II) to Cu(I). Thus, the copper-peroxide complex is able to bind DNA and to form a DNA-resveratrol-Cu(II) ternary complex.[30] These complexes would favor and give rise to internucleosomal DNA fragmentation, which is a hallmark of apoptosis.

Role of transporters in resveratrol stability

In order to exert its pleiotropic effects at the cellular and molecular level (e.g., cell cycle arrest, apoptosis), resveratrol must reach and penetrate into cells. To understand how resveratrol gains access to cells, various methods (fluorescence microscopy, flow cytometry, spectrofluorimetry) using the intrinsic fluorescence properties of resveratrol and its metabolites have been developed to follow their intracellular incorporation (Fig. 2). Indeed, on the basis of the spontaneous fluorescence upon UV excitation,[31] we have demonstrated by fluorescence microscopy that this polyphenol is essentially present in the cytoplasm and in the nucleolar region[32] (Fig. 2A). A similar localization was also reported for taxol,[33] a natural product responsible for cell cycle arrest and used as a chemotherapeutic agent. This intracellular localization may be related to the cell cycle

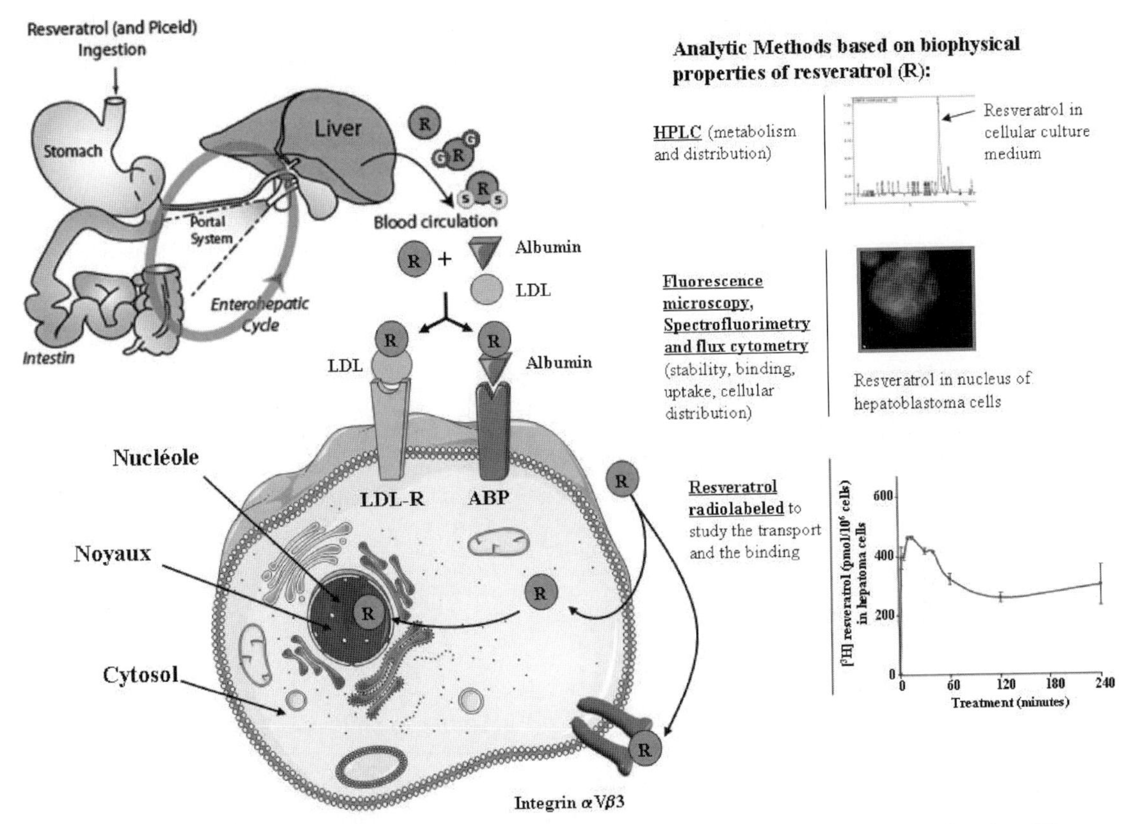

Figure 2. Fate of resveratrol and biophysical properties to analyze resveratrol metabolism and localization. Albumin and lipoproteins (LDL) are important carriers for resveratrol and are very likely to play essential roles in its distribution to the tissues and cells. Futhermore, the absorption of resveratrol could be due to both passive diffusion and complexation with the integrin. Resveratrol metabolites can be detected by HPLC, and the fluorescence properties of resveratrol are important to study the stability, binding, and its distribution.

perturbation previously described in HepG2 resveratrol-treated cells.[8] With spectrofluorimetry excitation spectra of resveratrol, its dimer, ε-viniferin, and their acetylated forms reveals two major peaks: one from 315 to 350 nm, corresponding to aromatic nuclei, and the second broader one from 400 to 450 nm, which corresponds to conjugated double bonds. As expected, ε-viniferin presents high excitation capacity in UV wavelenghts. Emission spectra at $\lambda_{ex.}$ of 330 nm display a large peak in the 350 to 450 nm range for both polyphenols. Their fluorescence intensities are related to the number of phenol groups and their acetylations. Flow cytometry analyses confirm a rapid fluorescence increase during the first 30 min, which is due to polyphenol uptake, and a high fluorescence after 1 h of treatment, which seems not to be due to resveratrol uptake or its metabolism. Indeed, the green emitted fluorescence (520 ± 10 nm) at $\lambda_{ex.}$ of 488 nm does not overlap resveratrol fluorescence and takes place after 24 h of resveratrol treatment. Interestingly, autofluorescence is higher for resveratrol triacetate at a low concentration.[31] Autofluorescence *in situ* measurement shows that resveratrol and related compounds induce deep changes in cell activity. These changes occur mainly by increasing NADPH cell content[31] and the number of green fluorescent cytoplasmic granular structures, which may be related to an induction of detoxifying enzyme mechanisms.

Fluorescent microscopic methods are also convenient for the study of resveratrol derivatives transport, such as resveratrol triacetate and ε-viniferin.[31] Indeed, the higher hydrophobicity provided by acetylation of the phenol groups enhances, as expected, the cellular uptake rate of resveratrol and ε-viniferin. In fact, resveratrol triacetate uptake occurs according to one of two processes: a passive one and a carrier-mediated one.[31] Acetylated forms are hydrolyzed by intracellular esterases and further conjugated, such as to the parent molecules. Furthermore, ε-viniferin, which has a lower antiproliferative effect toward cancer cells,[12,13,31] presents a similar uptake as resveratrol.

To quantify resveratrol uptake, the polyphenol was labeled with tritium in the ortho and para positions of benzenic rings.[32] The time-, dose-, and temperature-dependencies of tritiated resveratrol influx showed a passive diffusion and a carrier-mediated process in hepatoblastoma HepG2 cells.[32] These *in vitro* data are in agreement with the *in vivo* results of Bertelli *et al.*,[34] which have shown an accumulation of resveratrol in the liver of rats after oral administration. Interestingly, the uptake kinetics of resveratrol are similar in human hepatocytes without any toxicity.[32] These observations are relevant since various antineoplasic agents cause a hepatotoxicity that limits their efficacy in anticancer therapy. Indeed, in accordance with previous studies in other cell lines,[35,36] resveratrol appears to have specific cytotoxic effects toward hepatic tumor cells, as compared to normal human hepatocytes.

Numerous events at multiple stages can be implicated in the decrease of resveratrol accessibility to cells, notably its complexation with extracellular molecules, such as (1) serum proteins, (2) fatty acids, (3) lipoprotein, or (4) integrins.

Resveratrol interactions with serum proteins. Owing to its low water solubility, resveratrol must be bound to proteins and/or conjugated to remain at a high concentration in serum. The efficiency of a therapeutic substance is related to its capacity (selectivity and affinity) to bind to protein transporters.[37] Among serum proteins that could play a plasmatic carrier's role, albumin is a good candidate because it binds and carries out amphiphile molecules. Indeed, we have previously shown in human HepG2 cells that resveratrol is able to interact with plasmatic proteins such as albumin,[38] thanks to various binding assays with bovine serum albumin (BSA), radiolabeled resveratrol, gel filtration chromatography, and spectrofluorimetry. We showed that the rate of the passive transport of resveratrol is two-fold lower in serum-containing cell culture medium compared to serum-free medium.[32] Moreover, resveratrol uptake is dose-dependently decreased by the addition of BSA, thus suggesting a role of this major serum protein in its trapping. Recently, Bourassa *et al.*[39] showed that resveratrol binds BSA *via* hydrohilic and hydrophobic interactions with a number of bound resveratrol (n) being of 1.30 and with a binding constant K of 2.52 ($\pm$ 0.5) $\times$ 10^4 mol/L. Resveratrol binding altered BSA conformation with a major reduction of alpha helix and an increase in beta-sheet and turn structures, indicating a partial protein unfolding.[39] The interaction takes place mainly in the vicinity of tryptophanyl residues (Trp 212 and Trp 134) located in the protein domains I and II.[39] These studies are very important when these biochemical data are placed in an *in vivo* context. Indeed, the affinity and the binding of resveratrol with albumin can suggest that albumin could be a natural polyphenol reservoir (Fig. 2). Thus, albumin might play a pivotal role in the distribution and bioavailability of circulating resveratrol. Since resveratrol binds BSA, there is a high probability that the polyphenol binds to HSA, the major plasma protein in humans.[40] Indeed, HSA is able to bind resveratrol and maintain a high concentration in human serum;[15] however, HSA binds resveratrol only when its concentration is high. It seems that HSA has only one binding site for resveratrol, similar to hemoglobin (Hb).[15] The binding constant of resveratrol to HSA is greater than that of resveratrol to Hb, indicating that the affinity of HSA toward this polyphenol is higher than that of Hb. Molecular docking simulations of HSA–resveratrol and Hb–resveratrol show that resveratrol is located in the hydrophobic cavity between subdomain IB and IIA of HSA and in the central cavity of Hb.[15] Furthermore, thermodynamic analyses[15] suggest that hydrophobic interaction plays a major role in the binding of resveratrol to HSA and hydrogen bonding being the main force for binding of resveratrol to Hb. Thus, the distribution of resveratrol in humans will be affected by its binding affinity to HSA. Considering the concentration of HSA is about 40 g/L and Hb is about 140 g/L in plasma both proteins should play important roles in resveratrol transport. Furthermore, Hb can accumulate resveratrol when its concentration is high in serum. Also, HSA and Hb complexation with resveratrol derivatives, such as its acetylated form, could be examined. These results underline the importance of resveratrol binding to serum proteins and especially to albumin. The involvement of a carrier-mediated transport could allow a good uptake of resveratrol at low concentrations, despite its trapping to serum proteins. It may be hypothesized that resveratrol–albumin complexes should be retained by albumin membrane receptors and that resveratrol should be then delivered to the cell membranes, as described for fatty acid transport.

Resveratrol interactions with fatty acids. Fatty acids would ensure a lipophilic environment favorable to the binding of resveratrol. Indeed, spectrofluorimetric method reveals that resveratrol fluorescence exaltation is higher when lipids are bound to albumin.[38] Thus, fatty acids have a positive effect on the binding of resveratrol to BSA. Usually, fatty acids are used as vectors because they have a high affinity for the liver and an efficient cellular uptake resulting from specific interactions with transmembrane transporters (liver plasma membrane-fatty acid binding protein).

Resveratrol interactions with lipoproteins. Proteins other than albumin may also be implicated in the high-affinity fixation of resveratrol. Various red wine anti-oxidants such as resveratrol can interact with lipoproteins.[18,41] *In vitro* assays show that the concentrations of *trans*-resveratrol added to plasma increase with the order of their lipid content, that is, HDL < LDL < VLDL, and that resveratrol is more associated with lipoproteins than with lipoprotein-free proteins.[41] Our observations in colon cancer cells are in agreement with the involvement of lipoproteins in resveratrol transport (Fig. 2). Burkon and Somoza reported that *in vitro* more than 90% of free *trans*-resveratrol is bound to human plasma lipoproteins in a noncovalent manner.[42] This binding occurs also *in vivo*, as the presence of dietary polyphenolic compounds is detected in human LDL isolated from blood samples of healthy volunteers.[20] The involvement of LDL has been recently shown in 11 healthy male volunteers (aged 18–50), nonsmokers who consumed 250 mL of red wine.[20] The LDL samples obtained after 24 h reveal three metabolites (*trans*-resveratrol-3-*O*-glucuronide, *cis*-resveratrol-3-*O*-glucuronide and *cis*-resveratrol-3-*O*-glucoside) and free *trans*-resveratrol.[20] Similarly, when piceid is administrated to healthy volunteers the metabolites *trans*-resveratrol-3-sulfate, *trans*-resveratrol-3,4′-disulfate, *trans*-resveratrol-3,5-disulfate, and *trans*-resveratrol-C/O-diglucuronides were bound noncovalently to plasma protein.[42] Consequently, phenolic compound–LDL complexes would have a greater accessibility to lipid peroxyl radicals within LDL particles, and would be likely to exert their peroxyl scavenging activity in the arterial intima, where oxidation of LDL commonly occurs in microdomains sequestered from plasmatic antioxidants.[43]

Resveratrol interactions with cellular receptors. Although resveratrol uptake in hepatoma cells and hepatocytes involves mainly passive diffusion, resveratrol transport could be mediated by the binding to receptors. Recent reports show that resveratrol can interact with the integrin $\alpha_V\beta_3$ (Fig. 2), principally to the β3 monomer,[44] which is essential for transduction of the stilbene signal into the p53-dependent apoptosis of breast cancer cells[45] or that may contribute to its angiosuppressive activity.[46] Transmembrane receptors cannot explain the pleiotropic effects of resveratrol, as a part of polyphenol uptake involves a passive process. Other intracellular targets could contribute to its intracellular transport, such as the aryl hydrocarbon receptor (AhR). Resveratrol is shown to be a competitive antagonist of dioxin binding to AhR and promotes AhR translocation to the nucleus;[47] however, resveratrol's ability to bind AhR is controversial. Furthermore, resveratrol is able to bind estrogen receptor alpha and beta with comparable affinity, but with a 7,000-fold lower affinity than estradiol.[48] Molecular dynamics studies have shown that the binding of resveratrol to ER-alpha is stereoselective, with a weaker binding of the *cis* form as compared to the *trans* isomer.[49]

Resveratrol metabolites: stability and biological activities

Identification and pharmacokinetics studies of resveratrol metabolites in animal models and in humans show that resveratrol is efficiently absorbed upon oral administration, with detectable levels being recovered both in plasma and urine.[50] Rapid metabolism leads to the production of sulfates, glucuronides, and five distinct metabolites present in the urine: resveratrol monosulfate, two isomeric forms of resveratrol monoglucuronide, dihydroresveratrol monosulfate, and dihydroresveratrol monoglucuronide. The nature and quantity

Table 1. Increase of resveratrol production by plant or cells and extraction methods

Author, patent number, and publication date	Main results
Medina-Boliver and Dolan, M. US2010130623 2010–05-27	Methods for the *in vitro* optimization of polyphenol compounds
Stenhuus, B. *et al.* WO2009124879 2009–10-15	Use of yeast for the fermentation of plants in order to increase stilbene derivative extraction
Bru Martinez *et al.* WO2009106662 2009–09-03	Cyclodextrins are used to capture resveratrol in the cell culture supernatant in order to increase final *in vitro* production
Katz, *et al.* 2009 US2009035839 2009–02-05	Use of a recombinant microorganism for the production of resveratrol and derivatives
Yong, L.B. *et al.* KR20040086962 2004–10-13	Irradiation with UV light and physical aggression of the peanut leads to an increase of resveratrol production (66-fold)
Nam In, H. KR20020002247 2002–01-09	Production of a transgenic tomato able to produce a high content of resveratrol

of metabolites may differ among subjects owing to interindividual variability.[6,50,51] Once in the bloodstream, metabolites can be subjected to phase II metabolism with further conversions occurring in the liver, where entero-hepatic transport in the bile may result in some recycling back to the small intestine.[52] We have already shown in hepatic cells that resveratrol is highly conjugated after 4 h of incubation into mono- (3-sulfate-resveratrol and 4′-sulfate-resveratrol) and disulfate (3,4′-disulfate-resveratrol and 3,5-disulfate- resveratrol) derivatives, but no glucuronide conjugates could be found.[53] Resveratrol is also able to induce its own metabolism by increasing the activity of hepatic detoxifying enzymes of phase II.[53]

Resveratrol metabolites present a plasmatic half-life similar to resveratrol aglycone, but the peak levels of these metabolites were three- to eight-fold higher during 4 h before the urinary elimination phase.[51] The undeniable *in vivo* efficacy of resveratrol, despite its low bioavailability, could be explained by (1) the conversion of resveratrol sulfates and glucuronides back to resveratrol in target organs such as the liver;[54,55] (2) enterohepatic recirculation involving biliary secretion of resveratrol metabolites followed by deconjugation by gut microflora and then reabsorption;[56] and (3) by the activities of its metabolites. In the following sections we summarize the little information available on resveratrol metabolites biological effects in relation with their stability. These aspects are important to clarify in order to fully assess the pharmacokinetics and therapeutical properties of resveratrol.

Piceatannol

Piceatannol (3,5,3′,4′-tetrahydroxystilbene) is a polyphenol found in grapes and other plants[57,58] that differs from resveratrol by additional hydroxyl group in 3′ of benzenic ring (Fig. 1). In humans, piceatannol is produced as a major metabolite of resveratrol by CYP1B1 and CYP1A2.[59,60] Piceatannol can also be found as an oxidation product of resveratrol in aerated aqueous solutions submitted to gamma radiolysis to generate $HO^{\cdot}/O_2^{\cdot -}$ free radicals involved in oxidative stress *in vivo*.[61] Like resveratrol, piceatannol displays anticancer properties by inducing extrinsic and intrinsic pathways of cell death and by blocking cell cycle progression.[62–65]

Resveratrol glucuronides

The most abundant glucuronides metabolites of resveratrol are the *trans*-resveratrol-3-*O*-D-glucuronide and *trans*-resveratrol-4′-*O*-D-glucuronide.[50] In order to understand the important potential role of these glucuronides we have studied their biological activities and their stability. We have investigated the stability of resveratrol glucuronides by analyzing fluorescence spectra under various conditions, that is, by UV exposure darkness/sunlight or air exposure. Emission spectra of glucuronide metabolites are quickly modified by UV exposure. Contrary to resveratrol, the 3-*O* and 4′-*O*-D-glucuronide resveratrol metabolites were without effect on colon

Table 2. Optimization of chemical synthesis of resveratrol and production of new original analogues

Author, patent number, and publication date	Main results	Cancer application
Merritt, *et al.* US2010185006 2010–07-22	Original analogues with increased stability	Not reported
Yoshihiro, *et al.* US2010069486 2010–03-18	Mixture of stilbene derivative and platinium	Claimed and demonstrated *in vivo*
Cossio, F. P. *et al.* US2009298905 2009–12-03	Nitrogenated *trans*-stilbene analogues with an anticancer activity	Claimed
Srivinas, *et al.* US 2009136431 2009–05-28	Use of stilbene derivatives for the inhibition of histone deacetylase inhibitors	Claimed and demonstrated *in vitro*
Lee, R. M. *et al.* WO2008131320 2008–10-30	Original stilbene analogues acting directly on cancer cells but also on neovascularization	Claimed and demonstrated *in vitro*
Rimando, A. *et al.* WO2008070872 2008–06-12	Production of original stilbene derivatives and use for the prevention and treatment of colon cancer	Claimed and demonstrated *in vitro*
Zhu, L.S. CN100368388 2006–09-20	Optimization of synthesis with production of original analogues	Claimed and demonstrated *in vitro*[a]
Munekazu, *et al.* WO03020264 2003–03-13	Stilbene trimer and analogues with powerful activity in the prevention and treatment of cancer	Claimed and demonstrated *in vitro*[a]

[a]Chinese patent no translation available

cancer cell growth, as evidenced by cell proliferation and cell cycle analysis at a concentration of 60 μM (unpublished data). These observations are in accordance with others showing the absence of cytotoxic effect or antiviral activity toward cells infected with HIV-1.[66] However, even if they do not have by themselves biological activity, they may nevertheless constitute an *in vivo* reservoir of resveratrol mobilizable by the action of β-glucuronidase.

Resveratrol sulfates

The biological activity of resveratrol sulfates has been more studied than that of resveratrol glucuronides. Recently, five sulfate metabolites (mono-, di- or trisulfate derivatives) have been synthesized and their biological effects tested in a set of assays that are associated with cancer chemopreventive activity.[67] Among sulfates derivatives, two metabolites showed activity: the 3 and 4′-monosulfate. Indeed, resveratrol-4′-*O*-sulfate is able to bind the cyclooxygenase (COX) sites of the enzymes[67] and inhibit their activities, with nearly the same efficacy of resveratrol,[67,68] as well as NF-κB induction.[67] Resveratrol-3-*O*-sulfate shows a radical scavenger activity, inhibition of COX-1, and cytotoxicity.[67] In human colon cancer cells, resveratrol-3-*O*-sulfate exerts an antiproliferative effect via a cell cycle arrest. Overall, results show that the sulfate metabolites are less active than resveratrol with the exception of resveratrol 3-sulfate. In general, the sulfate metabolites activities decrease as the degree of sulfation increases. Regarding the stability of resveratrol and its metabolites, LC-MS-MS analysis show that the parent molecule is degraded to approximately 20% after an incubation time of 24 h at 37°C, whereas sulfate derivatives are stable.[67] We also observed a marked degradation of resveratrol 3-sulfate upon UV light exposure and certain instability of the molecule when left for 72 h at room temperature in the dark, or exposed to sunlight or air (unpublished data).

Optimization to produce stable resveratrol and galenics forms

We have previously shown that resveratrol solubility or stability can be improved by chemical modifications, but these modifications can also affect resveratrol production and delivery forms. Indeed, a recent series of patents have covered industrial production and new galenic forms in various industrial domains (patent numbers: CN101628 859, CN101591680, and CN101597214).

Table 3. Galenic preparation and mixture of compounds for the increase of stability and/or oral absorption

Author, patent number, and publication date	Main results
Finley, *et al.* US2010197801 2010–08-05	Discovery of a new mint oil by-product able to increase bioavailability of varieous compounds, including resveratrol
Kurhts, E. WO2010062824 2010–06-03	Use of a poloxamer solution to increase solubility and activity of stilbene derivatives
Polans, *et al.* WO 2010/059628 2010–05-27	Use of a poloxamer solution to increase solubility and activity of stilbene derivatives
Yue, Z. *et al.* CN101618021 2010–01-06	Chitosan is incorporated in a slow-release microsphere with a final increase of bioavailability
Peng, C. *et al.* CN10150154 2009–12-02	A new original soft capsule is able to increase oral absorption in humans
Ouyang, W. *et al.* CN101579291 2009 11-18	Preparation of a new nanoemulsion for various applications
Tamura, A. *et al.* JP2009173570 2009–08-06	New formulation and mixture for the increase of intestinal absorption
Rubin, D. WO2009089338 2009–07-16	A chewable carrier is able to increase mouth absorption of compounds, such as resveratrol
Lunsmann, W. *et al.* WO2009089011 2009–07-16	Novel formulations are reported with an increase of resveratrol concentration, leading to a decrease of ingested volume
Todd, Y.M. *et al.* WO2009082459 2009–07-02	Formulation of resveratrol and other drugs able to delay release of the compound *in vivo*, with an increase of bioavailability and activity
Sardi, W.F. WO2009039195 2009–03-26	Prevention of cancer by administration of resveratrol in combination with other drugs to increase "longevity gene products"
Bissery, M.C. *et al.* US2009023656 2009–01-22	Combination of stilbene and VEGF Trap for optimization of anticancer activity

Resveratrol production

Resveratrol can be obtained either by extraction methods or chemical synthesis. In food and the food complement industry, as well as in the field of cosmetics, optimization of resveratrol production and/or extraction methods is classically used and preferred to organic synthesis. The production of resveratrol by plants is increased either after physical chemical stress or after gene recombination.[69] Other approaches of industrial optimization processes using yeast fermentation or optimization of organic extraction are described in a recent series of patents from Chinese companies, confirming the major interest of this country in this research domain. Some examples of these approaches are presented in Table 1. In the pharmaceutical industry, in contrast, organic synthesis is preferred. Optimization of the synthetic process and creation of new structures demonstrating a more powerful biological activity are claimed (Table 2).

Composition and galenic forms of resveratrol

Resveratrol itself can be used in original formulation to increase solubility, stability, and/or intestinal absorption (Table 3). The recent patent from Polans *et al.* (see Table 3) indicates, for example, that the solubilization of resveratrol using a polymer of polyoxyethylene and polyoxypropylene used as surfactant in various other products leads to an increase in *in vitro* as well as *in vivo* activity in animal cancer models. In this patent, the authors reported the absence of some resveratrol metabolites confirming that maintenance of the structure integrity of the parent molecule is particularly important.

Conclusion

In this review, we have reported the current knowledge concerning resveratrol uptake, metabolism, stability of the parent molecule. This report highlights the fact that these events could influence the biological effects of resveratrol. It appears that resveratrol stability involves redox reactions and biotransformations that influence its antioxidant properties. The pharmacokinetics and metabolism of resveratrol represent other important issues, most notably, the putative effects of its metabolites on pathological models. For example, some metabolites, mainly sulfate-conjugated resveratrol, show biological effects in cellular models. How the modifications of stability, chemical structure, resveratrol metabolism, and interactions with other biomolecules could change the cellular and molecular targets needs further investigations.

Acknowledgments

This study was supported by the "Conseil Régional de Bourgogne," the "Ligue Bourguignone contre le Cancer" (especially Nièvre committee), and ANR Grant No. P008641.

Conflicts of interest

The authors declare no conflicts of interest.

References

1. Langcake, P. & R. Pryce. 1977. A new class of phytoalexins from grapevines. *Experentia*. **33:** 1151–1152.
2. Romero-Perez, A.I. *et al.* 1999. Piceid, the major resveratrol derivative in grape juices. *J. Agric. Food Chem.* **47:** 1533–1536.
3. Kris-Etherton, P.M. & C.L. Keen. 2002. Evidence that the antioxidant flavonoids in tea and cocoa are beneficial for cardiovascular health. *Curr. Opin. Lipidol.* **13:** 41–49.
4. Delmas, D., B. Jannin & N. Latruffe. 2005. Resveratrol: preventing properties against vascular alterations and ageing. *Mol. Nutr. Food Res.* **49:** 377–395.
5. Delmas, D. *et al.* 2006. Resveratrol as a chemopreventive agent: a promising molecule for fighting cancer. *Curr. Drug Targets*. **7:** 423–442.
6. Walle, T. *et al.* 2004. High absorption but very low bioavailability of oral resveratrol in humans. *Drug Metab. Dispos.* **32:** 1377–1382.
7. Belguendouz, L., L. Fremont & A. Linard. 1997. Resveratrol inhibits metal ion-dependent and independent peroxidation of porcine low-density lipoproteins. *Biochem. Pharmacol.* **53:** 1347–1355.
8. Delmas, D. *et al.* 2000. Inhibitory effect of resveratrol on the proliferation of human and rat hepatic derived cell lines. *Oncol. Rep.* **7:** 847–852.
9. Laza-Knoerr, A.L., R. Gref & P. Couvreur. 2010. Cyclodextrins for drug delivery. *J. Drug Target* **18:** 645–656.
10. Lopez-Nicolas, J.M. *et al.* 2006. Determination of stoichiometric coefficients and apparent formation constants for beta-cyclodextrin complexes of trans-resveratrol using reversed-phase liquid chromatography. *J. Chromatogr A.* **1135:** 158–165.
11. Riva, S. *et al.* 1998. Enzymatic modification of natural compounds with pharmacological properties. *Ann. N.Y. Acad. Sci.* **864:** 70–80.
12. Colin, D. *et al.* 2009. Effects of resveratrol analogs on cell cycle progression, cell cycle associated proteins and 5fluorouracil sensitivity in human derived colon cancer cells. *Int. J. Cancer* **124:** 2780–2788.
13. Marel, A.K. *et al.* 2008. Inhibitory effects of trans-resveratrol analogs molecules on the proliferation and the cell cycle progression of human colon tumoral cells. *Mol. Nutr. Food Res.* **52:** 538–548.
14. Jiang, Y.L. 2008. Design, synthesis and spectroscopic studies of resveratrol aliphatic acid ligands of human serum albumin. *Bioorg. Med. Chem.* **16:** 6406–6414.
15. Lu, Z. *et al.* 2007. Transport of a cancer chemopreventive polyphenol, resveratrol: interaction with serum albumin and hemoglobin. *J. Fluoresc.* **17:** 580–587.
16. Chen, X. *et al.* 2007. Stereospecific determination of cis- and trans-resveratrol in rat plasma by HPLC: application to pharmacokinetic studies. *Biomed. Chromatogr.* **21:** 257–265.
17. Camont, L. *et al.* 2009. Simple spectrophotometric assessment of the trans-/cis-resveratrol ratio in aqueous solutions. *Anal. Chim. Acta.* **634:** 121–128.
18. Blache, D. *et al.* 1997. Gas chromatographic analysis of resveratrol in plasma, lipoproteins and cells after in vitro incubations. *J. Chromatogr. B Biomed. Sci. Appl.* **702:** 103–110.
19. Basly, J.P. *et al.* 2000. Estrogenic/antiestrogenic and scavenging properties of (E)- and (Z)-resveratrol. *Life Sci.* **66:** 769–777.
20. Urpi-Sarda, M. *et al.* 2007. HPLC-tandem mass spectrometric method to characterize resveratrol metabolism in humans. *Clin Chem.* **53:** 292–299.
21. Zamora-Ros, R. *et al.* 2006. Diagnostic performance of urinary resveratrol metabolites as a biomarker of moderate wine consumption. *Clin Chem.* **52:** 1373–1380.
22. Orallo, F. 2006. Comparative studies of the antioxidant effects of cis- and trans-resveratrol. *Curr Med Chem.* **13:** 87–98.
23. Campos-Toimil, M. *et al.* 2007. Effects of trans- and cis-resveratrol on Ca2+ handling in A7r5 vascular myocytes. *Eur. J. Pharmacol.* **577:** 91–99.
24. Yanez, M. *et al.* 2006. Inhibitory effects of cis- and trans-resveratrol on noradrenaline and 5-hydroxytryptamine uptake and on monoamine oxidase activity. *Biochem. Biophys. Res. Commun.* **344:** 688–695.
25. Mazue, F. *et al.* 2010. Structural determinants of resveratrol for cell proliferation inhibition potency: experimental and docking studies of new analogs. *Eur. J. Med. Chem.* **45:** 2972–2980.
26. Sang, S. *et al.* 2007. Autoxidative quinone formation in vitro and metabolite formation in vivo from tea polyphenol (-)-epigallocatechin-3-gallate: studied by real-time mass

spectrometry combined with tandem mass ion mapping. *Free Radic. Biol. Med.* **43:** 362–371.

27. Yang, N.C., C.H. Lee & T.Y. Song. 2010. Evaluation of resveratrol oxidation in vitro and the crucial role of bicarbonate ions. *Biosci. Biotechnol. Biochem.* **74:** 63–68.
28. Long, L.H., A. Hoi & B. Halliwell. 2010. Instability of, and generation of hydrogen peroxide by, phenolic compounds in cell culture media. *Arch. Biochem. Biophys.* **501:** 162–169.
29. Jensen, J.S., C.F. Wertz & V.A. O'Neill. 2010. Preformulation Stability of trans-Resveratrol and trans-Resveratrol Glucoside (Piceid). *J. Agric. Food Chem.* **58:** 1685–1690.
30. Hadi, S.M. *et al.* 2010. Resveratrol Mobilizes Endogenous Copper in Human Peripheral Lymphocytes Leading to Oxidative DNA Breakage: a Putative Mechanism for Chemoprevention of Cancer. *Pharm. Res.* **27:** 979–988.
31. Colin, D. *et al.* 2008. Antiproliferative activities of resveratrol and related compounds in human hepatocyte derived HepG2 cells are associated with biochemical cell disturbance revealed by fluorescence analyses. *Biochimie.* **90:** 1674–1684.
32. Lancon, A. *et al.* 2004. Human hepatic cell uptake of resveratrol: involvement of both passive diffusion and carrier-mediated process. *Biochem. Biophys. Res. Commun.* **316:** 1132–1137.
33. Guy, R. *et al.* 1996. Fluorescent taxoids. *Chem. Biol.* **3:** 1021–1031.
34. Bertelli, A. *et al.* 1998. Plasma and tissue resveratrol concentrations and pharmacological activity. *Drugs Exp. Clin. Res.* **24:** 133–138.
35. Lu, J. *et al.* 2001. Resveratrol analog, 3,4,5,4′-tetrahydroxystilbene, differentially induces pro-apoptotic p53/Bax gene expression and inhibits the growth of transformed cells but not their normal counterparts. *Carcinogenesis* **22:** 321–328.
36. Gao, X. *et al.* 2002. Disparate in vitro and in vivo antileukemic effects of resveratrol, a natural polyphenolic compound found in grapes. *J. Nutr.* **132:** 2076–2081.
37. Khan, M.A., S. Muzammil & J. Musarrat. 2002. Differential binding of tetracyclines with serum albumin and induced structural alterations in drug-bound protein. *Int. J. Biol. Macromol.* **30:** 243–249.
38. Jannin, B. *et al.* 2004. Transport of resveratrol, a cancer chemopreventive agent, to cellular targets: plasmatic protein binding and cell uptake. *Biochem. Pharmacol.* **68:** 1113–1118.
39. Bourassa, P. *et al.* 2010. Resveratrol, genistein, and curcumin bind bovine serum albumin. *J. Phys. Chem. B.* **114:** 3348–3354.
40. Yang, F. *et al.* 2007. Effect of human serum albumin on drug metabolism: structural evidence of esterase activity of human serum albumin. *J. Struct. Biol.* **157:** 348–355.
41. Belguendouz, L., L. Fremont & M.T. Gozzelino. 1998. Interaction of transresveratrol with plasma lipoproteins. *Biochem. Pharmacol.* **55:** 811–816.
42. Burkon, A. & V. Somoza. 2008. Quantification of free and protein-bound trans-resveratrol metabolites and identification of trans-resveratrol-C/O-conjugated diglucuronides – two novel resveratrol metabolites in human plasma. *Mol. Nutr. Food Res.* **52:** 549–557.
43. Witztum, J.L. 1994. The oxidation hypothesis of atherosclerosis. *Lancet* **344:** 793–795.
44. Lin, H.Y. *et al.* 2006. Integrin alphaVbeta3 contains a receptor site for resveratrol. *Faseb. J.* **20:** 1742–1744.
45. Lin, H.Y. *et al.* 2008. Resveratrol is pro-apoptotic and thyroid hormone is anti-apoptotic in glioma cells: both actions are integrin and ERK mediated. *Carcinogenesis.* **29:** 62–69.
46. Belleri, M. *et al.* 2008. alphavbeta3 Integrin-dependent antiangiogenic activity of resveratrol stereoisomers. *Mol Cancer Ther.* **7:** 3761–3770.
47. Casper, R.F. *et al.* 1999. Resveratrol has antagonist activity on the aryl hydrocarbon receptor: implications for prevention of dioxin toxicity. *Mol Pharmacol.* **56:** 784–790.
48. Bowers, J.L. *et al.* 2000. Resveratrol acts as a mixed agonist/antagonist for estrogen receptors alpha and beta. *Endocrinology* **141:** 3657–3667.
49. Abou-Zeid, L.A. & A.M. El-Mowafy. 2004. Differential recognition of resveratrol isomers by the human estrogen receptor-alpha: molecular dynamics evidence for stereoselective ligand binding. *Chirality* **16:** 190–195.
50. Cottart, C.H. *et al.* 2010. Resveratrol bioavailability and toxicity in humans. *Mol. Nutr. Food Res.* **54:** 7–16.
51. Boocock, D.J. *et al.* 2007. Phase I dose escalation pharmacokinetic study in healthy volunteers of resveratrol, a potential cancer chemopreventive agent. *Cancer Epidemiol. Biomarkers Prev.* **16:** 1246–1252.
52. Crozier, A., I.B. Jaganath & M.N. Clifford. 2009. Dietary phenolics: chemistry, bioavailability and effects on health. *Nat Prod Rep.* **26:** 1001–1043.
53. Lancon, A. *et al.* 2007. Resveratrol in human hepatoma HepG2 cells: metabolism and inducibility of detoxifying enzymes. *Drug Metab. Dispos.* **35:** 699–703.
54. Wenzel, E. & V. Somoza. 2005. Metabolism and bioavailability of trans-resveratrol. *Mol. Nutr. Food Res.* **49:** 472–481.
55. Vitrac, X. *et al.* 2003. Distribution of [14C]-trans-resveratrol, a cancer chemopreventive polyphenol, in mouse tissues after oral administration. *Life Sci.* **72:** 2219–2233.
56. Marier, J.F. *et al.* 2002. Metabolism and disposition of resveratrol in rats: extent of absorption, glucuronidation, and enterohepatic recirculation evidenced by a linked-rat model. *J. Pharmacol. Exp. Ther.* **302:** 369–373.
57. Rimando, A.M. *et al.* 2004. Resveratrol, pterostilbene, and piceatannol in vaccinium berries. *J. Agric. Food Chem.* **52:** 4713–4719.
58. Roupe, K. *et al.* 2004. Determination of piceatannol in rat serum and liver microsomes: pharmacokinetics and phase I and II biotransformation. *Biomed. Chromatogr.* **18:** 486–491.
59. Potter, G.A. *et al.* 2002. The cancer preventative agent resveratrol is converted to the anticancer agent piceatannol by the cytochrome P450 enzyme CYP1B1. *Br. J. Cancer.* **86:** 774–778.
60. Piver, B. *et al.* 2004. Involvement of cytochrome P450 1A2 in the biotransformation of trans-resveratrol in human liver microsomes. *Biochem Pharmacol* **68:** 773–782.
61. Camont, L. *et al.* 2010. Liquid chromatographic/electrospray ionization mass spectrometric identification of the oxidation end-products of trans-resveratrol in aqueous solutions. *Rapid Commun Mass Spectrom.* **24:** 634–642.

62. Kimura, Y. & H. Okuda. 2000. Effects of naturally occurring stilbene glucosides from medicinal plants and wine, on tumour growth and lung metastasis in Lewis lung carcinoma-bearing mice. *J. Pharm. Pharmacol.* **52:** 1287–1295.
63. Wieder, T. *et al.* 2001. Piceatannol, a hydroxylated analog of the chemopreventive agent resveratrol, is a potent inducer of apoptosis in the lymphoma cell line BJAB and in primary, leukemic lymphoblasts. *Leukemia* **15:** 1735–1742.
64. Wolter, F. *et al.* 2002. Piceatannol, a natural analog of resveratrol, inhibits progression through the S phase of the cell cycle in colorectal cancer cell lines. *J. Nutr.* **132:** 298–302.
65. Liu, W.H. & L.S. Chang. 2010. Piceatannol induces Fas and FasL up-regulation in human leukemia U937 cells via Ca2+/p38alpha MAPK-mediated activation of c-Jun and ATF-2 pathways. *Int J Biochem Cell Biol.* **42:** 1498–1506.
66. Wang, L.X. *et al.* 2004. Resveratrol glucuronides as the metabolites of resveratrol in humans: characterization, synthesis, and anti-HIV activity. *J. Pharm. Sci.* **93:** 2448–2457.
67. Hoshino, J. *et al.* 2010. Selective synthesis and biological evaluation of sulfate-conjugated resveratrol metabolites. *J. Med. Chem.* **53:** 5033–5043.
68. Calamini, B. *et al.* 2010. Pleiotropic mechanisms facilitated by resveratrol and its metabolites. *Biochem J.* **429:** 273–282.
69. Soleas, G.J., E.P. Diamandis & D.M. Goldberg. 1997. Resveratrol: a molecule whose time has come? And gone? *Clin Biochem.* **30:** 91–113.

Ann. N.Y. Acad. Sci. ISSN 0077-8923

ANNALS OF THE NEW YORK ACADEMY OF SCIENCES
Issue: *Resveratrol and Health*

Chemoprevention in experimental animals

Nalini Namasivayam
Department of Biochemistry and Biotechnology, Annamalai University, Annamalainagar, Tamil Nadu, India

Address for correspondence: Nalini Namasivayam, Department of Biochemistry & Biotechnology, Annamalai University, Annamalainagar, Tamil Nadu, India. nalininam@yahoo.com

The potential cancer-preventive effects of resveratrol, evident from the data obtained by various studies, are summarized in this review. Resveratrol (*trans*-3,5,4′-trihydroxystilbene), a naturally occurring polyphenolic compound, was first isolated in 1940 as a constituent of the roots of white Hellebore (*Veratrum grandiflorum* O. Loes), and is now found to be present in various plants including grapes, berries, peanuts, and red wine. This review first briefly describes the current evidence on the link between resveratrol and cancer occurrence, based on epidemiological studies. Subsequently, investigations with resveratrol in animal models of colon carcinogenesis are presented, followed by a comprehensive compilation of resveratrol on cancer. In the second part, the article focuses on results from investigations on cancer-preventive mechanisms of resveratrol. Biological activities including antioxidant effects, modulation of carcinogen metabolism, anti-inflammatory potential, antioxidant properties, antiproliferative mechanisms by induction of apoptosis, and cell differentiation are discussed. Some novel information on its modulating effects on cell signaling pathway, metabolism studies, bioavailability, and cancer-preventive efficacy is also provided. Based on these findings, resveratrol may be used as a promising candidate for cancer chemoprevention.

Keywords: resveratrol; chemoprevention; experimental animals

Introduction

Cancer chemoprevention

Cancer chemoprevention can be defined as the prevention, inhibition, or reversal of carcinogenesis by administration of one or more chemical entities, either as individual drugs or as naturally occurring constituents of the diet.[1] The concept of chemoprevention, which is the use of natural or synthetic compounds to block, reverse, or prevent the development of cancers, has great appeal. There are at least two major mechanisms for cancer chemoprevention.[2,3] One is antimutagenesis. It includes the inhibition of the uptake of carcinogens, the formation/activation of carcinogens, the deactivation/detoxification of carcinogens, the blocking of carcinogen-DNA bindings, and the enhancement of fidelity of DNA repair. Another mechanism is antiproliferation/antiprogression. Examples are the modulation of hormone/growth factor activity, the modification of signal transduction, the inhibition of oncogene activity, the promotion of the cellular differentiation, the modulation of arachidonic acid metabolism, and the enhancement of apoptosis.[4]

Classes of chemopreventive agents

Absolute classification of chemopreventive agents is difficult because the precise mechanisms of action are not known for many compounds. In addition, many chemopreventive agents act through more than one mechanism, making it difficult to establish the most effective mode of action. On this basis, the chemopreventive agents are classified as (1) inhibitors of carcinogen formation; (2) blocking agent-inhibitors of tumor initiation; and (3) suppressing agent-inhibitors of tumor promotion/progression.[5]

Phytochemicals. Several studies have demonstrated that generous consumption of vegetables and fruits reduces the risk of colon cancer.[6] Although the nature of the constituents that are responsible for reduced risk has not been fully elucidated, it is clear that plant foods contain chemopreventive agents, including several micronutrients, such

doi: 10.1111/j.1749-6632.2010.05873.x

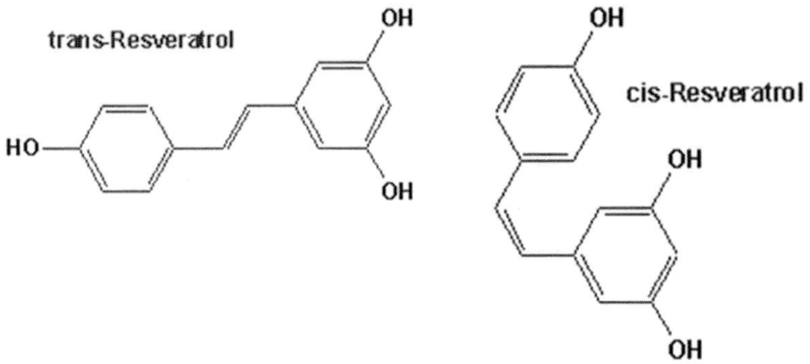

trans-Resveratrol
cis-Resveratrol

Figure 1. Structure of resveratrol.

as vitamins and minerals, and also nonnutrients, such as organosulfur compounds, polyphenols, and isoflavones, to cite a few. The diversity of these compounds is a positive feature, indicating that a variety of approaches to cancer prevention by these agents may be made so that the optimal selection will emerge.

History of resveratrol

Resveratrol (3,5,4′-trihydroxystilbene) is a naturally occurring phytoalexin produced by a variety of plants such as grapes (*Vitis vinifera*), peanuts (*Arachis hypogaea*), and mulberries in response to stress, injury, ultraviolet (UV) irradiation, and fungal (e.g., *Botrytis cinerea*) infection. Although phytoalexins have long been inferred to be important in the defense of plants against fungal infection, few reports show that they provide resistance to infection. Several plants, including grapevine, synthesize the stilbene-type phytoalexin, resveratrol, when attacked by pathogens.[7]

Chemistry. Resveratrol is found widely in nature, and a number of its natural and synthetic analogues, isomers, adducts, derivatives, and conjugates are known. It is an off-white powder with a melting point of 253–255° C and molecular weight of 228 kDa. Resveratrol is insoluble in water but dissolves in ethanol, carboxymethylcellulose, and dimethylsulphoxide. The stilbene-based structure of resveratrol consists of two phenolic rings linked by a styrene double bond to generate 3,5,4′-trihydroxy-*trans*-stilbene (Fig. 1). Although the presence of the double bond facilitates *trans*- and *cis*-isomeric forms of resveratrol [(E)- and (Z)-diastereoisomers, respectively], the *trans*-isomer is sterically the most stable form.[8]

Metabolism, pharmacokinetics, tissue distribution, and clearance. Numerous studies have examined the metabolism, pharmacokinetics, tissue distribution, and clearance of resveratrol.[9,10] Bertelli *et al.* studied the plasma kinetics and tissue bioavailability of this compound after oral administration in rats.[9] Resveratrol concentrations were measured in the plasma, heart, liver, intestine, and kidneys. Tissue concentrations of resveratrol showed a significant bioavailability and strong affinity for the liver and kidneys. The majority of the absorbed resveratrol was conjugated to yield resveratrol glucuronide (16.8%) and lesser amounts of resveratrol sulfate (3%). Only a small amount of resveratrol was absorbed and metabolized across the enterocyte of the jejunum and ileum. These findings suggest that resveratrol is most likely to be in the form of glucuronide conjugate after crossing the small intestine and entering the blood circulation. This may account for its significantly higher levels in the plasma, liver, intestine, and colon. Moreover, abundant *trans*-resveratrol-3-*O*-glucuronide and *trans*-resveratrol-3-*O*-sulfate were identified in rat urine. Virtually no unconjugated resveratrol was detected in the urine or serum samples.[10]

Biological and pharmacological effects. Several *in vivo* and *in vitro* studies describe various biological functions of resveratrol. Resveratrol is notable for its diverse biological actions in preclinical models at a very wide range of physiologically attainable and supraphysiological doses (Table 1).

Toxicity. To date, few studies have evaluated the toxicity of resveratrol in animals. A single dose of 2,000 mg resveratrol/kg body weight (bw) did not cause any detectable, toxicologically significant

Table 1. Biological effects of resveratrol

Therapeutic activities of resveratrol	References
Antibacterial and antifungicidal activities	Creasy and Coffee[81]
Antioxidant activity	Chanvitayapongs *et al.*[82]
Free radical scavenging activity	Belguendouz *et al.*[83]
Inhibition of lipid peroxidation	Frankel *et al.*[84]
Inhibition of eicosanoid synthesis	Kimura *et al.*[85]
Inhibition of platelet aggregation	Chung *et al.*[86]
Chelation of copper	Belguendouz *et al.*[83]
Anti-inflammatory property	Jang *et al.*[14]
Vasorelaxing activity	Chen and Pace-Asciak[87]
Modulation of lipids and lipoprotein metabolism	Frankel *et al.*[84]
Inhibition of rat gastric H^+, K^+-ATPase	Murakami *et al.*[88]
Inhibition of protein-tyrosine kinase and protein kinase C	Jayatilake *et al.*[89]

changes in rats.[11] Moreover, Crowell *et al.* have reported that oral administration of 300 mg resveratrol/kg bw for 28 days was nontoxic.[12]

Anticancer activity. Besides many other biological properties, resveratrol exhibits anticancer properties, as suggested by its ability to suppress cell proliferation in a wide variety of tumor cells, including lymphoid and myeloid cancers, multiple myeloma, cancers of the breast, prostate, stomach, colon, pancreas, and thyroid, melanoma, head and neck squamous cell carcinoma, ovarian carcinoma, and cervical carcinoma. In general, the anticancer activity of resveratrol has been attributed to the suppression of cell proliferation and induction of apoptosis.[13] Moreover, Jang *et al.* proved that resveratrol acts as an effective chemopreventive agent against mouse skin cancer model by blocking cancer development and formation at various stages, including initiation, promotion, and progression.[14] Recently, resveratrol was found to block the development of pre-neoplastic lesions in carcinogen-treated mouse mammary glands.[15] Although the anticarcinogenic function of resveratrol has been well-established to some degree, the mechanism by which resveratrol exerts its chemopreventive effects remain largely unknown. At present, many key molecular targets associated with anticancer activity has been revealed (Fig. 2).

Clinical trials. Many data are available from *in vitro* studies, experimental animal model studies, and few clinical trials in humans on the anticancer effects of resveratrol. Gautam *et al.* found that the *ex vivo* origin of contaminating tumor cells may reduce the incidence of relapse in patients undergoing bone marrow transplantation and demonstrated that resveratrol exhibits antileukemic activity.[16] A study by Wang *et al.* suggested that resveratrol significantly inhibits *in vivo* platelet aggregation induced by collagen.[17] Resveratrol also causes an increase in plasma adenosine levels and blood nucleosides in human subjects.[18]

Several reports show that the cancer preventive activities of resveratrol could be attributed to its ability to trigger apoptosis in carcinoma cells.[19,20] Resveratrol is metabolized by the enzyme cytochrome P450 1B1 (CYP1B1), found in a variety of different tumors, to form an antileukemic agent, piceatannol.[21,22] This observation provides a novel explanation for the cancer preventive property of resveratrol.

Cancer chemopreventive effect

Many studies have revealed the cancer chemopreventive and/or therapeutic potential of resveratrol. Jang *et al.* showed that resveratrol influences antiproliferative effects on human breast epithelial cells.[14] Carbo *et al.* have demonstrated that resveratrol administration to male Wistar rats inoculated with Yoshida AH-130 ascites hepatoma tumors resulted in a significant decrease in tumor cell content, and this response was found to be associated with a G2/M phase arrest and apoptosis.[23] Elattar and Virji[20,21] have shown that resveratrol induces significant dose-dependent inhibition in human oral squamous carcinoma cell (SCC-25) growth and DNA synthesis.[24,25] Resveratrol is also known to reduce the viability and DNA synthesizing capability of human promyelocytic leukemia (HL-60) cells via induction of apoptosis through the BCl-2 pathway.[26] Hsieh and Wu[9] investigated the effects of resveratrol on growth, induction of apoptosis, and modulation of prostate-specific gene

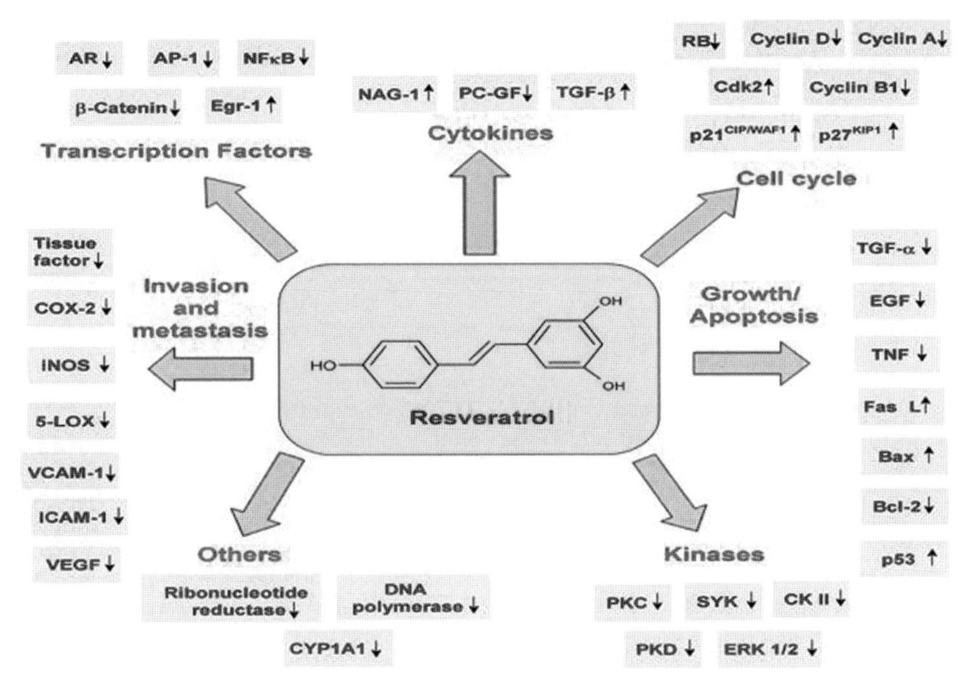

Figure 2. Molecular targets of resveratrol (Aggarwal *et al.*[90]).

expression using DU-145, PC-3, and JCA-1 human CaP cells.[27] This study suggests that resveratrol negatively modulates CaP cell growth by affecting mitogenesis as well as inducing apoptosis in a prostate cell type-specific manner. Resveratrol has also been shown to regulate PSA gene expression by an AR-independent mechanism.[27] In another study, Mitchell *et al.* demonstrated the inhibitory effects of resveratrol on androgen action in LNCaP cells.[28] This study observed that resveratrol represses different classes of androgen-regulated genes at the protein or mRNA level, including PSA, human glandular kallikrein-2, AR-specific coactivator ARA70, and the WAF1/p21.[28] In another study, Kampa *et al.* have shown that many antioxidant polyphenols present in wine, including resveratrol, inhibits the proliferation of human prostate cancer cell lines.[29]

Colon cancer chemopreventive effects

Resveratrol on pre-neoplastic changes

Reliable intermediate biomarkers for colon carcinogenesis need to be identified in order to use them to evaluate several agents for their carcinogenic or potential chemopreventive efficacy against colon tumors.[30] Aberrant crypt foci (ACF) are microscopic lesions that have been postulated to precede the development of adenomas and are considered as earliest premalignant lesions in colon carcinogenesis.[31] Furthermore, specificity studies showed that several other colon-specific carcinogens such as 1,2-dimethylhydrazine, 4-aminobiphenyl, *N*-nitroso-*N*-methylurea and 3-methylcholanthrene also induce ACF, indicating that development of these lesions in the colon is clearly related to the genotoxic events.[32] Multiplicity of ACF increases with time and appears to be a predictor of tumor outcome.[33] These observations justify the use of the colonic ACF assay as a useful tool for the evaluation of agents with potential chemopreventive properties in colon cancer prior to clinical studies.

Resveratrol was supplemented to 1,2-dimethylhydrazine-induced colon cancer rats to evaluate its effect on the preneoplastistic changes. The number of aberrant crypts (AC), AC/ACF (crypt multiplicity) and percentage of ACF inhibition in experimental groups were determined. Rats treated with the carcinogen showed 100% ACF incidence, in contrast to the complete lack of such lesions in control groups (Table 2). The number of ACF consisting of six or more aberrant crypts per rat in resveratrol-supplemented groups [group 6 (6.2 ± 1.4), group

Table 2. Resveratrol on aberrant crypt foci formation in rat colon

Groups	ACF incidence (%)	Total no. of ACF	No. of AC	Crypt multiplicity (AC/ACF)	Percent of ACF inhibition
DMH	10/10 (100)	100.3 ± 10.2^{a}	180.5 ± 14.5^{a}	1.8 ± 0.08^{a}	—
DMH + RES (I)	10/10 (100)	50.4 ± 5.3^{b}	85.6 ± 7.6^{b}	1.7 ± 0.08^{ab}	49.75
DMH + RES (PI)	10/10 (100)	39.4 ± 5.4^{c}	63.0 ± 20.9^{c}	1.6 ± 0.08^{b}	60.71
DMH + RES (EP)	10/10 (100)	$28.5 \pm 5.1^{d*}$	$37.0 \pm 14.3^{d*}$	$1.3 \pm 0.07^{c*}$	71.58

Data are presented as means ± SD of 10 rats in each group.
$^{a-d}P < 0.05$; values not sharing a common superscript letter are significantly different.
$^{*}P < 0.01$; values are significantly different as compared to DMH-alone–treated group.

5 (7.7 ± 1.0), and group 4 (8.2 ± 1.4)] were significantly ($P < 0.05$) lower than that of carcinogen-alone–treated rats (22.3 ± 2.4) (Table 3). The statistically significant ($P < 0.01$) reduction in the total number of ACF was higher in rats supplemented with resveratrol for the entire period (group 6).[34] Resveratrol supplementation for the entire period of the study caused a significant decrease in the total number of ACF, AC, and AC/ACF with increased percentage of inhibition.[34]

Total number of ACF, number of larger ACF and number of AC/ACF (crypt multiplicity) were used to evaluate the potency of colon cancer preventive agents. Accordingly, Sengottuvelan *et al.* also showed that resveratrol not only inhibited the growth of ACF by decreasing the total number of ACF consisting of various number of crypts (small, medium and large), but also inhibited its distribution in proximal, middle and distal regions of colon.[34] ACF develops as early as 2–4 weeks after carcinogen administration and appears predominantly in the medial colon during early time points. But as time progresses, ACF appears in the distal and proximal colon, and a proportion of ACF starts to exhibit focal expansion and may contain one to several crypts.[35] Resveratrol supplementation throughout the study period to colon cancer rats suppressed the formation of ACF in the distal colon, suggesting that resveratrol may intervene in the development of ACF at a later time point. These findings suggest that resveratrol suppresses early events (development of smaller ACFs) in colon carcinogenesis and also the formation of tumors.

In several studies, reduction in the number of multiple crypts (four or more), total, and regional distribution of ACF were used to define efficacy end points to predict the sensitivity and specificity of potential chemopreventive agents.[36] The expression of larger ACF (six or more aberrant crypts per focus) is considered more likely to progress to tumors.[37,38] In this study, resveratrol feeding showed a significant inhibition on the formation of larger ACF in the distal colon.

Hyperplasia was assessed by counting the number of cells per crypt column (crypt height). The number of cells in the crypt column was decreased on resveratrol supplementation. In this context, the decrease in crypt height might be correlated with a significant decrease in hyperplastic lesions.[39]

Although the mechanisms involved in the protective effects of resveratrol against ACF and tumor formation are not clearly understood, the inhibitory actions of resveratrol could be explained as follows: resveratrol is known to (i) affect bax and $p21^{CIP}$ expression in both ACF and surrounding mucosa; (ii) inhibit enzymes such as ribonucleotide reductase and DNA polymerases; (iii) modulate protein kinase C and cyclooxygenase-2 (COX-2) expression; (iv) inhibit ROS-mediated carcinogenesis; (v) inhibit tumor cell division; (vi) activate apoptotic cell death; and (vii) reduce carcinogen-induced luminal mutations.

Antioxidant activity of resveratrol

Cancer can be inhibited at different stages of its development. Induction of antioxidants and detoxifying enzymes by anticarcinogens appear to be a form of adaptation to metabolic stress.[40] An inverse relationship between the concentration of lipid peroxides and the rate of cell proliferation[41] and differentiation[42] is well documented. Moreover, a number of studies have demonstrated that tumor cells

Table 3. Resveratrol on ACF distribution in proximal, middle, and distal rat colon

ACF Distribution	DMH	DMH+RES (I)	DMH+RES (PI)	DMH+RES (EP)
Proximal colon	11.7 ± 0.9^a	6.7 ± 0.6^b	4.2 ± 1.3^c	1.7 ± 0.6^d
Small	9.2 ± 0.6	4.1 ± 0.3	3.2 ± 1.0	1.0 ± 0.4
Medium	1.5 ± 0.2	1.4 ± 0.2	0.7 ± 0.2	0.7 ± 0.2
Large	1.0 ± 0.1	1.2 ± 0.1	0.3 ± 0.1	–
Middle colon	30.1 ± 3.2^a	14.3 ± 1.6^b	12.3 ± 1.5^c	10.2 ± 1.6^d
Small	11.1 ± 1.2	8.3 ± 1.0	8.2 ± 1.0	2.8 ± 0.4
Medium	15.3 ± 1.9	5.0 ± 0.4	2.6 ± 0.3	5.3 ± 0.9
Large	3.7 ± 0.1	1.0 ± 0.2	1.5 ± 0.2	2.1 ± 0.3
Distal colon	58.5 ± 6.1^a	29.4 ± 3.1^b	22.9 ± 2.6^c	16.6 ± 2.9^d
Small	15.3 ± 1.6	13.3 ± 1.9	6.2 ± 0.3	8.3 ± 1.5
Medium	25.6 ± 3.5	10.1 ± 0.5	10.8 ± 1.6	4.2 ± 0.5
Large	17.6 ± 1.0	6.0 ± 0.7	5.9 ± 0.7	4.1 ± 0.9

Data are presented as means ± SD of 10 rats in each group.
$^{a-d}P < 0.01$; values not sharing a common superscript letter are significantly different.

have reduced levels of phospholipids and polyunsaturated fatty acids (PUFA). The low content of PUFA in tumor cells can be attributed to the loss or decreased activity of δ-6- and δ-6–5- desaturases,[43] lending support to the concept that the rate of lipid peroxidation is generally low in tumor cells.

The enzymes SOD and CAT and the glutathione system play key roles in the cellular defense against free radical damage.[44] Many data indicate that animal tumor cells lack complex enzyme systems, which normally exert protection by scavenging toxic oxygen species such as superoxide radicals, hydrogen peroxide, and lipid hydroperoxides.[45] The antioxidant activity against DMH-induced colon cancer showed a significant decrease in the activities of SOD, CAT, and GR, while the activities of glutathione dependent enzymes such as GPx and GST were significantly increased (almost doubled) in carcinogen-treated rats and served as markers of neoplastic tissues.[44] Colonic mucosal GSH (ubiquitous cellular reductant) levels were lowered in carcinogen-treated rats, which suggests that this tripeptide may be involved in the detoxification and possible repair mechanisms in the colonic mucosa.[45]

SOD, CAT, GSH, and GR replenishment (increase) upon resveratrol supplementation throughout the experimental period reflects a favorable balance between potentially harmful oxidants and protective antioxidants. Furthermore, elevated SOD and CAT activities can play an inhibitory role on cell transformation. CAT has been found to significantly decrease chromosomal aberrations and also delay or prevent the onset of spontaneous neoplastic transformation in mouse fibroblasts and epidermal keratinocytes.[46] The available reports suggest that the anticarcinogenic effect of resveratrol may be mediated by the induction of GSH because this endogenous tripeptide molecule can detoxify various carcinogens, serve as an intracellular antioxidant, and also regulate DNA and protein synthesis.[47]

Many findings report that decreased CAT activity in tumor cells is somehow compensated by an increase in GPx activity, which in turn prevents tumor cells from peroxidative attack. The overexpression of GST enhances the production of eicosanoids, another common attribute observed in many tumors.[47] Furthermore, the ratio among these antioxidant enzymes is important, as any imbalance will result in the accumulation of toxic free radicals that cause cell damage.[44] Previous findings firmly established GST inhibition as one of the major mechanisms to explain the chemopreventive efficacy of phytochemicals.[48] Reduction in GPx and GST activities on resveratrol supplementation shows that resveratrol may play a role in maintaining the balance between these antioxidant enzymes, which is in harmony with the previous reports.[48] However, the changes in GPx and GST enzyme activities might be due to the malignant state, and recovery of the enzyme activities could help to reverse malignancy.

Several studies have demonstrated that resveratrol exhibits a wide range of biological activities including anti-inflammatory and antitumor effects.[14] Polyphenols and flavonoids can prevent oxidative damage by their ability to scavenge reactive oxygen species such as hydroxyl radical and superoxide anion. The cytotoxic action of resveratrol against cancer cells may be through mobilization of endogenous copper and the consequent prooxidant action, which might be one of the mechanisms involved in ROS mediated tumor cell apoptosis and cancer chemoprevention.[49] Thus the significant increase in tissue (intestine and colon) lipid peroxidation products observed on chronic resveratrol supplementation may be correlated with its pro-oxidant property.[47]

Resveratrol on bacterial enzymes

Epidemiological studies and laboratory research have indicated a strong association between the metabolic activity of the intestinal microflora and cancer of the large bowel.[50] The activation of procarcinogens could be mediated enzymatically by intestinal bacteria and the activities of colonic bacterial enzymes are increased by dietary fat.[50] Increased expression of intestinal mucosal β-glucuronidase, β-glucosidase, and β-galactosidase in a population with a high risk of colon cancer is well documented. In addition, it has been reported that *E. coli*, a β-glucuronidase positive bacterium, increases the production of active carcinogenic metabolites in the colon and is thus responsible for colon carcinogenesis.[51]

Several other enzymes such as nitroreductase and sulphatases have also been implicated in the carcinogenic process, which are known to retoxify and release carcinogens in the gastrointestinal tract.[52] The carcinogens, on being reduced, may be converted into highly reactive intermediates, which in-turn can react with proteins and nucleic acids.[53] Fecal sulfatase activity should also be considered in the desulfation of conjugated toxins and in the degradation of sulfated mucins. Changes in the expression of sulfated molecules such as mucins and other glycoconjugates have been demonstrated in transformed colonic epithelial cells.[54] A change in mucinase activity is accompanied by a change in the rate of mucin degradation, leading to a shift in the balance between mucin secretion and degradation.[53,67] Enhanced degradation of the mucosal lining of colonic epithelial cells (mucin) ensures greater contact of the toxic carcinogen with the colonocytes. This may be accompanied by increased susceptibility of the colonic cells to being transformed.[53]

Measurement of colonic and fecal biotransforming enzyme activities in carcinogen exposed rats was found to be higher by several-fold compared to the control rats. The influence of diet on tumor development and carcinogen retoxifying enzymes have been evaluated for their influence on colon carcinogenesis.[55] The activities of these bacterial enzymes were significantly decreased following resveratrol supplementation, especially when 8 mg/kg body weight was supplemented throughout the study period.[56] Apart from antioxidant and antiproliferative properties, resveratrol is also known to modulate cytochrome P450 metabolic activity, thus preventing the activation of procarcinogenic substrates to carcinogens. This may just as well explain the protective action of resveratrol on carcinogen-induced colon carcinogenesis.

Resveratrol is also known to have strong antimicrobial activity.[11] Thus, the success of resveratrol supplementation is credited in part to its antibacterial effects. Furthermore, Onoue *et al.* suggested that reduction of bacterial enzyme activity is paralleled by a decrease in the frequency of colonic ACF.[49] Thus, one of the plausible explanations for the reduction in tumor incidence and ACF development may be associated with the reduced activities of fecal and colonic mucosal enzymes on resveratrol supplementation to carcinogen-treated rats. A strategy for colon-specific drug delivery is another way of releasing bioactive compounds/molecules into the colon. Resveratrol is readily absorbed and immediately glucuronidated and sulfated by intestinal cells and/or by the liver.[9] This conjugation makes resveratrol more hydrophilic and easily accessible to the intracellular targets. Nevertheless, hydrolytic enzymes such as β-glucuronidase and sulfatases are expressed at high levels in the extracellular space of certain bulky tumors, including ovarian cancers.[57,58] These enzymes may be capable of converting the resveratrol metabolites back to *trans*-resveratrol, providing tumor-selective bioactivation and a sufficient concentration of active drug to induce autophagy.

Antiproliferative activities of resveratrol

Ornithine decarboxylase (ODC) is found to be increasingly expressed in a variety of cancers and is considered to be important for carcinogenesis. ODC has been evaluated as an intermediate biomarker of cell proliferation in cancer chemopreventive studies.[59] Carcinogen exposure to rats resulted in increased expression of colonic ODC compared to the control group.[60] A significant decrease of ODC activity observed on resveratrol supplementation[61,74] could be, at least in part, related to the accumulation of cells at the S/G2 phase transition and suppression of the activity of nuclear transition factors, which are involved in a number of different signaling pathways associated with proliferation, differentiation, neuronal excitation, and cell death.

Colonic epithelial cell proliferation changes are considered to be indicators of increased risk of colon cancer.[62] Many studies showed that morphometric analysis of argyrophilic nucleolar organizing region-associated protein (AgNOR) was used as a marker for cell proliferation, which aids in the identification of cells with neoplastic potential. At 30 weeks, AgNOR expression was visualized as black dots in silver-stained histological sections. The number of AgNOR dots was consistently enhanced in non-lesional colonic crypts of carcinogen-treated rats, which reflects increased cell proliferation.[60]

Our data indicating a reduced number of AgNORs/nucleus during the entire period of resveratrol supplementation suggest that the mechanism of reduction of ACF number by resveratrol may be by reducing increased cell proliferation of the colonic epithelium.[60] Resveratrol downregulates the expression of cyclins D1 and D2, which are directly involved in cell cycle progression,[63] generally stimulated during malignancy,[64] and repressed by anticancer phytochemicals.[65] Resveratrol is also known to significantly reduce expression of transcription factors, including DP-1, involved in the control of cell proliferation.[66] In addition, resveratrol is known to possess antiproliferative and cell cycle-arresting properties *in vitro*, in several cancer cell lines.[13] Furthermore, resveratrol can inhibit several enzyme activities associated with cell proliferation and DNA replication.

Proapoptotic effects of resveratrol

Colon cancer reflects one or more disturbances in colonic tissue homeostasis. Changes in the rates of colonocyte proliferation, apoptosis, or both are involved in colon tumorigenesis. Indeed, a defect in an apoptosis mechanism is recognized as an important cause of carcinogenesis.[68] Dysregulation of proliferation alone is not sufficient for cancer formation; suppression of apoptotic signaling is also needed.[67] The proapototic effector molecule caspase 3 plays an important role in downstream apoptotic signaling by cleaving proteins vital for cell survival, by activating poly (ADP-ribose) polymerase (PARP), and by activating DNA fragmentation factors.[68] Following carcinogen administration, colonic expression of caspase 3 was downregulated. Several studies show that induction of apoptotic processes by activation of apoptosis-regulating molecules such as caspases can induce cancer cell death.[69] The mechanism responsible for the induction of caspase 3 by resveratrol is not known; however, it is possible that inhibition of cholesterol formation by resveratrol may alter the integrity of cell membranes (cell/mitochondrial) of tumor cells, thereby leading to activation of the proapoptotic effector molecules, including caspase 3. Dorrie *et al.* showed that resveratrol induced extensive apoptosis by depolarizing mitochondrial membranes and activating caspase cascade in several cancer cell lines.[70] Induction of p53 at the mRNA and protein levels is a commonly observed effect of resveratrol and may be considered a major cause of apoptosis.

Heat shock proteins (HSPs) are a family of chaperones induced by heat and other stresses; these proteins thus serve essential functions under stress (and nonstress) conditions. Several HSPs are differentially expressed and/or regulated during cell cycle and at various stages of development and differentiation.[71] The normal cellular functions of HSP70 and HSP27 are not completely understood. Over-expression of HSP70, an anti-apoptotic heat shock protein, may lead to accumulation of damaged cells and, therefore, enhanced cancer risk. Over-expression of HSP27 may increase cell tumorigenicity, possibly as a result of a drastic decrease in tumor cell apoptosis.[71] Furthermore, Lee *et al.* have shown that over-expression of HSP70 may be associated with abnormal p53 expression by tumor cells.[72] Accumulation of HSP27 in colon cancer cells reaching confluence is involved in their resistance to cytotoxic drugs. Moreover, over-expression of HSP70, with low levels of caspase 3 activity, is known to play an antiapoptotic role in malignant human tumors of various origins. It has been shown that HSP70 over-production can ameliorate apoptotic cell death and inhibit caspase 3 activation, leading to reduced apoptosis.[73]

Induction and activation of some HSPs in tumor cells may be controlled by several regulatory factors, including oncogenes such as c-myc, ras, and tumor suppressor genes (e.g., p53). In this regard, Sengottuvelan *et al.*[61] investigated the effect of resveratrol on HSP70 and HSP27 expression in carcinogen-induced rat colon carcinogenesis. An elevated expression of HSP70 and HSP27 in the colonic mucosa of carcinogen-exposed rats was observed at the end of 30 weeks.[61] Resveratrol supplementation suppressed the accumulation of HSP70 and HSP27 in carcinogen-treated rats. The inhibitory effect was most prominent when resveratrol was supplemented throughout the experimental period. This could be attributed to its antitumor activity, thus highlighting the chemotherapeutic potential of resveratrol, in addition to its chemopreventive activity.

Resveratrol on cell signaling and inflammatory markers

Protein kinase C (PKC), a serine/threonine-specific kinase known to exist as at least nine isoenzymes with different properties and subcellular localizations, appears to participate in signaling pathways involved in cell proliferation, differentiation, apoptosis, and malignant transformation.[74] The changes in PKC distribution, followed by a decrease in cytosolic kinase expression in carcinogen-exposed rats, is consistent with an apparent down-regulation of PKC, which has been noted in several cultured cells treated with tumor promoters. Kahl-Rainer *et al.* have suggested that PKC may diffuse from the cytoplasm to the nucleus of the cell and thereby transmit signals from the cytosolic side of the plasma membrane to the nucleus and induce malignant transformation.[75] Doi *et al.* have also recently speculated that altered regulation of PKC, or a process closely linked to this phenomenon, may be important in the process of multi-stage carcinogenesis.[76]

1,2-Dimethylhydrazine (DMH) is known to stimulate colonic epithelial cell proliferation, and it is possible that the PKC alterations noted in the previous experiments are secondary to proliferative changes induced by DMH. The signaling events leading to apoptotic cell death upon exposure to antiproliferative agents has emerged as a target candidate for chemopreventive agents. Sengottuvelan *et al.*[77,78] have shown that resveratrol suppresses the translocation and overexpression of membrane PKC in carcinogen-exposed rats. This action of resveratrol may be due to the suppression of colonic mucosal turnover of phosphoinositides and diacylglycerol mass or to its modulatory action on PKC enzyme expression.[77,78] This inhibitory effect could also be explained, in part, by the antiproliferative and antioxidant property of resveratrol as other phenolic antioxidants inhibit phorbol ester-mediated activation of PKC.

The enzyme cyclooxygenase-2 (COX-2), an inducible early-response protein, plays an important role in inflammation and in carcinogenesis.[79] Assay of COX-2 expression can be used to monitor the process of carcinogenesis, and suppression of COX-2 expression has become a target for cancer chemoprevention. It is known that COX-2 activity is elevated in carcinogen-induced rodents and in human colorectal tumors. Sengottuvelan *et al.* showed that the expression pattern of COX-2 was elevated in DMH-treated rats.[78] Over-expression of COX-2 can result in the inhibition of programmed cell death in prostate cancer cells. Evidence from *in vitro* and animal model studies suggests that the COX-2 inhibition may suppress carcinogenesis by affecting/promoting a number of pathways such as angiogenesis, tumor invasion, and apoptosis. Resveratrol supplementation inhibited the over-expression of COX-2 in carcinogen-treated rats. Several studies have shown that inhibition of COX-2 activity in different cancer models is effective in the treatment of cancer initiation, promotion, and progression.[80] Reduced expression of COX-2 by the colonic cells may be due to the anti-oxidant, anti-inflammatory, or immunomodulatory effects of resveratrol. In addition, decreased COX-2 expression on resveratrol supplementation may also be attributed to the ability of resveratrol to inhibit cell proliferation.

Conclusion

Many phenolic compounds present in food and vegetables are known to possess potent and desirable biological activities against cancer and cardiovascular disease. The most universal property is related to their functions as antioxidants, manifested by their ability to trap free radicals. There are several mechanisms that may account for the biological properties of resveratrol, such as anticancer, anti-hyperlipidemic, and anti-inflammatory effects. The possible signal transduction pathways inhibited by resveratrol may include (1) scavenging ROS; (2) inhibition of cell proliferation; (3) induction of apoptosis; or (4) inhibition of protein kinases. Based on the findings that signal transduction may be affected by resveratrol, further studies are needed to determine how effective resveratrol or its analogues are in preventing or treating inflammation, cardiovascular disease, and cancer.

Conflicts of interest

The author declares no conflicts of interest.

References

1. Morse, M.A. & G.D. Stoner. 1993 Cancer chemoprevention: principles and prospects. *Carcinogenesis* **14:** 1737–1746.
2. Kelloff, G.J. 1999. Perspectives on cancer chemoprevention research and drug development, *Adv. Cancer Res.* **78:** 199–334.
3. Shureiqi, P.R. & D.E. Brenner. 2000.Chemoprevention: general perspective. *Crit. Rev. Oncol. Hematol.* **33:** 157–167.
4. Kohli, M., J. Yu, C. Seaman, A. Bardelli, K.W. Kinzler, B. Vogelstein, C. Lengauer, L. Zhang. 2004. SMAC/Diablo-dependent apoptosis induced by nonsteroidal antiinflammatory drugs (NSAIDs) in colon cancer cells. Proc. *Natl. Acad. Sci. USA* **101:** 16897–16902.
5. Stoner, G.D., M.A. Morse & G.J. Kelloff. 1997. Perspectives in cancer chemoprevention. *Environ. Health Perspect.* **105:** 945–954.
6. Safe, S., M.J. Wargovich, C.A. Lamartiniere, H. Mukhtar. 1999. Symposium on mechanisms of action of naturally occurring anticarcinogens. *Toxicol. Sci.* **52:** 1–8.
7. Fremont, L. 2000. Biological effects of resveratrol. *Life Sci.* **66:** 663–673.
8. Trela, B. & A. Waterhouse. 1996. Resveratrol: isomeric molar absorptivities and stability. *J. Agric. Food Chem.* **44:** 1253–1257.
9. Bertelli, A.A., L. Giovannini, R. Stradi, *et al.* 1998. Evaluation of kinetic parameters of natural phytoalexin in resveratrol orally administered in wine to rats. *Drugs Exp. Clin. Res.* **24:** 51–55.
10. Yu, C., Y.G. Shin, A. Chow, *et al.* 2002. Human, rat, and mouse metabolism of resveratrol. *Pharm. Res.* **19:** 1907–1914.
11. Ashby, J., H. Tinwell, W. Pennie, A.N. Brooks, P.A. Lefevre, N. Beresford, & J.P. Sumpter. 1999. Partial and weak oestrogenicity of the red wine constituent resveratrol: Consideration of its superagonist activity in MCF-7 cells and its suggested cardiovascular protective effects. *J. Appl. Toxicol.* **19:** 39–45.
12. Crowell, J.A., P.J. Korytko, R.L. Morrissey, *et al.* 2004. Resveratrol-associated renal toxicity. *Toxicol. Sci.* **82:** 614–619.
13. Hsieh, T.C. & J.M. Wu. 1999. Differential effects on growth, cell cycle arrest, and induction of apoptosis by resveratrol in human prostate cancer cell lines. *Exp. Cell. Res.* **249:** 109–115.
14. Jang, M., L. Cai, G.O. Udeani, *et al.* 1997. Cancer chemopreventive activity of resveratrol, a natural product derived from grapes. *Science* **275:** 218–220.
15. Bhat, K.P., D. Lantvit, K. Christov, *et al.* 2001. Estrogenic and antiestrogenic properties of resveratrol in mammary tumor models. *Cancer Res.* **61:** 7456–7463.
16. Gautam, S.C., Y.X. Xu, M. Dumaguin, *et al.* 2000. Resveratrol selectively inhibits leukemia cells: a prospective agent for ex vivo bone marrow purging. *Bone Marrow Transplant.* **25:** 639–645.
17. Wang, Z., J. Zou, Y. Huang, *et al.* 2002. Effect of resveratrol on platelet aggregation in vivo and in vitro. *Chin. Med. J.* **115:** 378–380.
18. Goldberg, D.M., J. Yan & G.J. Soleas. 2003. Absorption of three wine-related polyphenols in three different matrices by healthy subjects. *Clin. Biochem.* **36:** 79–87.
19. Kuo, P.L., L.C. Chiang & C.C. Lin. 2002. Resveratrol induced apoptosis is mediated by p53-dependent pathway in Hep G2 cells. *Life Sci.* **72:** 23–34.
20. Mahyar-Roemer, M., A. Katsen, P. Mestres & K. Roemer. 2001. Resveratrol induces colon tumor cell apoptosis independently of p53 and precede by epithelial differentiation, mitochondrial proliferation and membrane potential collapse. *Int. J. Cancer* **94:** 615–622.
21. Potter, G.A., L.H. Patterson, E. Wanogho, *et al.* 2002. The cancer preventive agent resveratrol is converted to the anticancer agent piceatannol by the cytochrome P450 enzyme CYP1B1. *Br. J. Cancer* **86:** 774–778.
22. Chang, T.K., J. Chen & W.B. Lee. 2001. Differential inhibition and inactivation of human CYP1 enzymes by trans-resveratrol: evidence for mechanism-based inactivation of CYP1A2. *J. Pharmacol. Exp. Ther.* **299:** 874–882.
23. Carbo, N., P. Costelli, F.M. Baccino, *et al.* 1999. Resveratrol, a natural product present in wine, decreases tumour growth in a rat tumour model. *Biochem. Biophys. Res. Commun.* **254:** 739–743.
24. Elattar, T.M. & A.S. Virji. 1999. Modulating effect of resveratrol and quercetin on oral cancer cell growth and proliferation. *Anticancer Drugs* **10:** 187–193.
25. Elattar, T.M. & A.S. Virji. 1999. The effect of red wine and its components on growth and proliferation of human oral squamous carcinoma cells. *Anticancer Res.* **19:** 5407–5414.
26. Surh, Y.J., Y.J. Hurh, J.Y. Kang, *et al.* 1999. Resveratrol, an antioxidant present in red wine, induces apoptosis in human promyelocytic leukemia (HL-60) cells. *Cancer Lett.* **140:** 1–10.
27. Hsieh, T.C. & J.M. Wu. 2000. Grape-derived chemopreventive agent resveratrol decreases prostate-specific antigen (PSA) expression in LNCaP cells by an androgen receptor (AR)-independent mechanism. *Anticancer Res.* **20:** 225–228.
28. Mitchell, S.H., W. Zhu & C.Y. Young. 1999. Resveratrol inhibits the expression and function of the androgen receptor in LNCaP prostate cancer cells. *Cancer Res.* **59:** 5892–5895.

29. Kampa, M., A. Hatzoglou, G. Notas, *et al.* 2000. Wine antioxidant polyphenols inhibit the proliferation of human prostate cancer cell lines. *Nutr. Cancer* **37:** 223–233.
30. Wargovich, M.J., C. Harris, C.D. Chen, *et al.* 1992. Growth kinetics and chemoprevention of aberrant crypts in the rat colon. *J. Cell Biochem. Suppl.* **16G:** 51–54
31. Bird, R.P. 1987. Observation and quantification of aberrant crypts in the murine colon treated with a colon carcinogen: preliminary findings. *Cancer Lett.* **37:** 147–151
32. Bilbin, M., B. Tudek & H. Czeczot. 1992. Induction of aberrant crypts in the colons of rats by alkylating agents. *Acta Biochim. Pol.* **39:** 113–137
33. McLellan, E.A., A. Medline & R.P. Bird. 1991. Dose response and proliferative characteristics of aberrant crypt foci: putative preneoplastic lesions in rat colon. *Carcinogenesis* **12:** 2093–2098
34. Sengottuvelan, M., R. Senthilkumar & N. Nalini. 2006. Modulatory influence of dietary resveratrol during different phases of 1,2-dimethylhydrazine induced mucosal lipid-peroxidation, antioxidant status and aberrant crypt foci development in rat colon carcinogenesis. *Biochim. Biophys. Acta* **1760:** 1175–1183.
35. Bird, R.P. 1995. Role of aberrant crypt foci in understanding the pathogenesis of colon cancer. *Cancer Lett.* **93:** 55–71.
36. Bird, R.P. & C.K. Good. 2000. The significance of aberrant crypt foci in understanding the pathogenesis of colon cancer. *Toxicol. Lett.* **112–113:** 395–402.
37. Whiteley, L.O. & D.M. Klurfeld. 2000. Are dietary fiber-induced alterations in colonic epithelial cell proliferation predictive of fiber's effect on colon cancer? *Nutr. Cancer* **36:** 131–149
38. Wattenberg, L.W. 1985. Chemoprevention of cancer. *Cancer Res.* **45:** 1–8.
39. Das, U. 2002. A radical approach to cancer. *Med. Sci. Monit.* **8:** 79–92.
40. Navarro, J., E. Obrador, J. Carretero, *et al.* 1999. Changes in glutathione status and the antioxidant system in blood and in cancer cells associated with tumour growth in vivo. *Free Radic. Biol. Med.* **26:** 410–418.
41. Sun, Y. 1990. Free radicals, antioxidant enzymes, and carcinogenesis. *Free Radic. Biol. Med.* **8:** 583–599.
42. Masotti, L., E. Casali & T. Galeotti. 1988. Lipid peroxidation in tumour cells. *Free Radic. Biol. Med.* **4:** 377–386.
43. Meister, A. & M.E. Anderson. 1983. Glutathione. *Annu. Rev. Biochem.* **52:** 711–760.
44. Jones, G.M., K.K. Sanford, R. Parshad, *et al.* 1985. Influence of added catalase on chromosome stability and neoplastic transformation of mouse cells in culture. *Br. J. Cancer* **52:** 583–590.
45. Kulkarni, A.A. & A.P. Kulkarni. 1995. Retinoids inhibit mammalian glutathione transferases. *Cancer Lett.* **91:** 185–189.
46. Azmi, A.S., S.H. Bhat & S.M. Hadi. 2005. Resveratrol-Cu(II) induced DNA breakage in human peripheral lymphocytes: implications for anticancer properties. *FEBS Lett.* **579:** 3131–3135.
47. Sengottuvelan, M., R. Senthilkumar & N. Nalini. 2006. Modulatory influence of dietary resveratrol during different phases of 1,2-dimethylhydrazine induced mucosal lipid-peroxidation, antioxidant status and aberrant crypt foci development in rat colon carcinogenesis. *Biochim. Biophys. Acta* **1760:** 1175–1183.
48. Reddy, B.S., A. Engle, S. Katsifis, *et al.* 1989. Biochemical epidemiology of colon cancer: effect of types of dietary fiber on fecal mutagens, acid, and neutral sterols in healthy subjects. *Cancer Res.* **49:** 4629–4635.
49. Onoue, M., S. Kado, Y. Sakaitani, *et al.* 1997. Specific species of intestinal bacteria influence the induction of aberrant crypt foci by 1,2-dimethylhydrazine in rats. *Cancer Lett.* **113:** 179–186.
50. Gorbach, S.L. & B.R. Goldin. 1990. The intestinal microflora and the colon cancer connection. *Rev. Infect. Dis.* **12:** 252–261.
51. Kinouchi, T., K. Kataoka, K. Miyanishi, *et al.* 1993. Biological activities of the intestinal microflora in mice treated with antibiotics or untreated and the effects of the microflora on absorption and metabolic activation of orally administered glutathione conjugates of K-region epoxides of 1-nitropyrene. *Carcinogenesis* **14:** 869–874.
52. Nieuw Amerongen, A.V., J.G. Bolscher, E. Bloemena & E.C. Veerman. 1998. Sulfomucins in the human body. *Biol. Chem.* **379:** 1–18.
53. Shiau, S.Y. & Y.O. Ong. 1992. Effects of cellulose, agar and their mixture on colonic mucin degradation in rats. *J. Nutr. Sci. Vitaminol.* **38:** 49–55.
54. Goldin, B.R. & S.L. Gorbach. 1984. Alterations of the intestinal microflora by diet, oral antibiotics, and Lactobacillus: decreased production of free amines from aromatic nitro compounds, azo dyes, and glucuronides. *J. Natl. Cancer Inst.* **73:** 689–695.
55. Freeman, H.J. 1986. Effects of differing purified cellulose, pectin, and hemicellulose fiber diets on fecal enzymes in 1,2-dimethylhydrazine-induced rat colon carcinogenesis. *Cancer Res.* **46:** 5529–5532.
56. Sengottuvelan, M. & N. Nalini. 2006. Dietary supplementation of resveratrol suppresses colonic tumour incidence in 1,2-dimethylhydrazine-treated rats by modulating biotransforming enzymes and aberrant crypt foci development. *Br. J. Nutr.* **96:** 145–153.
57. Schumacher, U., E. Adam & U. Zangemeister-Wittke, R. Gossrau. 1996. Histochemistry of therapeutically relevant enzymes in human tumours transplanted into severe combined immunodeficient (SCID) mice: nitric oxide synthase-associated diaphorase, beta-D-glucuronidase and non-specific alkaline phosphatase. *Acta Histochem.* **98:** 381–387.
58. Okuda, T., H. Saito, A. Sekizawa, *et al.* 2001. Steroid sulfatase expression in ovarian clear cell adenocarcinoma: immunohistochemical study. *Gynecol. Oncol.* **82:** 427–434.
59. Tanaka, T., T. Kojima, M. Suzui & H. Mori. 1993. Chemoprevention of colon carcinogenesis by the natural product of a simple phenolic compound protocatechuic acid: suppressing effects on tumor development and biomarkers expression of colon tumorigenesis. *Cancer Res.* **53:** 3908–3913.
60. Sengottuvelan, M., P. Viswanathan & N. Nalini. 2006. Chemopreventive effect of trans-resveratrol–a phytoalexin against colonic aberrant crypt foci and cell proliferation in 1,2-dimethylhydrazine induced colon carcinogenesis. *Carcinogenesis* **27:** 1038–1046.

61. Sengottuvelan, M., K. Deeptha & N. Nalini. 2009. Influence of dietary resveratrol on early and late molecular markers of 1,2-dimethylhydrazine-induced colon carcinogenesis. *Nutrition* **25:** 1169–1176.
62. Bostick, R.M., L. Fosdick, G.A. Grandits, *et al.* 1997. Colorectal epithelial cell proliferative kinetics and risk factors for colon cancer in sporadic adenoma patients. *Cancer Epidemiol. Biomarkers Prev.* **6:** 1011–1019.
63. Wolter, F., B. Akoglu, A. Clausnitzer & J. Stein. 2001. Downregulation of the cyclin D1/Cdk4 complex occurs during resveratrol-induced cell cycle arrest in colon cancer cell lines. *J. Nutr.* **131:** 2197–2203.
64. Suzuki, R., H. Kuroda, H. Komatsu, *et al.* 1999. Selective usage of D-type cyclins in lymphoid malignancies. *Leukemia* **13:** 1335–1342.
65. Carlson, B., T. Lahusen, S. Singh, *et al.* 1999. Downregulation of cyclin D1 by transcriptional repression in MCF-7 human breast carcinoma cells induced by flavopiridol. *Cancer Res.* **59:** 4634–4641.
66. Choubey, D. & J.U. Gutterman. 1997. Inhibition of E2F-4/DP-1-stimulated transcription by p202. *Oncogene* **15:** 291–301.
67. Bedi, A., P.J. Pasricha, A.J. Akhtar, *et al.* 1995. Inhibition of apoptosis during development of colorectal cancer. *Cancer Res.* **55:** 1811–1816.
68. Creagh, E.M. & S.J. Martin. 2001. Caspases: cellular demolition experts. *Biochem. Soc Trans.* **29:** 696–702.
69. Kong, A.N., R. Yu, V. Hebbar, *et al.* 2001. Signal transduction events elicited by cancer prevention compounds. *Mutat. Res.* **480–481:** 231–241.
70. Dorrie, J., H. Gerauer, Y. Wachter & S.J. Zunino. 2001. Resveratrol induces extensive apoptosis by depolarizing mitochondrial membranes and activating caspase-9 in acute lymphoblastic leukemia cells. *Cancer Res.* **61:** 4731–4739.
71. Morimoto, R.I. 1991. Heat shock: the role of transient inducible responses in cell damage, transformation, and differentiation. *Cancer Cells* **3:** 295–301.
72. Lee, S.K., Z.H. Mbwambo, H. Chung, *et al.* 1998. Evaluation of the antioxidant potential of natural products. *Comb. Chem. High Throughput Screen* **1:** 35–46.
73. Samali, A. & T.G. Cotter. 1996. Heat shock proteins increase resistance to apoptosis. *Exp. Cell Res.* **223:** 163–170.
74. Goodnight, J., H. Mischak & J.F. Mushinski. 1994. Selective involvement of protein kinase C isozymes in differentiation and neoplastic transformation. *Adv. Cancer Res.* **64:** 159–209.
75. Kahl-Rainer, P., J. Karner-Hanusch, W. Weiss & B. Marian. 1994. Five of six protein kinase C isoenzymes present in normal mucosa show reduced protein levels during tumor development in the human colon. *Carcinogenesis* **15:** 779–782.
76. Doi, S., D. Goldstein, H. Hug & I.B. Weinstein. 1994. Expression of multiple isoforms of protein kinase C in normal human colon mucosa and colon tumors and decreased levels of protein kinase C beta and eta mRNAs in the tumors. *Mol. Carcinog.* **11:** 197–203
77. Sengottuvelan, M., K. Deeptha & N. Nalini. 2009. Resveratrol ameliorates DNA damage, prooxidant and antioxidant imbalance in 1,2-dimethylhydrazine induced rat colon carcinogenesis. *Chem. Biol. Interact.* **181:** 193–201.
78. Sengottuvelan, M., K. Deeptha & N. Nalini. 2009. Resveratrol attenuates 1,2-dimethylhydrazine (DMH) induced glycoconjugate abnormalities during various stages of colon carcinogenesis. *Phytother. Res.* **23:** 1154–1158.
79. Sheng, H., C.S. Williams, J. Shao, P. Liang, R.N. DuBois, & R.D. Beauchamp. 1998. Induction of cyclooxygenase-2 by activated Ha-ras oncogene in Rat-1 fibroblasts and the role of mitogen-activated protein kinase pathway. *J. Biol. Chem.* **273:** 22120–22127.
80. Tuynman, J. B., M.P. Peppelenbosch, & D.J. Richel. 2004. COX-2 inhibition as a tool to treat and prevent colorectal cancer. *Crit. Rev. Oncol. Hematol.* **52:** 81–101.
81. Creasy, L.L. & M. Coffee. 1988. Phytoalexin production potential of grape berries. *J. Am. Soc. Hort. Sci* **113:** 230–234.
82. Chanvitayapongs, S., B. Draczynska-Lusiak & A.Y. Sun. 1997 Amelioration of oxidative stress by antioxidants and resveratrol in PC12 cells. *Neuroreport* **8:** 1499–1502.
83. Belguendouz, L., L. Fremont & A. Linard. 1997. Resveratrol inhibits metal ion-dependent and independent peroxidation of porcine low-density lipoproteins. *Biochem. Pharmacol.* **53:** 1347–1355.
84. Frankel, E.N., A.L. Waterhouse & J.E. Kinsella. 1993. Inhibition of human LDL oxidation by resveratrol. *Lancet* **341:** 1103–1104.
85. Kimura, Y., H. Okuda & S. Arichi. 1985. Effects of stilbenes on arachidonate metabolism in leukocytes. *Biochim. Biophys. Acta* **834:** 275–278.
86. Chung, M.I., C.M. Teng, K.L. Cheng, *et al.* An antiplatelet principle of Veratrum formosanum. *Planta. Med.* **58:** 274–276.
87. Chen, C.K. & C.R. Pace-Asciak. 1996. Vasorelaxing activity of resveratrol and quercetin in isolated rat aorta. *Gen. Pharmacol.* **27:** 363–366.
88. Murakami, K., S.Y. Chan & A. Routtenberg. 1986. Protein kinase C activation by cis-fatty acid in the absence of Ca2+ and phospholipids. *J. Biol. Chem.* **261:** 15424–15429.
89. Jayatilake, G.S., H. Jayasuriya, E.S. Lee, *et al.* 1993. Kinase inhibitors from Polygonum cuspidatum. *J. Nat. Prod.* **56:** 1805–1810.
90. Aggarwal, B.B., A. Bhardwaj, R.S. Aggarwal, *et al.* 2004. Role of resveratrol in prevention and therapy of cancer:preclinical and clinical studies *Anticancer Res.* **24:** 2783–2840.

Ann. N.Y. Acad. Sci. ISSN 0077-8923

ANNALS OF THE NEW YORK ACADEMY OF SCIENCES
Issue: *Resveratrol and Health*

Resveratrol modulates astroglial functions: neuroprotective hypothesis

André Quincozes-Santos and Carmem Gottfried
Research Group in Neuroglial Plasticity, Department of Biochemistry, Institute of Health's Basic Science, Federal University of Rio Grande do Sul, Porto Alegre, Rio Grande de Sul, Brazil

Address for correspondence: Carmem Gottfried, Ph.D., Laboratório de Plasticidade Neuroglial, Departamento de Bioquímica, Instituto de Ciências Básicas da Saúde, Universidade Federal do Rio Grande do Sul, Rua Ramiro Barcelos 2600 anexo, 90035-003 Porto Alegre-RS, Brazil. cgottfried@ufrgs.br

Resveratrol, a redox active compound present in grapes and wine, has a wide range of biological effects, including cardioprotective, chemopreventive, and anti-inflammatory activities. The central nervous system is a target of resveratrol, which can pass the blood–brain barrier and induce neuroprotective effects. Astrocytes are one of the most functionally diverse groups of cells in the nervous system, intimately associated with glutamatergic metabolism, transmission, synaptic plasticity, and neuroprotection. In this review, we focus on the resveratrol properties and response to oxidative insult on important astroglial parameters involved in brain plasticity, such as glutamate uptake, glutamine synthetase activity, glutathione content, and secretion of the trophic factor S100B.

Keywords: astrocytes; resveratrol; glutamatergic metabolism; neuroprotective; oxidative stress

Resveratrol: from plants to mammal brain-targeting

Plants

Many natural components of diet have been investigated in recent years, in particular the antioxidants that have been shown to cause numerous biological effects in different cell types and tissues.[1–3] The redox active compound resveratrol (3,4′,5-trihydroxy-*trans*-stilbene) is one of the most studied antioxidants; it was first found in roots of white hellebore and later in roots of *Polygonum cuspidatum*.[4] In plants, resveratrol is a phytoalexin found mainly in grapes, grape juice, wine, and berries.[5,6] In ancient medicine Hippocrates made observations on the medicinal properties of wine; Galen also reported that preparations of wine and herbs could be used as antidotes to poisons.[7] Today, red wine has gained particular attention especially owing to the French paradox that describes, particularly in southern France, an inverse correlation between intake of a diet rich in lipids (and wine) and a low incidence of heart disease.[8–10] Among the wines with the highest concentration of resveratrol, wines from the south of Brazil stand out; due to the high humidity of the soil, they naturally have a higher amount of the phytoalexin resveratrol.[11]

Mammals

Since the first reported detection of resveratrol in grapevines in 1976, a plethora of beneficial effects have been described in mammals, including cardioprotective, chemopreventive, and anti-inflammatory activities.[5,6,10,12–15] Many papers locate the most diverse actions of resveratrol in its direct antioxidant and scavenger effects or by its ability to modulate and improve cellular antioxidant defenses.[6,13] One of the signaling pathways modulated by resveratrol is that of the sirtuin protein family (SIRT1–7 in mammals). This is a conserved family of NAD^+-dependent deacetylases (class III histone deacetylases) that exerts effects related to lifespan extension in diverse species.[16] In mammalian cells, resveratrol induces SIRT1-dependent effects that are consistent with improved cellular function and implicated to play a role in a number of age-related human diseases. The effects of resveratrol on sirtuins may explain its positive effect on longevity.[13,17]

doi: 10.1111/j.1749-6632.2010.05857.x

Brain-targeting

As mentioned, the central nervous system (CNS) is a target of resveratrol because this polyphenol can pass the blood–brain barrier.[18] Among resveratrol's neuroprotective roles are benefits described in animal models of Alzheimer's and Parkinson's diseases[19] and ischemia.[20]

Resveratrol, oxidative stress, and brain pathology

From epidemiological studies, resveratrol is recognized as a component that offers many health benefits: it may protect cell constituents against oxidative damage and, therefore, limit the risk of diseases such as atherosclerosis and cancer by directly acting on reactive oxygen species (ROS) or by stimulating endogenous defense systems.[6,9,21–23] Oxidative stress has strong implications for many human diseases and has been connected with neurodegenerative disorders.[24] Brain cells have the capacity to produce peroxides, particularly hydrogen peroxide (H_2O_2), in large amounts.[25] H_2O_2 concentrations of up to 100 μM have been reported for brain in a microdialysis study.[26] In this context, the defense of glial cells against peroxide-mediated oxidative damage would likely be essential for maintaining brain functions.

Astroglial plasticity

The last 25 years have seen an exponential increase in knowledge of the neuroglial plasticity.[27] Astroglial cells have been implicated in numerous ways in brain metabolism, especially by the fact as they influence neuronal function, particularly at the level of synapses.[27–31] Numerous studies demonstrated that astrocytes play a significant role in neurodegenerative disorders[32–34] and exert a fundamental protective function against oxidative stress because of their effects on the metabolism of the antioxidant glutathione (GSH) and the defence against ROS.[35]

Primary astrocytes and C6 astroglial cell cultures are good models to study glial function, signaling pathways, and mechanisms of peroxide disposal by brain cells.[36–41] In such cultures, however, the influence of other types of brain cells on the antioxidant potential is lacking, in contrast to the *in vivo* situation. Nevertheless, in spite of the fact that comparison of the *in vitro* results with the *in vivo* condition is limited, mainly because astrocyte cells in cultures are two-dimensional and the astroglia *in situ* exist in a three-dimensional matrix, an enormous amount of molecular information has been learned from the study of astroglial cultures (primary and lineage cells), particularly as pertains to the molecular mechanisms that underlie glutamate metabolism, much of which is applicable *in vivo*.

Glutamate is the major excitatory neurotransmitter in the CNS and plays an important role in neural plasticity and neurotoxicity.[42] The modulation of extracellular glutamate determines its physiological and excitotoxic actions. The main mechanism responsible for the maintenance of low-extracellular concentrations of glutamate is performed by a family of glutamate transporter proteins, which use the electrochemical gradients across the plasma membranes as driving forces for uptake.[27] In astrocytes, glutamate is converted into glutamine by the enzyme glutamine synthetase (EC 6.3.1.2).[43] Glutamine is released by astrocytes and taken up by neurons to be again converted to glutamate; this system is called glutamate–glutamine cycle.[44] The interaction between presynaptic and postsynaptic neurons together with astrocytes characterizes the tripartite synapse.[45] Glutamate uptake is also important for maintaining levels of GSH, the major antioxidant of the brain. GSH is a tripeptide formed by amino acids cysteine, glutamate, and glycine, where the sulfhydryl group (SH) of cysteine serves as a proton donor and is responsible for the biological antioxidant effect of GSH.[44] Moreover, GSH secreted from astrocytes serves as the basis for the synthesis of GSH neuronal.[43,46] A large variety of neurological and psychiatric disorders, including depression, anxiety disorders, schizophrenia, chronic pain, epilepsy, and

Table 1. Effects of resveratrol on glutamate uptake

	C6 astroglial culture		Primary astrocyte culture[51]	Acute hippocampal slice[50]
	I[38]	II[38]		
RSV	⇧	=	⇧	⇧
H_2O_2	=	⬇	⬇	⬇
RSV + H_2O_2	⇧	⬇⬇	⇧	⇧

RSV, resveratrol; I (1 mM H_2O_2/30 min); II (0.1 mM H_2O_2/6 h). Arrows indicate increase or decrease of glutamate uptake compared to control conditions. Two arrows indicate potentiating effect.

Alzheimer's and Parkinson's diseases, demonstrate pathophysiology impairments in the glutamatergic system.[47]

Resveratrol and astroglial plasticity

Resveratrol modulates glutamate metabolism

As astroglial cells are responsible for the uptake of extracellular glutamate, our group has studied the effect of resveratrol on glutamate metabolism in primary culture, cell lines, and acute hippocampal slices.[38,48–51] First, we demonstrated in C6 astroglial cells that resveratrol increased glutamate uptake with doses ranging from 0.1 to 250 μM.[48] Afterward, in agreement with these results, we demonstrated that resveratrol increases glutamate uptake in both primary astrocyte culture and hippocampal slices. In all designed studies the increase in glutamate uptake was around 50% compared to the control condition, except for an opposite effect obtained with the highest concentration of resveratrol (250 μM), which decreased glutamate uptake in primary cell culture, indicating a hormetic phenomenon.[52,53] The concept of chemical hormesis states that chemicals are able to display opposite effects at low and higher levels.[54]

Astroglial effects of resveratrol are influenced by redox condition

The influence of redox condition of the milieu on the effect of resveratrol, summarized in Tables 1–3, was undertaken by two models of oxidative insult:[38] (I) higher concentration of hydrogen peroxide (1 mM), but short-time of exposure (30 min/acute); and (II) lower concentrations of hydrogen peroxide (0.1 mM) and longest time of incubation (up to 6 h). We observed an interesting dual effect of 100 μM resveratrol mostly protecting cells against H_2O_2-induced damage in model I and potentiating it in model II, suggesting a pro-oxidant effect. In both models, H_2O_2 decreased glutamate uptake, and resveratrol completely prevented this effect in model I and strongly potentiated H_2O_2 insult in model II (Table 1).[38] The beneficial effect of resveratrol on glutamate metabolism mediated by modulation of important astroglial cell activities is promising; however, this dual effect of resveratrol observed *in vitro* needs to be extended to *in vivo* conditions, under different animal models of stress, to be better clarified.

Table 2. Effects of resveratrol on glutamine synthetase activity

	C6 astroglial culture	
	I[38]	II[38]
RSV	⇧	⇧
H_2O_2	=	⇩
RSV + H_2O_2	⇧	⇩⇩

RSV, resveratrol; I (1 mM H_2O_2/30 min); II (0.1 mM H_2O_2/6 h). Arrows indicate increase or decrease of glutamate uptake compared to control conditions. Two arrows indicate potentiating effect.

As the glutamatergic system is involved in several brain pathologies,[55,56] the modulation of glutamate uptake by resveratrol may represent an important pharmacological opportunity. Our studies suggest that resveratrol itself can also be influenced by the surrounding redox environment.

The effect of resveratrol against glutamate excitotoxicity shown in this review may explain the efficacy of resveratrol in protecting brain disorders such as Alzheimer's and Parkinson's diseases, stroke, and ischemia injury. In addition, resveratrol has been able to protect organotipic hippocampal culture against ischemia.[57] Thus, resveratrol may represent new therapeutic potential in protecting brain disorders involving glutamate and oxidative stress.

Resveratrol modulates major glutamate destinations in astrocyte

As resveratrol increased glutamate uptake by astrocyte, we have been investigating two major destinations of glutamate in glial cells: (1) the conversion of glutamate to glutamine, by measuring the activity of the astrocyte marker enzyme GS; and (2) the amount of GSH, the main antioxidant defense of the CNS. The glutamate–glutamine cycle is defined as carrying glutamine from astrocytes to neurons and glutamate in the opposite direction.[58,59] After uptake by astrocytes, glutamate is converted to glutamine, which in turn is returned to neurons to be reconverted into glutamate.[59] GSH synthesis is a mainly astrocytic process, and astroglial GSH exported to the extracellular space is essential for providing neurons with the GSH precursors.[60]

Resveratrol was able to increase the activity of the enzyme GS in both C6 astroglial cells[38,48] and

Table 3. Effects of resveratrol on glutathione content

	C6 astroglial culture		Primary astrocyte culture[51]	Acute hippocampal slice[50]
	I[38]	II[38]		
RSV	↑	=	↑	↑
H_2O_2	↓	↓	↓	↓
RSV + H_2O_2	↑	↓↓	↑	↑

RSV, resveratrol; I (1 mM H_2O_2/30 min); II (0.1 mM H_2O_2/6 h). Arrows indicate increase or decrease of glutamate uptake compared to control conditions. Two arrows indicate potentiating effect.

primary culture of astrocytes,[49] indicating an important role in glutamate–glutamine cycle (Table 2). Glutamine levels are related with cellular redox variations and have been decreased under catabolic stress.[44] ROS appears to be a key pleiotropic modulator that may be involved in different pathways leading to modifications of macromolecules such as proteins and lipids.[24,61] The activity of the enzyme GS was impaired by oxidative stress, and resveratrol was able to prevent this effect. As glutamine is an important source of glutamate, it also helps to maintain GSH levels after injuries in CNS.[62,63] There are many models of study for understanding the pathophysiology of Alzheimer's disease, and one of them involves the administration of intracerebroventricular streptozotocin.[64] In this model there is a decrease in GSH levels, and resveratrol was able to restore the amount of this antioxidant, displaying an important *in vivo* effect of resveratrol in dementia.

Resveratrol increased intracellular GSH in astroglial cell culture[38] and hippocampal slices.[49] However, under oxidative insult resveratrol also displays a dual effect that depends on the redox condition of the milieu (Table 3), similar to the effect observed with glutamate uptake. In an intense (1 mM H_2O_2) and acute (30 min) oxidative insult, resveratrol prevented H_2O_2-induced GSH decrease, but after 6 h of oxidative insult resveratrol displayed a pro-oxidant effect, potentiating GSH decrease.[38]

In summary, we have demonstrated that resveratrol may have important role in neuroprotection by increasing glutamate uptake, GS activity, and GSH levels. Neurons are unable to take up extracellular GSH, but they can make use of cysteinyl-glicine and cysteine, two molecules derived from extracellular GSH. Thus, the neurons

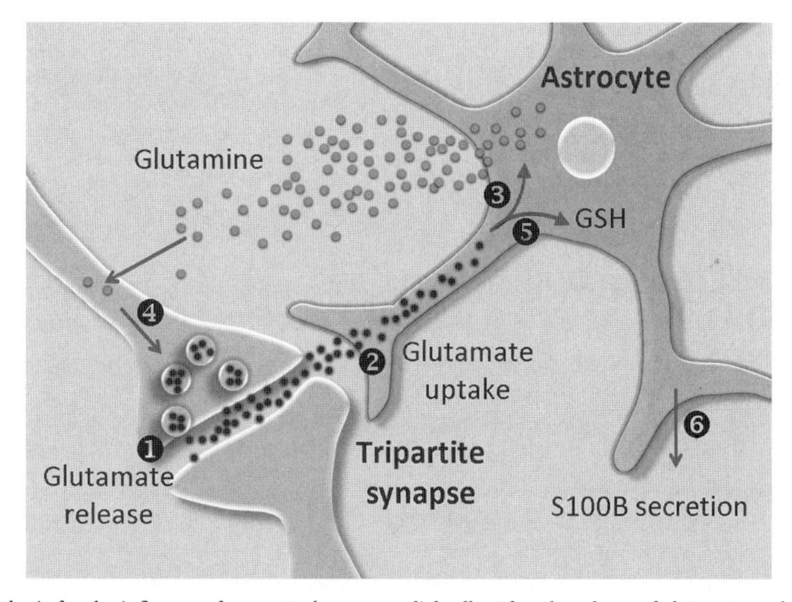

Figure 1. Hypothesis for the influence of resveratrol on neuroglial cells. After the release of glutamate at the synaptic cleft 1, resveratrol may improve glutamate uptake by astrocytes 2. This process stimulates the enzyme GS to convert glutamate into glutamine 3, which in turn is able to be released into the extracellular fluid, which is taken up by neurons and reconverted into glutamate 4. Additionally, resveratrol may stimulate another important fate of glutamate in astrocytes, particularly the synthesis of the tripeptide L-γ-glutamyl-L-cysteinyl-glycine or glutathione (GSH) 5, and promote the secretion of the trophic factor S100B 6.

need GSH from astrocytes to synthesize it, and resveratrol can modulate the glutamate–glutamine cycle through GS activity and GSH levels.

Hypothesis for the resveratrol on the tripartite synapse

There has been a lack of studies demonstrating the effect of polyphenolic compounds on neuroglial communication and signaling. Astrocytes are one of the most functionally diverse groups of cells in the nervous system, intimately associated with glutamatergic metabolism and transmission, S100B secretion, and, thus, with synaptic plasticity and neuroprotection.[27,28,33,65] S100B is a trophic factor produced and secreted by astrocytes involved in neuronal survival and activity during brain injury and recovery.[66–68] Emerging evidence indicates that signaling between perisynaptic astrocytes and neurons at the tripartite synapse plays an important role when neural circuits are formed and refined.[45] Given the role of glutamate in CNS injury, it is important to develop strategies to reduce glutamate-mediated excitotoxicity in neurological disorders. Among neural cells, astrocytes are more resistant to oxidative stress and provide a protective role for neurons, mainly due to their higher GSH content. Hence, resveratrol modulation of glutamate metabolism *in vitro* is an important key to clarifing how effectively this polyphenol acts *in vivo*. As resveratrol was able to induce *in vitro* a significant increase in glutamate uptake, GS activity, GSH levels, and S100B secretion, this indicates that astrocytes may be targets of resveratrol *in vivo* to improve brain pathologies (Fig. 1).

Conclusions and future directions

In spite of the vast progress made in our understanding of resveratrol's effects on the brain, our knowledge in this area is still rudimentary because the majority of the experiments have been performed in cell culture or in brain slices. The number of unanswered questions generated by this scenario highlights the relevance of further studies regarding the effect of resveratrol on neuroglial plasticity. It is important that this knowledge be translated into effective treatments for neural pathologies, including Parkinson's and Alzheimer's diseases. The results found *in vitro* need to be extended to studies *in vivo* with different doses and redox conditions to get a clear picture of the effect of resveratrol on human health.

Acknowledgments

This work was supported by the Conselho Nacional de Desenvolvimento Científico e Tecnológico (CNPq), Coordenação de Aperfeiçoamento de Pessoal de Nível Superior (CAPES), FINEP/Rede IBN 01.06.0842-00, and INCT-EN National Institute of Science and Technology for Excitotoxicity and Neuroprotection.

Conflicts of interest

The authors declare no conflicts of interest.

References

1. Joseph, J. *et al.* 2009. Nutrition, brain aging, and neurodegeneration. *J. Neurosci.* **29:** 12795–12801.
2. Xia, E.Q. *et al.* 2010. Biological activities of polyphenols from grapes. *Int. J. Mol. Sci.* **11:** 622–646.
3. Halliwell, B. 2007. Dietary polyphenols: good, bad, or indifferent for your health? *Cardiovasc. Res.* **73:** 341–347.
4. Nonomura, S., H. Kanagawa & A. Makimoto. 1963. [Chemical Constituents of Polygonaceous Plants. I. Studies on the Components of Ko-J O-Kon. (Polygonum Cuspidatum Sieb. Et Zucc.)]. *Yakugaku Zasshi.* **83:** 988–990.
5. Pervaiz, S. 2003. Resveratrol: from grapevines to mammalian biology. *FASEB J.* **17:** 1975–1985.
6. Fremont, L. 2000. Biological effects of resveratrol. *Life Sci.* **66:** 663–673.
7. Johnson, H. 2004. In *The Story of Wine*. Mitchell Beazley Publishers, Ed. Octopus Publishing Group. London, UK.
8. Renaud, S. & M. de Lorgeril. 1992. Wine, alcohol, platelets, and the French paradox for coronary heart disease. *Lancet* **339:** 1523–1526.
9. Dore, S. 2005. Unique properties of polyphenol stilbenes in the brain: more than direct antioxidant actions; gene/protein regulatory activity. *Neurosignals* **14:** 61–70.
10. Soleas, G.J., E.P. Diamandis & D.M. Goldberg. 1997. Resveratrol: a molecule whose time has come? And gone? *Clin. Biochem.* **30:** 91–113.
11. Souto, A.A. *et al.* 2001. Determination of trans-resveratrol concentrations in Brazilian red wines by HPLC. *J. Food Compos. Anal.* **14:** 441–445.
12. Delmas, D., B. Jannin & N. Latruffe. 2005. Resveratrol: preventing properties against vascular alterations and ageing. *Mol. Nutr. Food Res.* **49:** 377–395.
13. Baur, J.A. & D.A. Sinclair. 2006. Therapeutic potential of resveratrol: the in vivo evidence. *Nat. Rev. Drug Discov.* **5:** 493–506.
14. Das, S. & D.K. Das. 2007. Anti-inflammatory responses of resveratrol. *Inflamm Allergy Drug Targets.* **6:** 168–173.
15. Dudley, J. *et al.* 2009. Resveratrol, a unique phytoalexin present in red wine, delivers either survival signal or death signal to the ischemic myocardium depending on dose. *J. Nutr. Biochem.* **20:** 443–452.

16. Baur, J.A. 2010. Resveratrol, sirtuins, and the promise of a DR mimetic. *Mech. Ageing Dev.* **131:** 261–269.
17. Albani, D., L. Polito & G. Forloni. 2009. Sirtuins as novel targets for Alzheimer's disease and other neurodegenerative disorders: experimental and genetic evidence. *J. Alzheimers Dis.* **19:** 11–26.
18. Baur, J.A. *et al.* 2006. Resveratrol improves health and survival of mice on a high-calorie diet. *Nature* **444:** 337–342.
19. Sarkar, F.H. *et al.* 2009. Cellular signaling perturbation by natural products. *Cell Signal* **21:** 1541–1547.
20. Della-Morte, D. *et al.* 2009. Resveratrol pretreatment protects rat brain from cerebral ischemic damage via a sirtuin 1-uncoupling protein 2 pathway. *Neuroscience* **159:** 993–1002.
21. Jang, M. *et al.* 1997. Cancer chemopreventive activity of resveratrol, a natural product derived from grapes. *Science* **275:** 218–220.
22. Virgili, M. & A. Contestabile. 2000. Partial neuroprotection of in vivo excitotoxic brain damage by chronic administration of the red wine antioxidant agent, trans-resveratrol in rats. *Neurosci. Lett.* **281:** 123–126.
23. de la Lastra, C.A. & I. Villegas. 2007. Resveratrol as an antioxidant and pro-oxidant agent: mechanisms and clinical implications. *Biochem. Soc. Trans.* **35:** 1156–1160.
24. Halliwell, B. 2006. Oxidative stress and neurodegeneration: where are we now? *J. Neurochem.* **97:** 1634–1658.
25. Dringen, R., P.G. Pawlowski & J. Hirrlinger. 2005. Peroxide detoxification by brain cells. *J. Neurosci. Res.* **79:** 157–165.
26. Hyslop, P.A. *et al.* 1995. Measurement of striatal H2O2 by microdialysis following global forebrain ischemia and reperfusion in the rat: correlation with the cytotoxic potential of H2O2 in vitro. *Brain Res.* **671:** 181–186.
27. Magistretti, P.J. 2006. Neuron-glia metabolic coupling and plasticity. *J. Exp. Biol.* **209:** 2304–2311.
28. Araque, A. 2008. Astrocytes process synaptic information. *Neuron. Glia Biol.* **4:** 3–10.
29. Araque, A., G. Carmignoto & P.G. Haydon. 2001. Dynamic signaling between astrocytes and neurons. *Annu. Rev. Physiol.* **63:** 795–813.
30. Nishiyama, H. *et al.* 2002. Glial protein S100B modulates long-term neuronal synaptic plasticity. *Proc. Natl. Acad. Sci. USA* **99:** 4037–4042.
31. Perea, G. & A. Araque. 2009. GLIA modulates synaptic transmission. *Brain Res. Rev.* **63:** 93–102.
32. Salmina, A.B. 2009. Neuron-glia interactions as therapeutic targets in neurodegeneration. *J. Alzheimers Dis.* **16:** 485–502.
33. Wang, D.D. & A. Bordey. 2008. The astrocyte odyssey. *Prog. Neurobiol.* **86:** 342–367.
34. Markiewicz, I. & B. Lukomska. 2006. The role of astrocytes in the physiology and pathology of the central nervous system. *Acta Neurobiol. Exp. (Wars)* **66:** 343–358.
35. Schulz, J.B. *et al.* 2000. Glutathione, oxidative stress and neurodegeneration. *Eur. J. Biochem.* **267:** 4904–4911.
36. Quincozes-Santos, A. *et al.* 2007. Resveratrol attenuates oxidative-induced DNA damage in C6 Glioma cells. *Neurotoxicology* **28:** 886–891.
37. Van Eldik, L.J. & D.B. Zimmer. 1987. Secretion of S-100 from rat C6 glioma cells. *Brain Res.* **436:** 367–370.
38. Quincozes-Santos, A. *et al.* 2009. The janus face of resveratrol in astroglial cells. *Neurotox. Res.* **16:** 30–41.
39. Cechin, S.R., P.R. Dunkley & R. Rodnight. 2005. Signal transduction mechanisms involved in the proliferation of C6 glioma cells induced by lysophosphatidic acid. *Neurochem. Res.* **30:** 603–611.
40. Desagher, S., J. Glowinski & J. Premont. 1996. Astrocytes protect neurons from hydrogen peroxide toxicity. *J. Neurosci.* **16:** 2553–2562.
41. Aksenova, M.V. *et al.* 2005. Cell culture models of oxidative stress and injury in the central nervous system. *Curr. Neurovasc. Res.* **2:** 73–89.
42. Danbolt, N.C. 2001. Glutamate uptake. *Prog. Neurobiol.* **65:** 1–105.
43. Hertz, L. 2006. Glutamate, a neurotransmitter—and so much more. A synopsis of Wierzba III. *Neurochem. Int.* **48:** 416–425.
44. Mates, J. M. *et al.* 2002. Glutamine and its relationship with intracellular redox status, oxidative stress and cell proliferation/death. *Int. J. Biochem. Cell Biol.* **34:** 439–458.
45. Araque, A. *et al.* 1999. Tripartite synapses: glia, the unacknowledged partner. *Trends Neurosci.* **22:** 208–215.
46. Banerjee, R., V. Vitvitsky & S.K. Garg. 2008. The undertow of sulfur metabolism on glutamatergic neurotransmission. *Trends Biochem. Sci.* **33:** 413–419.
47. Marino, M.J. & P.J. Conn. 2006. Glutamate-based therapeutic approaches: allosteric modulators of metabotropic glutamate receptors. *Curr. Opin. Pharmacol.* **6:** 98–102.
48. dos Santos, A.Q. *et al.* 2006. Resveratrol increases glutamate uptake and glutamine synthetase activity in C6 glioma cells. *Arch. Biochem. Biophys.* **453:** 161–167.
49. de Almeida, L.M. *et al.* 2007. Resveratrol increases glutamate uptake, glutathione content, and S100B secretion in cortical astrocyte cultures. *Cell. Mol. Neurobiol.* **27:** 661–668.
50. de Almeida, L.M. *et al.* 2008. Resveratrol protects against oxidative injury induced by H_2O_2 in acute hippocampal slice preparations from Wistar rats. *Arch. Biochem. Biophys.* **480:** 27–32.
51. Vieira de Almeida, L.M. *et al.* 2008. Protective effects of resveratrol on hydrogen peroxide induced toxicity in primary cortical astrocyte cultures. *Neurochem. Res.* **33:** 8–15.
52. Kouda, K. & M. Iki. 2010. Beneficial effects of mild stress (hormetic effects): dietary restriction and health. *J. Physiol. Anthropol.* **29:** 127–132.
53. Day, R.M. & Y.J. Suzuki. 2005. Cell proliferation, reactive oxygen and cellular glutathione. *Dose Response* **3:** 425–442.
54. Kendig, E.L., H.H. Le & S.M. Belcher. 2010. Defining hormesis: evaluation of a complex concentration response phenomenon. *Int. J. Toxicol.* **29:** 235–246.
55. Coyle, J.T. & P. Puttfarcken. 1993. Oxidative stress, glutamate, and neurodegenerative disorders. *Science* **262:** 689–695.
56. Trotti, D., N.C. Danbolt & A. Volterra. 1998. Glutamate transporters are oxidant-vulnerable: a molecular link between oxidative and excitotoxic neurodegeneration? *Trends Pharmacol. Sci.* **19:** 328–334.
57. Zamin, L.L. *et al.* 2006. Protective effect of resveratrol against oxygen-glucose deprivation in organotypic hippocampal

slice cultures: involvement of PI3-K pathway. *Neurobiol. Dis.* **24:** 170–182.

58. Benjamin, A.M. & J.H. Quastel. 1975. Metabolism of amino acids and ammonia in rat brain cortex slices in vitro: a possible role of ammonia in brain function. *J. Neurochem.* **25:** 197–206.
59. Westergaard, N., U. Sonnewald & A. Schousboe. 1995. Metabolic trafficking between neurons and astrocytes: the glutamate/glutamine cycle revisited. *Dev. Neurosci.* **17:** 203–211.
60. Pope, S.A., R. Milton & S.J. Heales. 2008. Astrocytes protect against copper-catalysed loss of extracellular glutathione. *Neurochem. Res.* **33:** 1410–1418.
61. Halliwell, B. & M. Whiteman. 2004. Measuring reactive species and oxidative damage in vivo and in cell culture: how should you do it and what do the results mean? *Br. J. Pharmacol.* **142:** 231–255.
62. McKenna, M.C. 2007. The glutamate-glutamine cycle is not stoichiometric: fates of glutamate in brain. *J. Neurosci. Res.* **85:** 3347–3358.
63. Hertz, L. & H.R. Zielke. 2004. Astrocytic control of glutamatergic activity: astrocytes as stars of the show. *Trends Neurosci.* **27:** 735–743.
64. Sharma, M. & Y.K. Gupta. 2002. Chronic treatment with trans resveratrol prevents intracerebroventricular streptozotocin induced cognitive impairment and oxidative stress in rats. *Life Sci.* **71:** 2489–2498.
65. Donato, R. *et al.* 2009. S100B's double life: intracellular regulator and extracellular signal. *Biochim. Biophys. Acta* **1793:** 1008–1022.
66. Donato, R. 2001. S100: a multigenic family of calcium-modulated proteins of the EF-hand type with intracellular and extracellular functional roles. *Int. J. Biochem. Cell Biol.* **33:** 637–668.
67. Donato, R. *et al.* 2009. S100B's double life: Intracellular regulator and extracellular signal. *Biochim. Biophys. Acta* **1793:** 797–811.
68. Goncalves, C.A., M. Concli Leite & P. Nardin. 2008. Biological and methodological features of the measurement of S100B, a putative marker of brain injury. *Clin Biochem.* **41:** 755–763.

Ann. N.Y. Acad. Sci. ISSN 0077-8923

ANNALS OF THE NEW YORK ACADEMY OF SCIENCES
Issue: *Resveratrol and Health*

Resveratrol and apoptosis

Hung-Yun Lin,[1] Heng-Yuan Tang,[1] Faith B. Davis,[1] and Paul J. Davis[1,2]
[1]Ordway Research Institute, Albany, New York. [2]Albany Medical College, Albany, New York

Address for correspondence: Hung-Yun Lin, Ph.D., Signal Transduction Laboratory, Ordway Research Institute, 150 New Scotland Avenue, Albany, NY 12208. hlin@ordwayresearch.org

Resveratrol is a naturally occurring stilbene with desirable cardioprotective and anti-cancer properties. We have demonstrated the existence of a plasma membrane receptor for resveratrol near the arginine–glycine–aspartate (RGD) recognition site on integrin $\alpha_v\beta_3$ that is involved in stilbene-induced apoptosis of cancer cells. Resveratrol treatment *in vitro* causes activation and nuclear translocation of mitogen-activated protein kinase (ERK1/2), consequent phosphorylation of Ser-15 of p53, and apoptosis. An RGD peptide blocks these actions of resveratrol. By a PD98059-inhibitable process, resveratrol causes inducible COX-2 to accumulate in the nucleus where it complexes with pERK1/2 and p53. Chromatin immunoprecipitation reveals binding of nuclear COX-2 to promoters of certain p53-responsive genes, including *PIG3* and *Bax*. NS-398, a specific pharmacologic inhibitor of COX-2, prevents resveratrol-induced complexing of nuclear ERK1/2 with COX-2 and with pSer-15-p53 and subsequent apoptosis; cyclooxygenase enzyme activity is not involved. Molecular steps in the pro-apoptotic action of resveratrol in cancer cells include induction of intranuclear COX-2 accumulation relevant to activation of p53. Epidermal growth factor, estrogen, and thyroid hormone act downstream of ERK1/2 to prevent resveratrol-induced apoptosis.

Keywords: resveratrol; integrin $\alpha_v\beta_3$; ERK1/2; p53; COX-2; apoptosis

Introduction

Resveratrol, a phytoalexin that occurs naturally in grape skin and several medicinal plants, has anti-cancer and other biologic properties, such as suppression of atherogenesis.[1] The stilbene resveratrol has chemopreventive action in a mouse model of skin cancer;[2] it also has limited anti-estrogenic activity[3] that, in our experience, does not modulate the anti-cancer activity of the compound. The mechanism of the antitumor effects of resveratrol is incompletely understood, but in some tumor cell lines it is apparent that the end result of resveratrol action is apoptosis.[4–10]

In studies of the transduction of the resveratrol signal into apoptosis, we have identified an initiation site, a receptor, for resveratrol on a cell surface protein, an integrin. The integrins are heterodimeric structural plasma membrane glycoproteins whose extracellular domains bind matrix proteins and other extracellular factors.[11] The intracellular domain of certain integrins, such as integrin $\alpha_v\beta_3$, may activate ERK1/2.[11] Integrin antagonist small peptides, designed to mimic the integrin-matrix protein adhesion recognition sequence RGD (Arg–Gly–Asp), have displayed efficacy in the treatment of cancer. In contrast, RGE (Arg–Gly–Glu) peptides do not bind to the integrin and are ineffective in physiological models.[12] In this review paper we will discuss the transduction of the resveratrol signal by $\alpha_v\beta_3$, including ERK1/2 activation and p53 phosphorylation, into p53-dependent apoptosis. Induction by resveratrol of nuclear accumulation of COX-2 also facilitates p53-dependent apoptosis. Finally, we discuss modulation by an endogenous growth factor and nonpeptide hormones of resveratrol-induced apoptosis. Such effects may affect clinical trials of resveratrol.

Mechanisms of resveratrol-induced apoptosis

Integrin $\alpha_v\beta_3$ bears a receptor for resveratrol

Using radiolabeled resveratrol, we recently determined that the stilbene binds dissociably to integrin

doi: 10.1111/j.1749-6632.2010.05846.x

$\alpha_v\beta_3$, leading to activation of the mitogen-activated protein kinase (MAPK; ERK1/2) signal transduction cascade.[13] A structural protein of the plasma membrane that is essential to the interactions of cancer cells and blood vessel cells with specific extracellular matrix proteins, $\alpha_v\beta_3$ integrin has been shown by us to be a receptor for several small molecules, such as thyroid hormone analogues[12,14] and dihydrotestosterone (DHT)[15] as well as resveratrol. Identification of a putative receptor site for resveratrol in cancer cells—one that initiates transduction via ERK1/2 of the resveratrol signal downstream into p53-dependent apoptosis[4–9]—supports the credibility and specificity of the compound as a potential therapeutic agent. The receptor may also be a vehicle for *in vitro* structure-activity studies of resveratrol analogues.

With regard to cancer cells, the roles of integrins are more complex than the transduction of outside-in signals that originate from integrin-matrix protein interactions. For example, dysregulation of the β3 integrins has been implicated in cancer pathogenesis. Tumor growth and associated angiogenesis, particularly as mediated by vascular endothelial growth factor, are enhanced in β3-null mice.[16] Integrin β3 overexpression, in contrast, can suppress tumor growth of a human glioma model in rats.[17] Such results suggest that promotion of integrin β3 expression in cancer cells may be a therapeutic goal in the setting of cancer, but this topic is beyond the scope of the current review.

Activation of ERK1/2 and p53

Extracellular signal-regulated kinases 1 and 2 (ERK1/2) are MAPK isoforms that are inducible components of normal cellular signal transduction processes. Pathways of ERK1/2 activation can be triggered by a variety of stimuli and have been particularly well characterized in growth factor-stimulated cells or in the setting of inflammation. ERK1/2 is directly activated by MAPK-kinase (MEK1/2). Upstream activators in the MEK–MAPK pathway are Raf-1 kinase and, at the cell membrane, Ras.[18]

Resveratrol activates MAPK at low concentrations (1 pM–10 μM), but higher concentrations (50–100 μM) of resveratrol can inhibit this signal transducing kinase in cancer cells.[19] We have shown that resveratrol induces ERK1/2 activation in prostate,[4,5] breast,[6,7,12] glial,[8] head and neck cancer cells,[9] and ovarian cancer cells.[10] The activation of ERK1/2 by resveratrol may be blocked by either *H-ras* antisense oligonucleotides or a MAPK kinase (MEK) inhibitor, PD 98059.[4–6] Constitutively active MAPK is required for maintenance of the malignant state, but short-term activation of MAPK may direct cells to apoptosis.[20] Studies of Dong's group indicate that resveratrol also activates other kinases such as p38 kinase and cJun N-terminal kinase.[21]

Resveratrol induces p53-dependent and p53-independent apoptosis.[4–9,22] In this review, we focus on resveratrol-induced p53-dependent apoptosis. p53 is an oncogene suppressor protein that is present at low levels in normal cells.[23] In response to a variety of stresses, including DNA damage,[23,24] levels of cellular p53 activity rise, the result of a posttranslational mechanism that stabilizes the protein.[25] p53 is involved in several aspects of cell cycle arrest including apoptosis, genome integrity, and DNA repair.[23,24] When cells are treated with ionizing or UV radiation, phosphorylation of p53 occurs at several serine and threonine residues,[25] as does acetylation of several lysine residues.[25] Activated p53 binds to DNA by an allosteric mechanism dependent upon both phosphorylation and acetylation of the protein.

Serine phosphorylation of wild-type p53 at different sites has different biological consequences.[23,25] Phosphorylation of p53 at N-terminal residues may block MDM2 binding,[5] thus promoting stabilization of p53, and may be a factor that facilitates or is required for the acetylation of p53 at a C-terminal lysine.[26] Acetylation has been reported to increase sequence-specific DNA binding of p53 *in vitro*, and acetylation is essential for recruitment of the protein to coactivators-–such as CBP (CREB-binding protein)/p300—involved in histone acetylation.[27] These co-activators contain intrinsic histone acetyltransferase activity. Acetylation of the p53 protein appears to be a property of CBP/p300 (PCAF) and certain other p53 coactivators.[28] We have also reported that resveratrol promotes acetylation of p53.[6]

Recent studies suggest that activation of certain promoter regions requires phosphorylation of specific p53 residues.[29] A link between N-terminal phosphorylation of p53 and C-terminal acetylation of the protein has been described in *in vitro* studies.[26–28] That is, serine phosphorylation at residues 15 and/or 33 prompts acetylation of p53 at Lys-382 (by p300) or at Lys-320 by p300/CBP-associated factor (PCAF), another p53 coactivator. CBP/p300 and PCAF are coactivators for a variety of transcription factors in addition to p53.

Resveratrol induces phosphorylation of p53 at N-terminal serine 15 (Ser-15)[4–10] and C-terminal serine 392 (Ser-392)[6] in several human cancer cell lines and induces acetylation of p53.[6] Activation of ERK1/2 is involved in resveratrol-induced serine phosphorylation of p53 in cancer cells.[4–9] Interestingly, resveratrol induces Ser-15 phosphorylation of a mutant p53 in prostate cancer DU145 cells,[4,5] which restores wild-type p53 DNA binding activity and induces apoptosis.

COX-2 translocation to the nucleus

Cyclooxygenase-2 (COX-2), the rate-limiting enzyme in prostaglandin synthesis, is induced in many cells by inflammatory mediators. Two cyclooxygenase (COX) isoforms have been identified.[30,31] COX-1 is a constitutively expressed form of the enzyme and is ubiquitous in its intracellular distribution, whereas COX-2 is inducible and is present in inflammatory foci and neovasculature[30] but may also be constitutively expressed in tumors. Constitutive expression of *COX-2* in cells and animal models is associated with tumor cell growth and metastasis, enhanced cellular adhesion and inhibition of apoptosis.[30,31] COX-2 utilization of arachidonic acid also perturbs the level of intracellular free arachidonic acid and subsequently affects several cellular functions.[31] COX-2 has been shown to localize in the endoplasmic reticulum (ER), Golgi complex, and nuclear envelope (NE).[32,33] Catalytically active COX-1 and COX-2 are localized in the NE and ER of PGE_2-releasing cells.[30] More recent studies have suggested that for functional coupling and PGE_2 biosynthesis, cytosolic PLA_2, COXs, and PGEs appear to be localized in the perinuclear region.[34]

Resveratrol induces nuclear accumulation of COX-2 in a variety of tumor cells, including human breast cancer, glioma, head and neck squamous cell cancer cells, ovarian and prostate cancer cells. A specific COX-2 inhibitor, NS398, and siRNA of COX-2 block resveratrol-induced Ser-15 phosphorylation and apoptosis in cancer cells.[6–9] These results suggest that inducible COX-2 plays an important role in p53-dependent apoptosis in cancer cells. Two other laboratories have recently presented evidence that COX-2 can be pro-apoptotic[35,36] and Hinz *et al.* have suggested that COX-2 inhibitors could be deleterious in management of certain tumors because of the pro-apoptotic action of the protein.[37] Pharmacologic inhibition of COX-2 has resulted in conflicting results. A selective COX-2 inhibitor, NS-398, significantly enhanced genotoxic stress-induced apoptosis in several types of $p53^{+/+}$ normal human cells, through a caspase-dependent pathway.[38] However, others have shown that NS-398 in high doses decreases caspase-3 activity in human osteosarcoma cells.[39] Han *et al.* have reported that COX-2 expression is inducible by wild-type p53 and by DNA damage, and involves p53-mediated activation of the Ras/Raf/MAPK signal transduction cascade.[38] Other studies have indicated that overexpression of COX-2 may induce an antiproliferative effect that is related to p53 and p21 expression.[34,40]

Our belief is that constitutive COX-2 expression supports tumor growth and is anti-apoptotic, whereas inducible COX-2—induced, for example, by resveratrol and localized largely to the cell nucleus—is pro-apoptotic via a mechanism that very specifically depends upon phosphorylation of the Ser-15 of p53. Such a mechanism may be exploitable in the management of specific cancers. We speculate that clinical anti-cancer regimens may ultimately be designed to target sequentially the constitutively expressed and inducible pools of COX-2, particularly in the management of tumors in which pharmacologic COX-2 inhibition produces cell cycle arrest, rather than apoptosis, and when resveratrol or other agents capable of inducing COX-2 result in apoptosis.[6–9,40–42] Recently, studies of nitric oxide (NO)-releasing acetylsalicylic acid (ASA) have revealed that ASA inhibits colon cancer growth by increasing COX-2 expression.[41] Overexpression of COX-2 inhibits platelet-derived growth factor-induced mesangial cell proliferation through the induction of p53, as well as of p21.[34] These observations imply functional interactions of COX-2 and p53.

COX-2-nucleoprotein complexes bind promoters

Studies by Parfenova *et al.*[33] indicate that nuclear COX-2 in vascular endothelial cells is physically associated with the nuclear matrix that spatially organizes chromatin. In contrast to similar studies in nontumor cells, we have shown that resveratrol-induced nuclear accumulation of COX-2 in tumor cell nuclei may be associated with the generation of complexes of COX-2 and ERK1/2 that are relevant to p53-dependent apoptosis.[7–9] Inhibition of COX-2 in colon cancer cells by a selective COX-2 inhibitor increased the nuclear localization of active p53,[40] while treatment of human brain tumor cells with the nonsteroidal anti-inflammatory agent, flurbiprofen, increased the level of COX-2 expression, and fostered complexing of COX-2 with p53, leading to tumor growth suppression.[42] Other studies indicate that cytosolic COX-2 may associate directly with p53 and this complex formation may reduce the activity of p53.[43] Electrophilic prostaglandins produced by COX-2 are shown to inhibit wild-type p53 activity by covalently binding p53 in cytosol.[44]

The biochemical steps between formation of the COX-2-MAPK (ERK1/2) complex and activation of p53 in resveratrol-treated cells are now clear.[7–9] In fact, recent studies have demonstrated that post-translational modification of COX-2 in the form of tyrosine phosphorylation regulates COX-2 activity in cerebral endothelial cells.[45] Studies from our laboratory also indicate that resveratrol-induced COX-2 associates with activated ERK1/2 in the nucleus of cancer cells.[8,9] Activated nuclear ERK1/2 has been shown to form complexes with transcriptionally active proteins, such as receptors for nonpeptide hormones, signal transducing and activator of transcription (STAT) proteins, and the oncogene suppressor protein, p53.[46] Activated ERK1/2 in this context serves to serine phosphorylate (activate) the proteins with which it is associated. We therefore considered the possibility that, following its activation by resveratrol and consequent translocation to the nucleus, ERK1/2 may play a role in the resveratrol-induced p53-dependent apoptosis. Either inhibition of ERK1/2 activation pharmacologically by PD98050 or inhibition of COX-2 by NS398 blocked the nuclear complexing of COX-2 with phosphorylated p53 and with activated ERK1/2.[7–9]

The induction of COX-2 has been shown to be either ERK1/2- or p38-dependent.[7,8,10,47] Interestingly, in addition to the p38 and ERK1/2 signal transduction pathways, the PI3K/Akt cascade may also be involved in the functions of COX-2. The PI3K/Akt pathway is activated by COX-2 or its product, PGE_2.[48] There are several transcriptional factors and co-activators such as Akt, NF-κB, and STAT proteins residing in cytoplasm that translocate to the nucleus after posttranslational modification.[46,49] COX-2 is also subject to posttranslational modification,[9,45] but it is not clear yet whether activated ERK1/2 plays a role in the posttranslational modification/phosphorylation of nuclear COX-2. Mechanisms underlying the translocation of inducible COX-2 to the nucleus remain undefined. However, COX-2 may form complexes with caveolin-1 (Cav-1)[50] and with a member of the ubiquitin and small ubiquitin-like modifier (SUMO) family, which regulate the function of a variety of transcription factors by altering their intracellular targeting, their interaction with specific partners, and/or their stability.[10]

The finding that nuclear COX-2 can bind to the promoter region of one or more genes[9,10] suggests that the COX-2 may be transcriptionally active or serve as a co-factor (corepressor or coactivator) for transactivator proteins, and that a view of this protein exclusively as a critical enzyme in the production of prostaglandins may be too limited. It is clear that more information is needed about the putative transcriptional role of this protein. It will also be important to determine whether constitutively expressed COX-2 protein has transcriptional roles. Chromatin immunoprecipitation studies that we have recently conducted with antibody to COX-2 have revealed increased COX-2 binding to the promoter regions of certain p53-responsive genes, such as *Bad* and *PIG3*, in resveratrol-treated cells.[9,10]

Independence of COX-2 enzymatic activity

Resveratrol has been reported to inhibit COX-2 expression induced by carcinogens in mouse skin, human breast, and rat esophageal epithelial cells.[51–53] In addition, resveratrol also is an inhibitor of COX-2 enzyme activity.[53] Zahner *et al.*[35] have shown that hyper-expression of *COX-2* can induce p53 gene expression and that this effect of COX-2 is wholly independent of the enzymatic activity of COX-2 and the production of PGE_2. The COX-2 specific inhibitor NS398, which not only inhibits COX-2 enzymatic activity but also reduces abundance of COX-2, inhibits Ser-15 phosphorylation of p53 and p53-dependent apoptosis in cancer cells treated with resveratrol.[7–9] NS398 does not affect ERK1/2 activation in resveratrol-treated cells.[7,8] These studies indicate that NS398 inhibits resveratrol-induced COX-2 in the nuclei where COX-2 forms complexes with SUMO-1, pERK1/2, and p53. COX-2 may lose the capability to form complexes with other nucleoproteins and therefore may not potentiate p53-dependent apoptosis in resveratrol-treated cancer cells when NS398 is present.

Mechanisms of inhibition of resveratrol-induced apoptosis

Growth factors

Certain endogenous polypeptide growth factors, including EGF, insulin-like growth factor, and fibroblast growth factor, have been implicated in the development and progression of cancers.[5,54] These factors are capable of rapidly activating ERK1/2 in several cell lines.[5] EGF inhibits the action of resveratrol on both ERK1/2 activation and the induction of apoptosis in both androgen-responsive and unresponsive prostate cancer cells. These results suggest that increased ambient EGF levels would oppose any clinical actions that resveratrol may have. Resveratrol abrogation of a PKC-mediated ERK1/2 activation response in prostate cancer cells correlates with isozyme-selective PKC-α inhibition.[55] EGF inhibits ERK1/2 activation, p53 phosphorylation, and apoptosis caused by resveratrol via a PKC-α-dependent pathway. Resveratrol does not influence Ang II-mediated transactivation of EGF-receptor, but potently inhibits EGF-induced phosphorylation of Akt kinase. This suggests that resveratrol acts downstream of EGF-receptor transactivation.[56]

Resveratrol inhibits activities of several transcription factors. The actions of EGF and TGF-α on transcription are NF-κB-mediated.[57] Signal transducer and activator of transcription 3 (STAT3) is constitutively active in tumor cells.[58] STAT3 is also activated by growth factors and Src kinases.[59] STAT3 activation has been linked with chemoresistance and radioresistance and chemopreventive agents have been shown to suppress STAT3 activation. Thus inhibitors of STAT3 activation have potential for both prevention and therapy of cancer. Plant polyphenols such as curcumin, resveratrol, flavopiridol, indirubin, magnolol, piceatannol, parthenolide, EGCG, and cucurbitacin inhibit STAT3 activation.[60] The proto-oncogene proteins c-Fos and c-Jun participate in a plethora of signaling pathways. Growth factor-induced c-*fos* and c-jun expression occurs very rapidly and transiently.[61] EGF-induced immediate-early gene expression is involved in cell proliferation[62] and this process may interfere with resveratrol-induced c-*fos* and c-*jun* expression.[4]

Resveratrol inhibits tumorigenesis in mouse skin through interference with pathways of reactive oxidants and possibly by modulating the expression of c-fos and TGF-β1.[51] Resveratrol can suppress proliferation of Ishikawa cells through down-regulation of EGF.[63] EGF increases the complexing of MDM2 and p53 in cytosol, thus blocking the function of p53 and increasing p53 degradation.[4] Further, resveratrol decreases, while E_2 and EGF increase, cell migration.[64]

Resveratrol also inhibits the expression of several autocrine growth stimulators, including prostate cancer cell-derived growth factor and transforming growth factor-α (TGF-α).[65] In addition, resveratrol significantly elevates the expression of the growth inhibitor TGF-β2 mRNA, without changes in TGF-β1 and TGF-β3 expression. These data suggest that resveratrol inhibits proliferation by altering autocrine growth modulator pathways in cancer cells.[66]

Nonpeptide hormones

Physiologic concentrations of thyroid hormones (L-thyroxine, T_4, and 3,5,3′-triiodo-L-thyronine, T_3) bind to a plasma membrane receptor on integrin $\alpha_v\beta_3$ and cause proliferation *in vitro* of several tumor cell lines.[8,14] While this cell surface thyroid hormone receptor is on the same integrin as that upon which the resveratrol receptor is located,[12] and both sites can lead to activation of ERK1/2, the binding sites are discrete,[8,67] do not appear to interact, and have distinctive downstream consequences. However, RGD peptide can block the binding site of both thyroid hormone and resveratrol. Tumor cell proliferation induced by thyroid hormone is ERK1/2-requiring and an ERK1/2 inhibitor, PD 98059, is effective in decreasing the proliferative action of the hormone.[14,68] T_3, but not T_4, can nongenomically activate the phosphatidylinositol 3-kinase (PI3K) cascade.[68] Downstream of ERK1/2 in the transduction of the resveratrol and thyroid hormone signals is a functional divergence that accounts for stimulation of cancer cell proliferation by thyroid hormone and induction of apoptosis by resveratrol. Resveratrol increases nuclear abundance of COX-2, whereas T_4 alone has no effect on nuclear COX-2. However, thyroid hormone inhibits formation of nuclear complexes between pERK1/2 and COX-2 in resveratrol-treated cells and also inhibits resveratrol-induced p53 phosphorylation.

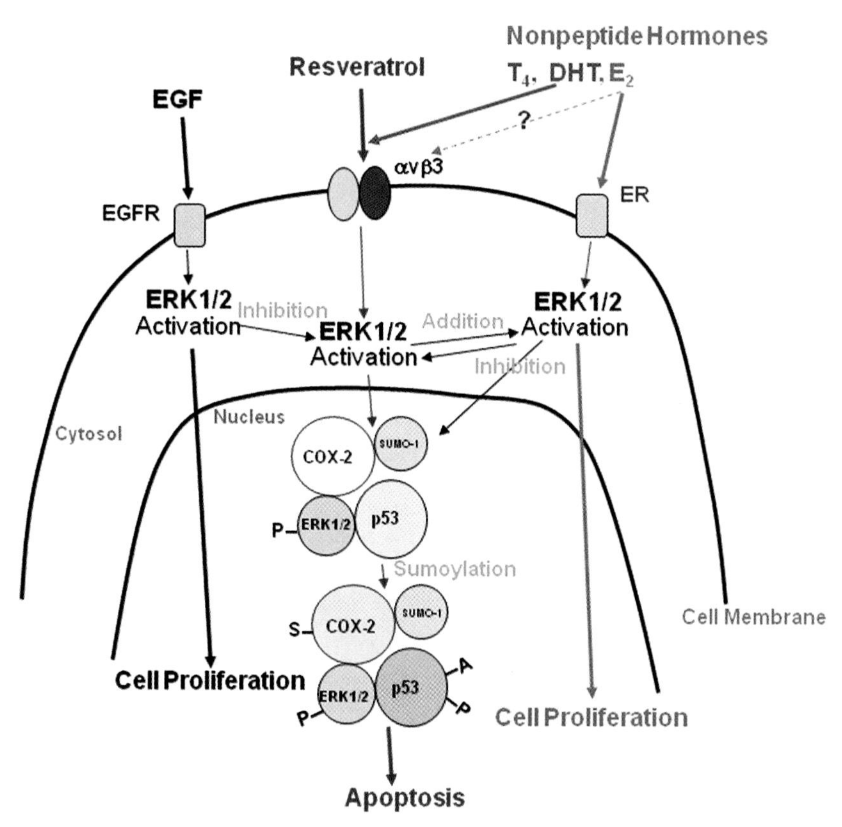

Figure 1. Signal transduction pathways involved in resveratrol-induced apoptosis and in inhibitory effects of nonpeptide hormones and EGF on actions of resveratrol. Resveratrol binds to integrin $\alpha_v\beta_3$ on the cancer cell surface and activates ERK1/2, which is essential for resveratrol-induced expression of COX-2. The newly synthesized COX-2 is associated with SUMO-1, and phosphoERK1/2 translocates to the nucleus where the complex binds to p53, in the course of which COX-2 is sumoylated and p53 is phosphorylated and acetylated. The COX-2 and p53 complex binds to promoters of p53-responsive genes and initiates transcription. Thyroid hormone and DHT each bind to discrete sites on integrin $\alpha_v\beta_3$ to activate ERK1/2 and cell proliferation. Estrogen (E2) binds to the cell surface estrogen receptor (ER) and integrin $\alpha_v\beta_3$ may play an accessory role. The activation of ERK1/2 by resveratrol and nonpeptide hormone are separate processes that do not interfere with each other. However, the activation of ERK1/2 by nonpeptide hormones is essential to the inhibition of apoptosis induced by resveratrol. EGF binds to EGF receptor to activate ERK1/2. The activation of ERK1/2 by EGF directly inhibits resveratrol-induced ERK1/2. Therefore, the inhibitory effect of nonpeptide hormones on resveratrol action occurs downstream of ERK1/2 but that of EGF occurs directly at ERK1/2 activation. A, acetylation; P, phosphorylation; and S, sumoylation.

How this signal transduction step (nuclear complexing of COX-2 and activated ERK1/2, upstream of activation of p53) is differentially affected by T_4 and resveratrol is not yet clear. Among the possible explanations is that the pools of activated ERK1/2 that result from the transduction of resveratrol and thyroid hormone signals are discrete.[69]

In addition to thyroid hormone, estradiol (E_2) can affect the action of resveratrol. E_2 activates ERK1/2 rapidly via a cell surface receptor.[70] A downstream effect of this action of estrogen on ERK1/2 activation may be phosphorylation of Ser-118 of ER-α and enhanced transcriptional activity of the receptor.[71] ERK1/2 activation induced by estrogen in breast cancer cells has been linked to cell proliferation.[72] It also has been shown that E_2 may increase cell migration via crosstalk with epidermal growth factor receptor (EGFR) signaling.[73] Both resveratrol and E_2 stimulate ERK1/2 activation and ERK1/2 activation is additive in the presence of E_2 and resveratrol. However, E_2 inhibits resveratrol-induced phosphorylation of serines 6, 15, and 392 in p53, as well as acetylation of p53, despite the enhancement by E_2 of ERK1/2 activity.

While there is congruence at ERK1/2 of signals generated by resveratrol and E_2, there is, downstream of ERK1/2, a divergence of effects of these two agents on nuclear transcription by p53. This also suggests the existence of discrete intracellular pools of ERK1/2, separately regulated by resveratrol and E_2. The antiapoptotic effect of E_2 in the presence of resveratrol involves rapid stimulation of signaling cascades, including the ERK1/2 pathway. The ER inhibitor ICI 182780 suppresses both E_2-induced ERK1/2 activation and the inhibitory effect of E_2 on resveratrol-induced serine phosphorylation of p53. Although resveratrol has been shown by others to have estrogen-like activity in a transcriptional assay in cells expressing predominantly ERα,[74] ICI 182780 does not diminish resveratrol-induced activation of ERK1/2 and serine phosphorylation of p53, and in fact appeared to enhance these effects of the stilbene. Others have also reported that activities of resveratrol are not affected by ICI 182780.[75] E_2 inhibits posttranslational modifications of p53 that are essential for p53-dependent gene transcription by ER-α-dependent signal transduction pathways. These results have implications for the possible use of resveratrol therapeutically in cancer treatment, as they suggest that resveratrol may be most effective in an estrogen-depleted setting.

Conclusions

Via a plasma membrane receptor identified on integrin $\alpha_v\beta_3$, resveratrol induces ERK1/2- and p53-dependent apoptosis in a variety of cancer cells. The promotion of apoptosis by the stilbene also depends upon a nuclear pool of resveratrol-inducible COX-2 that complexes with ERK1/2 and p53 and may be transcriptionally active. The desirable pro-apoptotic activity of resveratrol is opposed by physiologic concentrations of thyroid hormone. The latter has a discrete receptor on the same integrin and also activates ERK1/2. Downstream of ERK1/2, however, thyroid hormone interferes with formation of resveratrol-induced, p53-requiring nucleoprotein complexes and blocks apoptosis. Estrogen may also modulate resveratrol-induced apoptosis. Finally, the polypeptide growth factor, EGF, may interfere with the action of resveratrol. This may occur through crosstalk between the integrin receptor for resveratrol and the EGF receptor on the cancer cell surface. There is an inhibitory effect on ERK1/2 activation when resveratrol is co-incubated with growth factors but not with hormones. A schematic of the signals induced by resveratrol, hormones and growth factors is shown in Figure 1. In the intact animal test model and in the clinic, we postulate that normal circulating levels of growth factors and hormones may blunt an oncologic response to resveratrol. The clinical utility of resveratrol in the setting of cancer may be enhanced by pharmacologic blockade of the cell surface receptor for thyroid hormone and receptors for several other endogenous factors that are capable of suppressing the pro-apoptotic action of the stilbene.

Acknowledgments

Research presented in this review paper was supported in part by an endowment established by M. Frank and Margaret Domiter Rudy and funding from Charitable Leadership Foundation. Partial studies have been presented at the Graduate Institute of Medical Sciences, Taipei Medical University, Taipei, Taiwan, and the 1st International Conference of Resveratrol and Health in Denmark. Authors appreciate the invitation of Dr. Wen-Sen Lee and Ms. Ling Chang (Taipei Medical University, Taiwan) and Dr. Ole Vang (Roskilde University, Denmark).

Conflicts of interest

The authors declare no conflicts of interest.

References

1. Delmas, D., B. Jannin & N. Latruffe. 2005. Resveratrol: preventing properties against vascular alterations and ageing. *Mol. Nutr. Food Res.* **49:** 377–395.
2. Jang, M., L. Cai, G.O. Udeani, *et al.* 1997. Cancer chemopreventive activity of resveratrol, a natural product derived from grapes. *Science* **275:** 218–220.
3. Mueller, S.O., S. Simon, K. Chae, *et al.* 2004. Phytoestrogens and their human metabolites show distinct agonistic and antagonistic properties on estrogen receptor alpha (ERalpha) and ERbeta in human cells. *Toxicol. Sci.* **80:** 14–25.
4. Lin, H.Y., A. Shih, F.B. Davis, *et al.* 2002. Resveratrol-induced serine phosphorylation of p53 causes apoptosis in a mutant p53 prostate cancer cell line. *J. Urol.* **168:** 748–755.
5. Shih, A., S.L. Zhang, J.H. Cao, *et al.* 2004. Inhibitory effect of EGF on resveratrol-induced apoptosis in prostate cancer cells is mediated by protein kinase C-alpha. *Mol. Cancer Ther.* **11:** 1355–1364.
6. Zhang, S.L., F.B. Davis, J.H. Cao, *et al.* 2004. Oestrogen inhibits resveratrol-induced posttranslational modification of p53 and apoptosis in breast cancer cells. *Br. J. Cancer* **91:** 178–185.
7. Tang, H.Y., A. Shih, J.H. Cao, *et al.* 2006. Inducible COX-2 facilitates p53-dependent apoptosis in human breast cancer cells. *Mol. Cancer Ther.* **5:** 2034–2204.
8. Lin, H.Y., H.Y. Tang, T. Keating, *et al.* 2008. Resveratrol is pro-apoptotic and thyroid hormone is anti-apoptotic in

glioma cells: both actions are integrin- and ERK-mediated. *Carcinogenesis* **29:** 62–69.
9. Lin, H.Y., M.Z. Sun, H.Y. Tang, *et al.* 2008. Resveratrol causes COX-2- and p53-dependent apoptosis in head and neck squamous cell cancer cells. *J. Cell. Biochem.* **104:** 2131–2142.
10. Lin, C., D. Crawford, S. Lin, *et al.* 2010. Inducible COX-2-dependent apoptosis in human ovarian cancer cells. *Carcinogenesis* In press.
11. Plow, E.F., T.A. Haas, L. Zhang, *et al.* 2000. Ligand binding to integrins. *J. Biol. Chem.* **275:** 21785–21788.
12. Bergh, J.J., H.Y. Lin, L. Lansing, *et al.* 2005. L-Thyroxine induces mitogen-activated protein kinase activation via binding to integrin alpha(V)beta3. *Endocrinology* **146:** 2864–2871.
13. Lin, H.Y., L. Lansing, F.B. Davis, *et al.* 2005. Integrin $\alpha V\beta 3$ contains a receptor site for resveratrol. *FASEB J.* **20:** 1742–1744.
14. Davis, F.B., H.Y. Tang, A. Shih, *et al.* 2006. Acting via a cell surface receptor, thyroid hormone is a growth factor for glioma cells. *Cancer Res.* **66:** 7270–7275.
15. Lin, H.Y., M. Sun, C. Lin, *et al.* 2009. Androgen-induced human breast cancer cell proliferation is mediated by discrete mechanisms in estrogen receptor-α-positive and -negative breast cancer cells. *J. Steroid Biochem. Mol. Biol.* **113:** 182–188.
16. Hodivala-Dilke, K.M., K.P. McHugh, D.A. Tsakiris, *et al.* 1999. Beta3-integrin-deficient mice are a model for Glanzmann thrombasthenia showing placental defects and reduced survival. *J. Clin. Invest.* **103:** 229–238.
17. Kanamori, M., S.R. Vanden Berg, G. Bergers, *et al.* 2004. Integrin beta3 overexpression suppresses tumor growth in a human model of gliomagenesis: implications for the role of beta3 overexpression in glioblastoma multiforme. *Cancer Res.* **64:** 2751–2758.
18. Malarkey, K., C.M. Belham, A. Paul, *et al.* 1995. The regulation of tyrosine kinase signalling pathways by growth factor and G-protein-coupled receptors. *Biochem. J.* **309:** 361–375.
19. Miloso, M., A.A. Bertelli, G. Nicolini, *et al.* 1999. Resveratrol-induced activation of the mitogen-activated protein kinase in human neuroblastoma SH-SY5Y cells. *Neurosci. Lett.* **264:** 141–144.
20. Lassus, P., P. Roux, O Zugasti, *et al.* 2000. Extinction of rac1 and Cdc42Hs signalling defines a novel p53-dependent apoptotic pathway. *Oncogene* **19:** 2377–2385.
21. Dong Z. 2003. Molecular mechanism of the chemopreventive effect of resveratrol. *Mutat. Res.* **523–524:** 145–150.
22. Mahyar-Roemer, M. & K. Roemer. 2001. p21 Waf1/Cip1 can protect human colon carcinoma cells against p53-dependent and p53-independent apoptosis induced by natural chemopreventive and therapeutic agents. *Oncogene* **20:** 3387–3398.
23. Lavin, M.F. & N. Gueven. 2006. The complexity of p53 stabilization and activation. *Cell Death Differ.* **13:** 941–950.
24. Agarwal, M.L., W.R. Taylor, M.V. Chernov, *et al.* 1998. The p53 network. *J. Biol. Chem.* **273:** 1–4.
25. Colman, M.S., C.A. Afshari & J.C. Barrett. 2000. Regulation of p53 stability and activity in response to genotoxic stress. *Mutat. Res.* **462:** 179–188.
26. Kapoor, M., R. Hamm, W. Yan, *et al.* 2000. Cooperative phosphorylation at multiple sites is required to activate p53 in response to UV radiation. *Oncogene* **19:** 358–364.
27. Berlev, N.A., L. Liu, N.H. Chehab, *et al.* 2001. Acetylation of p53 activates transcription through recruitment of coactivators/histone acetytransferases. *Mol. Cell* **8:** 1243–1254.
28. Dornan, D., H. Shimizu, N.D. Perkins, *et al.* 2003. DNA-dependent acetylation of p53 by the transcription coactivator p300. *J. Biol. Chem.* **278:** 13431–13441.
29. Appella, E. & C.W. Anderson. 2001. Post-translational modifications and activation of p53 by genotoxic stresses. *Eur. J. Biochem.* **268:** 2764–2772.
30. Morita, I., M. Schindler, M.K. Regier, *et al.* 1995. Different intracellular locations for prostaglandin endoperoxide H synthase-1 and -2. *J. Biol. Chem.* **270:** 10902–10908.
31. Smith, W.L., R.M. Garavito & D.L. DeWitt. 1996. Prostaglandin endoperoxide H synthases (cyclooxygenases)-1 and -2. *J. Biol. Chem.* **271:** 33157–33160.
32. Murakami, M., K. Nakashima, D. Kamei, *et al.* 2003. Cellular prostaglandin E2 production by membrane-bound prostaglandin E synthase-2 via both cyclooxygenases-1 and -2. *J. Biol. Chem.* **278:** 37937–37947.
33. Parfenova, H., V.N. Parfenov, B.V. Shlopov, *et al.* 2001. Dynamics of nuclear localization sites for COX-2 in vascular endothelial cells. *Am. J. Physiol. Cell Physiol.* **281:** C166–C178.
34. Ueno, N., M. Murakami, T. Tanioka, *et al.* 2001. Coupling between cyclooxygenase, terminal prostanoid synthase, and phospholipase A2. *J Biol. Chem.* **276:** 34918–34927.
35. Zahner, G., G. Wolf, R. Reinking, *et al.* 2002. Cyclooxygenase-2 overexpression inhibits platelet-derived growth factor-induced mesangial cell proliferation through induction of the tumor suppressor gene p53 and the cyclin-dependent kinase inhibitors p21waf-1/cip-1 and p27kip-1. *J. Biol. Chem.* **277:** 9763–9771.
36. Yuan, B., K. Ohyama, T. Bessho, *et al.* 2006. Contribution of inducible nitric oxide synthase and cyclooxygenase-2 to apoptosis induction in smooth chorion trophoblast cells of human fetal membrane tissues. *Biochem. Biophys. Res. Commun.* **341:** 822–827.
37. Hinz, B., R. Ramer, K. Eichele, *et al.* 2004. Up-regulation of cox-2 expression is involved in R(+)-methanandamide-induced apoptosis of human neuroglioma cells. *Mol. Pharmacol.* **66:** 1643–1651.
38. Han, J.A., J.I. Kim, P.P. Ongusaha, *et al.* 2002. p53 mediated induction of Cox-2 counteracts p53- or genotoxic stress-induced apoptosis. *EMBO J.* **21:** 5635–5644.
39. Moalic-Juge, S., B. Liagre, R. Duval, *et al.* 2002. The anti-apoptotic property of NS-398 at high dose can be mediated in part through NF-kappaB activation, hsp70 induction and a decrease in caspase-3 activity in human osteosarcoma cells. *Int. J. Oncol.* **20:** 1255–1262.
40. Swamy, M.V., C.R. Herzog & C.V. Rao. 2003. Inhibition of COX-2 in colon cancer cell lines by celecoxib increases the nuclear localization of active p53. *Cancer Res.* **63:** 5239–5242.
41. Williams, J.L., N. Nath, J. Chen, *et al.* 2002. Growth inhibition of human colon cancer cells by nitric oxide

(NO)-donating aspirin is associated with cyclooxygenase-2 induction and beta-catenin/T-cell factor signaling, nuclear factor-kappaB, and NO synthase 2 inhibition: implications for chemoprevention. *Cancer Res.* **63:** 7613–7618.
42. King, J.G. Jr. & K. Khalili. 2001. Inhibition of human brain tumor cell growth by the anti-inflammatory drug, flurbiprofen. *Oncogene* **20:** 6864–6870.
43. Corcoran, C.A., Q. He, Y. Huang, *et al.* 2005. Cyclooxygenase-2 interacts with p53 and interferes with p53-dependent transcription and apoptosis. *Oncogene* **24:** 1634–1640.
44. Moos, P.J., K. Edes & F.A Fitzpatrick. 2000. Inactivation of wild-type p53 tumor suppressor by electrophilic prostaglandins. *Proc. Natl. Acad. Sci. USA* **97:** 9215–9220.
45. Parfenova, H., L. Balabanova & C.W. Leffler. 1998. Posttranslational regulation of cyclooxygenase by tyrosine phosphorylation in cerebral endothelial cells. *Am. J. Physiol.* **274:** C72–C81.
46. Shih, A., H.Y. Lin, F.B. Davis, *et al.* 2001. Thyroid hormone promotes serine phosphorylation of p53 by mitogen-activated protein kinase. *Biochemistry* **40:** 2870–2878.
47. Cheng, H.F. & R.C. Harris. 2002. Cyclooxygenase-2 expression in cultured cortical thick ascending limb of Henle increases in response to decreased extracellular ionic content by both transcriptional and post-transcriptional mechanisms. Role of p38-mediated pathways. *J. Biol. Chem.* **277:** 45638–45643.
48. Markowitz, S.D. 2007. Aspirin and colon cancer–targeting prevention? *N. Engl. J. Med.* **356:** 2195–2198.
49. Bai D., L. Ueno & P.K. Vogt. 2009. Akt-mediated regulation of NFkappaB and the essentialness of NFkappaB for the oncogenicity of PI3K and Akt. *Int. J Cancer.* **125:** 2863–2870.
50. Perrone G., M. Zagami, V. Altomare, *et al.* 2007. COX-2 localization within plasma membrane caveolae-like structures in human lobular intraepithelial neoplasia of the breast. *Virchows Arch.* **451:** 1039–1045.
51. Jang, M. & J.M. Pezzuto. 1998. Effects of resveratrol on 12-O-tetradecanoylphorbol-13-acetate-induced oxidative events and gene expression in mouse skin. *Cancer Lett.* **134:** 81–89.
52. Banerjee, S., C. Bueso-Ramos & B.B. Aggarwal. 2002. Suppression of 7,12-dimethylbenz(a)anthracene-induced mammary carcinogenesis in rats by resveratrol: role of nuclear factor-kappaB, cyclooxygenase 2, and matrix metalloprotease 9. *Cancer Res.* **62:** 4945–4954.
53. Li, Z.G., T. Hong, Y. Shimada, *et al.* 2002. Suppression of N-nitrosomethylbenzylamine (NMBA)-induced esophageal tumorigenesis in F344 rats by resveratrol. *Carcinogenesis* **23:** 1531–1536.
54. Djakiew, D. 2000. Dysregulated expression of growth factors and their receptors in the development of prostate cancer. *Prostate* **42:** 150–160.
55. Stewart, J.R. & C.A. O'Brian. 2004. Resveratrol antagonizes EGFR-dependent Erk1/2 activation in human androgen-independent prostate cancer cells with associated isozyme-selective PKC alpha inhibition. *Invest. New Drugs* **22:** 107–117.
56. Haider, U.G., T.U. Roos, M.I. Kontaridis, *et al.* 2005. Resveratrol inhibits angiotensin II- and epidermal growth factor-mediated Akt activation: role of Gab1 and Shp2. *Mol. Pharmacol.* **68:** 41–48.
57. Chew, C.S., K. Nakamura & A.C. Petropoulos. 1994. Multiple actions of epidermal growth factor and TGF-alpha on rabbit gastric parietal cell function. *Am. J. Physiol.* **267:** G818–G826.
58. Fletcher, S., J.A. Drewry, V.M. Shahani, *et al.* 2009. Molecular disruption of oncogenic signal transducer and activator of transcription 3 (STAT3) protein. *Biochem. Cell Biol.* **87:** 825–833.
59. Yang, X., D. Qiao, K. Meyer, *et al.* 2009. Signal transducers and activators of transcription mediate fibroblast growth factor-induced vascular endothelial morphogenesis. *Cancer Res.* **69:** 1668–1677.
60. Aggarwal, B.B., G. Sethi, K.S. Ahn, *et al.* 2006. Targeting signal-transducer-and-activator-of-transcription-3 for prevention and therapy of cancer: modern target but ancient solution. *Ann. N.Y. Acad. Sci.* **1091:** 151–169.
61. Chen, D.B. & J.S. Davis. 2003. Epidermal growth factor induces c-fos and c-jun mRNA via Raf-1/MEK1/ERK-dependent and -independent pathways in bovine luteal cells. *Mol. Cell Endocrinol.* **200:** 141–154.
62. Scapoli, L., M.E. Ramos-Nino, M. Martinelli, *et al.* 2004. Src-dependent ERK5 and Src/EGFR-dependent ERK1/2 activation is required for cell proliferation by asbestos. *Oncogene* **23:** 805–813.
63. Kaneuchi, M., M. Sasaki, Y. Tanaka, *et al.* 2003. Resveratrol suppresses growth of Ishikawa cells through down-regulation of EGF. *Int. J. Oncol.* **23:** 1167–1172.
64. Azios, N.G. & S.F. Dharmawardhane. 2005. Resveratrol and estradiol exert disparate effects on cell migration, cell surface actin structures, and focal adhesion assembly in MDA-MB-231 human breast cancer cells. *Neoplasia* **7:** 128–140.
65. Aggarwal, B.B., A. Bhardwaj, R.S. Aggarwal, *et al.* 2004. Role of resveratrol in prevention and therapy of cancer: preclinical and clinical studies. *Anticancer Res.* **24:** 2783–2840.
66. Serrero, G. & R. Lu. 2001. Effect of resveratrol on the expression of autocrine growth modulators in human breast cancer cells. *Antioxid. Redox Signal.* **3:** 969–979.
67. Lin, H.Y., H.Y. Tang, A. Shih, *et al.* 2007. Thyroxine induces MAPK activation and blocks resveratrol-induced apoptosis in human thyroid cancer cell lines. *Steroids* **72:** 180–187.
68. Lin, H.Y., M. Sun, H.Y. Tang, *et al.* 2009. L-Thyroxine vs. 3,5,3′-triiodo-L-thyronine and cell proliferation: activation of mitogen-activated protein kinase and phosphatidylinositol 3-kinase. *Am. J. Physiol. Cell Physiol.* **296:** C980–C991.
69. Davis, P.J., F.B. Davis & V. Cody. 2005. Membrane receptors mediating thyroid hormone action. *Trends Endocrinol. Metab.* **16:** 429–435.
70. Wade, C.B. & D.M. Dorsa. 2003. Estrogen activation of cyclic adenosine 5′-monophosphate response element-mediated transcription requires the extracellularly regulated kinase/mitogen-activated protein kinase pathway. *Endocrinology* **144:** 832–838.
71. Kato, S., H. Endoh, Y. Masuhiro, *et al.* 1995. Activation of the estrogen receptor through phosphorylation by mitogen-activated protein kinase. *Science* **270:** 1491–1494.

72. Santen, R.J., R.X. Song, R. McPherson, *et al.* 2002. The role of mitogen-activated protein (MAP) kinase in breast cancer. *J. Steroid Biochem. Mol. Biol.* **80:** 239–256.
73. Azios, N.G. & S.F. Dharmawardhane. 2005. Resveratrol and estradiol exert disparate effects on cell migration, cell surface actin structures, and focal adhesion assembly in MDA-MB-231 human breast cancer cells. *Neoplasia* **7:** 128–140.
74. Bhat, K.P., D. Lantvit, K. Christov, *et al.* 2001. Estrogenic and antiestrogenic properties of resveratrol in mammary tumor models. *Cancer Res.* **61:** 7456–7463.
75. Cho, D.I., N.Y. Koo, W.J. Chung, *et al.* 2002. Effects of resveratrol-related hydroxystilbenes on the nitric oxide production in macrophage cells: structural requirements and mechanism of action. *Life Sci.* **71:** 2071–2082.

Issue: *Resveratrol and Health*

Chemopreventive effects of resveratrol and resveratrol derivatives

Thomas Szekeres,[1] Philipp Saiko,[1] Monika Fritzer-Szekeres,[1] Bob Djavan,[2] and Walter Jäger[3]
[1]Institute for Laboratory Medicine, Department for Medical and Chemical Laboratory Diagnostics, Medical University of Vienna, General Hospital of Vienna, Vienna, Austria. [2]Department of Urology, New York University, New York, New York. [3]Department of Clinical Pharmacy and Diagnostics, Faculty of Life Sciences, University of Vienna, Vienna, Austria

Address for correspondence: Thomas Szekeres, M.D., Ph.D., Institute for Laboratory Medicine, Medical University of Vienna, General Hospital of Vienna, 5H, Waehringer Guertel 18–20, A-1090 Vienna, Austria. thomas.szekeres@meduniwien.ac.at

Resveratrol is considered to have a number of beneficial effects. Recently, our group modified the molecule and synthesized a number of compounds with different biochemical effects. Polymethoxy and polyhydroxy derivatives of resveratrol were shown to inhibit tumor cell growth in various cell lines and inflammation pathways (cyclooxygenases activity), in part more effectively than resveratrol itself. One lead compound (hexahydroxystilbene, M8) turned out to be the most effective inhibitor of tumor cell growth and of cyclooxygenase 2 activity. M8 was then studied in two different human melanoma mouse models. This novel resveratrol analog was able to inhibit melanoma tumors in a primary tumor model alone and in combination with dacarbacine, an anticancer compound that is used for melanoma treatment. We also tested the development of lymph node metastasis in a second melanoma model and again M8 successfully inhibited the tumor as well as the size and weight of lymph node metastasis. Hydroxylated resveratrol analogs therefore represent a novel class of anticancer compounds and promising candidates for *in vivo* studies.

Keywords: resveratrol; resveratrol metabolism; free radicals; antitumor effects; polyhydroxyphenols; melanoma; ribonucleotide reductase

Introduction

Resveratrol (3,4′,5-trihydroxy-*trans*-stilbene, Fig. 1) is considered to be the ingredient of red wine responsible for the so-called French paradox.[1] Indeed, the incidence of heart infarction is 40% lower in France than in the rest of the Western world, despite similar demographics. One reason for this effect might be the drinking habits in France and the preventive effects of wine ingredients like alcohol and/or hydroxyphenolic compounds such as gallic acid or resveratrol. Resveratrol was identified to be one important ingredient of red wine. It is an excellent free radical scavenger with numerous beneficial effects including the prevention of blood vessel disease. For instance, platelet inhibition or inhibition of low density lipoprotein (LDL) cholesterol oxidation by resveratrol might be reasons for these beneficial effects. In addition to anti-inflammatory effects, resveratrol was identified as a chemopreventive agent against malignant transformation.[2–4] Potter *et al.* identified polyhydroxy metabolites of resveratrol and speculated that the beneficial effects might also be caused by metabolites and intermediates.[5] Our group has previously studied various other polyhydroxyphenols, such as didox and trimidox and their biochemical and cytotoxic effects, in various tumor cell systems and in animals, elucidating mechanisms responsible for the biochemical interactions of these compounds. Elford *et al.*, who invented didox and trimidox (polyhydroxy-substituted benzohydroxamic acid derivates), have found that the number and position of hydroxyl groups on the phenol ring significantly influences the activity of the compounds.[7,8] These findings were confirmed by our experiments, and we speculated that similar effects might be true for resveratrol and resveratrol

doi: 10.1111/j.1749-6632.2010.05864.x

Resveratrol (3,5,4'-trihydroxy-*trans*-stilbene)

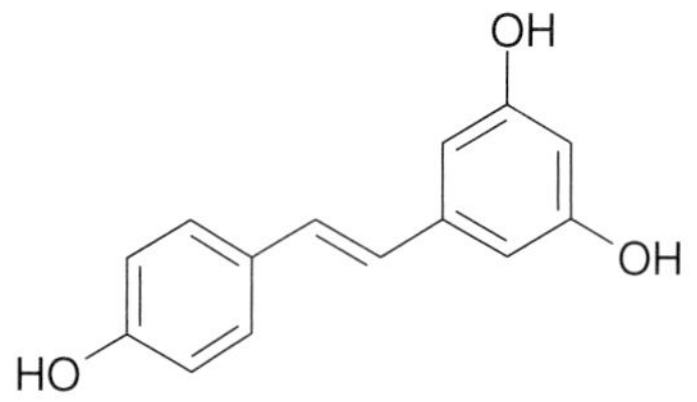

Piceatannol (3,5,3',4'-tetrahydroxy-*trans*-stilbene)

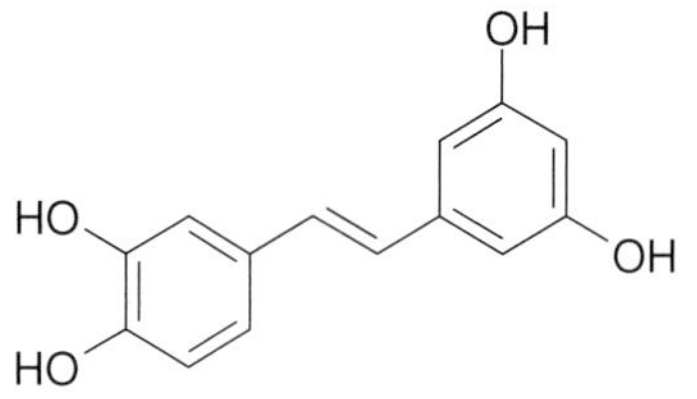

M8 (3,4,5,3',4',5'-hexahydroxy-*trans*-stilbene)

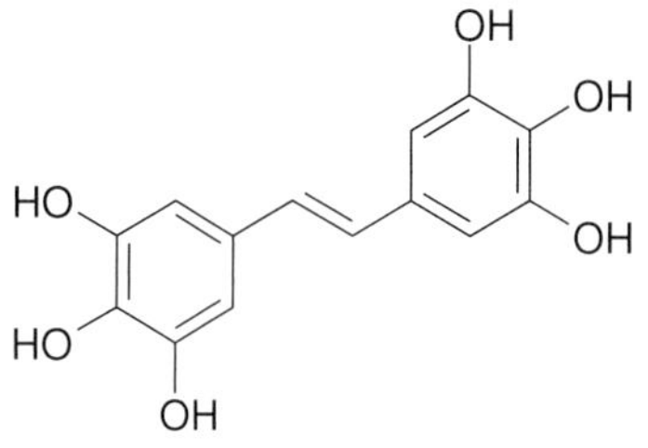

Figure 1. Structures of resveratrol, piceatannol, and hexahydroxystilbene (M8).

analogs. Resveratrol is also a molecule consisting of hydroxyphenolic moieties and shares important biochemical targets with didox or trimidox. For instance, despite other biochemical effects it was shown that not only didox and trimidox but resveratrol is a highly active inhibitor of the enzyme ribonucleotide reductase (RR), a key enzyme of DNA synthesis in malignant cells.[9] Inhibition of this enzyme can target tumor cells and cause inhibition of rapidly proliferating tumor cell growth, whereas non-malignant cells proliferate at a lower rate and do not need RR activity for survival. The enzyme needs a free tyrosyl radical for activity, and compounds with free radical scavenging activity can inhibit enzyme activity. This leads to the inhibition of the formation of deoxynucleoside triphosphates (dNTP) in tumor cells. As dNTPs are needed for DNA synthesis, inhibition of RR can cause growth arrest and apoptosis of tumor cells. Therefore, this enzyme is considered to be an excellent target for anticancer agents.

Metabolism of resveratrol

Several studies conducted on humans have reported a very low oral bioavailability based on extensive biotransformation. The metabolic pattern of resveratrol is complex; Boocock *et al.* identified two monosulfates, one disulfate, two monoglucuronides, and one glucuronide-sulfate in human plasma.[10] In contrast, Burkon *et al.* identified seven metabolites, namely 3-sulfate, 3,4′-disulfate, 3,5-disulfate, 3-glucuronide, 4′-glucuronide, and two diglucuronides (Fig. 2).[11] These discrepancies may be explained by a strong influence of applied dose on the metabolic profile of resveratrol. *In vitro* studies conducted by our lab using human intestinal Caco-2 cells and the isolated perfused rat liver showed that sulfation prevailed in the lower resveratrol concentrations, but when the applied dose was raised, sulfate formation dropped dramatically. As a consequence of the observed inhibition of resveratrol sulfation, conjugation with glucuronic acid was the main metabolic pathway in higher resveratrol concentrations.[12,13]

Although extensively metabolized, many preclinical studies have already proved the anticancer activity of resveratrol *in vivo*, and several phase I/II clinical trials of oral resveratrol in cancer prevention and therapy are already on the way. This discrepancy between observed *in vivo* action and extensive biotransformation may be explained by enzymatic

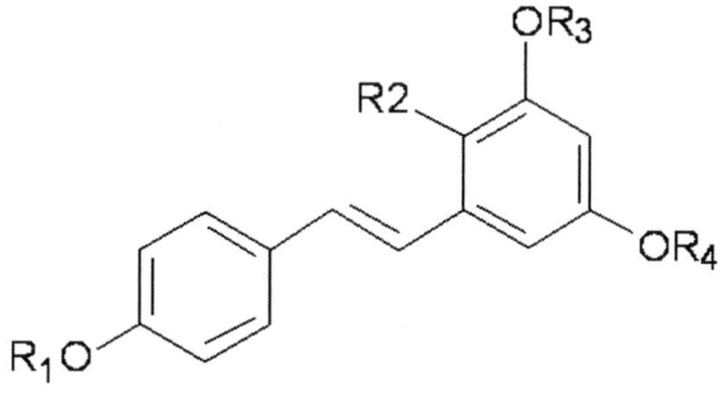

Figure 2. Major resveratrol metabolites identified in humans.

Table 1. Structures of the resveratrol analogs 1–12[18]

Compd.	Pos. 3 (= R1)	Pos. 4 (= R2)	Pos. 5 (= R3)	Pos. 3′ (= R4)	Pos. 4′ (= R5)	Pos. 5′ (= R6)
1	$-OCH_3$	−H	$-OCH_3$	−H	$-OCH_3$	−H
2	$-OCH_3$	$-OCH_3$	$-OCH_3$	−H	$-OCH_3$	−H
3	$-OCH_3$	−H	$-OCH_3$	$-OCH_3$	−H	$-OCH_3$
4	$-OCH_3$	−H	$-OCH_3$	$-OCH_3$	$-OCH_3$	−H
5	$-OCH_3$	$-OCH_3$	$-OCH_3$	$-OCH_3$	−H	$-OCH_3$
6	$-OCH_3$	$-OCH_3$	$-OCH_3$	$-OCH_3$	$-OCH_3$	$-OCH_3$
7	−OH	−H	−OH	−H	−OH	−H
8	−OH	−OH	−OH	−H	−OH	−H
9	−OH	−H	−OH	−OH	−H	−OH
10	−OH	−H	−OH	−OH	−OH	−H
11	−OH	−OH	−OH	−OH	−H	−OH
12	−OH	−OH	−OH	−OH	−OH	−OH

R4 R5 R6 R1 R2 R3

7: resveratrol (3,5,4′-trihydroxy-*trans*-stilbene), **10**: piceatannol (3,5,3′4′-tetrahydroxy-*trans*-stilbene, **12**: M8 (3,4,5,3′,4′,5′-hexahydroxy-*trans*-stilbene).

hydrolysis of the resveratrol conjugates in the tissue, enterohepatic recirculation after deconjugation in the gut, or possible biological activity of the metabolites themselves. Indeed, very recent data showed that resveratrol conjugates are biological active.[10,14] Resveratrol-3-sulfate but not the 3-glucuronide exhibited biological activities known to be mediated by parent resveratrol: induction of quinone reductase 1, free radical scavenging, inhibition of cyclooxygenase 1 or 2 isoenzymes (COX-1/2), and inhibition of alpha-induced NF-κB activity or activation of SIRT1.[15,16] Resveratrol-3-glucuronide, on the other hand, showed higher antioxidant activity than the parent compound itself.[17] These recent studies indicate that the resveratrol conjugates may act through different molecular targets and therefore contribute *in vivo* to the diversity of health benefits previously attributed only to resveratrol itself.

Analogs of resveratrol

In the attempt to improve the beneficial effects of resveratrol, including the induction of programmed cell death in tumor cells and the inhibition of inflammation by inhibiting cyclooxygenase activity, we synthesized a number of polyhydroxy and polymethoxy substituted analogs of resveratrol (Table 1).[18] Cyclooxygenase is the enzyme responsible for prostaglandin synthesis from arachidonic acid. Two isoenzymes are known: COX-1 and COX-2. COX-1 is constitutively expressed and its inhibition is responsible for side effects of various anti-inflammatory compounds, and COX-2 inhibition is responsible for the anti-inflammatory and pain relieving effects. Resveratrol, like aspirin, inhibits both forms of cyclooxygenases. Looking for analogs of resveratrol that might exhibit COX-2 selectivity, a quantitative structure–activity relationship study was performed in order to identify various structural parameters of the molecules with inhibitory effects on COX-1 and COX-2 inhibition (Table 2).[18]

The methoxy derivatives were weak inhibitors of COX-1 with 2–55-fold higher IC_{50}-values (the drug concentration that causes 50% inhibition of enzyme activity) than resveratrol.[18] On the other hand,

Table 2. Inhibitory effect of 1–12 and the reference compound celecoxib on COX-1 and COX-2 activity[18]

Compound	IC_{50} (μM)		Selectivity index COX-1/COX-2
	COX-1	COX-2	
1	1.228	1.667	0.74
2	9.099	7.797	1.17
3	27.783	1.575	17.64
4	2.834	0.796	3.56
5	7.247	0.514	14.11
6	11.348	0.355	31.97
7	0.535	0.996	0.54
8	2.072	0.04537	45.67
9	0.00998	0.00171	5.83
10	4.713	0.0113	417.08
11	0.01027	0.00138	7.44
12	0.748	0.00104	719.23
Celecoxib	19.026	0.03482	546.41

7: resveratrol (3,5,4′-trihydroxy-*trans*-stilbene), **10**: piceatannol (3,5,3′4′-tetrahydroxy-*trans*-stilbene, **12**: M8 (3,4,5,3′,4′,5′-hexahydroxy-*trans*-stilbene).

hydroxylated resveratrol analogs, especially tetra and pentahydroxystilbenes, were more potent inhibitors, with lower IC_{50}-values.[18] All hydroxylated analogs were more potent against COX-2 than resveratrol.[18] Based on the IC_{50} values for COX-1 and COX-2, the relative IC_{50} ratios of COX-1/COX-2 were calculated. Resveratrol showed a weak COX-2 inhibition with a selectivity index (COX-1/COX-2) of 0.5.[18] However, hexahydroxystilbene (M8) exhibited the highest selectivity index for COX-2 (selectivity index = 719); it was even more selective towards COX-2 than Celebrex® (selectivity index = 546), a commercially available highly selective COX-2 inhibitor.[18] Methoxylated resveratrol analogs did not show any COX selectivity and inhibited COX-2 at much higher concentration than hydroxylated analogs.[18] Very similar effects were seen when free radical scavenging capacities were investigated using our compounds.[19]

Free radical scavenging experiments with O_2 (electron spin resonance) and 2,2-diphenyl-1-picrylhydrazyl (photometry) revealed that tetrahydroxystilbene and hexahydroxystilbene exerted a more than 6600-fold higher anti-radical activity than resveratrol.[19] Furthermore, in HL-60 human leukemia cells, hydroxystilbenes with ortho-hydroxyl groups exhibited a more than three-fold higher cytotoxic activity than hydroxystilbenes with other substitution patterns.[19] Oxidation of ortho-hydroxystilbenes in a microsomal model system resulted in the existence of ortho-semiquinones.[19]

Biochemical effects of resveratrol and resveratrol analogs

Resveratrol was shown to cause a number of biochemical effects. Anti-inflammatory effects are described above. One important mechanism of action seems to be the free radical scavenging activity. Inhibition of RR was mentioned in the introduction; resveratrol causes an imbalance of dNTPs, which are precursors of DNA synthesis. In cancer cells, resveratrol can, for instance, induce apoptosis and downregulation of either NF-κB or Bcl2. All these biochemical effects play a role in the antineoplastic effects of resveratrol and resveratrol analogs.[20–28] Our group first synthesized a number of analogs and then tested these compounds alone and in combination.[18–28] One important drug combination is facilitating combination effects on dNTP synthesis, with the effects of a second group of antimetabolites to achieve additive and synergistic cytotoxic effects. First, we combined resveratrol and resveratrol analogs with cytosine arabinoside (Ara-C), which is converted to its triphosphate (Ara-CTP) for activity.[23,27,28] Inhibition of RR depletes dNTP synthesis and decreases their pool sizes, with the consequence that more Ara-C can be phosphorylated to its active metabolite, which is then responsible for synergism.[27] We could show that resveratrol, but also piceatannol, the monohydroxylated resveratrol metabolite as well as M8 were capable of enhancing the effects caused by Ara-C.[23,27,28] Inhibition of DNA synthesis, such as other biochemical effects, also causes programmed cell death or apoptosis.

Hexahydroxystilbene (M8) was also shown to induce apoptosis at concentrations much lower than resveratrol (Fig. 3), which is in line with other experiments that showed the superior activity of M8.[18,19]

Structure–activity relationship

Structure–activity studies revealed that increasing the number of OH groups and their position (ortho) on the phenol ring could increase the free radical scavenging capacity of the compounds.[19]

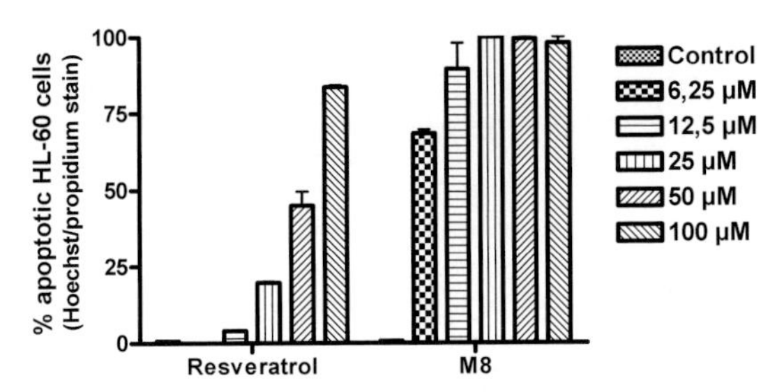

Figure 3. Induction of apoptosis by resveratrol and M8 in human HL-60 promyelocytic leukemia cells.

We believe that the increased cytotoxicity of ortho-hydroxystilbenes is related to the presence of ortho-semiquinones formed during metabolism or autoxidation.[19] In addition, cytotoxic activity could be enhanced, which was shown by lowering the IC_{50} values of the compounds active against a number of human tumor cell lines, including leukemia, prostate, colon, breast, and melanoma tumor cell lines.[19–29] Polyhydroxy resveratrol derivatives, in particular the most effective hexahydroxy compound (M8), significantly inhibited the activity of the enzyme RR, which is considered to be a very important target for antitumor activity due to significantly increased enzyme activity in human tumor cells.[28]

We therefore selected M8 as lead compound for further investigations. M8 was used for *in vivo* experiments in human melanoma models.[29,30]

In vivo effects of hexahydroxystilbene

As mentioned, M8 (hexahydroxystilbene) turned out to be the most effective resveratrol analog synthesized by our group.[18,19] It provided the lowest IC_{50} values in tumor cell lines, showed highly selective cyclooxygenase 2 inhibitory activity at very low concentrations of the compound, and it was the most effective free radical scavenger out of a large group of polymethoxy and polyhydroxystilbene analogs.[18,19] The compound was then examined in animal models. Two questions were addressed: is the compound toxic *in vivo* and is it proven to be active against *in vivo* tumors as well?[29,30] We first investigated the single agent activity and mechanism of action of M8 in a human melanoma severe combined immunodeficiency (SCID) mouse xenotransplantation model.[29] This tumor entity is still very difficult to treat and in the advanced stage only very few options for treatment are left. Therefore, there is a need for additional treatment options.

Mice suffering from palpable tumor disease after inoculation of human melanoma 518A2 cells were treated with i.p. doses of 2.5 and 5 mg/kg/day for a period of 4 weeks.[29] Both doses were well tolerated and significantly reduced the weight of the primary tumor.[29] Then, M8 was tested in combination with dacarbacine (DTIC), one of the few standard chemotherapeutic regimens used in human melanoma.[29] After establishment of palpable tumors, the animals were treated with 2.5 mg/kg/day M8 for 14 days and, in addition, DTIC (80 mg/kg i.p.) was administered on days 4 and 6.[29] The combination yielded highly significant synergistic effects.[29] Three out of six animals treated with the combination were tumor free after 14 days of treatment; the other three animals displayed small residual tumors.[29] These promising results prompted us to further investigate the effects of M8 in a melanoma metastasis tumor model *in vivo*.[30] In a CB17 scid mouse model with M24met cells, the *in vivo* effects of M8 were studied. This model allows for the investigation of the effects on the primary tumor as well as the treatment effects on distant lymph node metastasis. Treatment of animals after the development of palpable tumors with 2.5 and 5 mg/kg M8 for 10 days caused a significant reduction of the primary tumor, which was accompanied by a reduction of $Ki\text{-}67^{+}$ cells in the tumor.[30] As the development of metastasis is a critical step for tumor spreading and disease progression, the effects of M8 on lymph-node metastasis were studied in this model. The above-mentioned treatment regimen was continued for another 10 days, and then lymph nodes were evaluated. Lymph nodes of the control group weighed 0.75 mg whereas treatment with 2.5 or 5 mg/kg M8 reduced weight significantly to 0.3 mg.[30] These results were confirmed when lymph node volumes were calculated. The number of tumor cells in the lymph nodes could also be reduced significantly by ~50% by treatment with M8.[30] Inguinal stage of metastasis was reached in 56% of untreated animals, whereas none of the treated animals displayed inguinal tumor metastasis.[30] We therefore conclude that treatment leads to a reduction of distant lymph node metastasis, which underlines that M8 might be a very promising approach for the treatment of human melanomas. Further beneficial effects might be seen with drug combinations using M8 together with conventional chemotherapeutic regimens.

Conclusion

We conclude that resveratrol, as well as polyhydroxy analogs of resveratrol, are able to alter various biochemical pathways, and thus they might be used not only as preventive agents against blood vessel disease but also in the treatment of inflammation and cancer.

Conflicts of interest

The authors declare no conflicts of interest.

References

1. Renaud, S. & M. de Lorgeril. 1992. Wine, alcohol, platelets, and the French paradox for coronary heart disease. *Lancet.* **339:** 1523–1526.
2. Pace-Asciak C.R., S. Hahn, E.P. Diamandis, *et al.* 1995. The red wine phenolics trans-resveratrol and quercetin block human platelet aggregation and eicosanoid synthesis: implications for protection against coronary heart disease. *Clin. Chim. Acta.* **235:** 207–219.
3. Frankel E.N., A.L. Waterhouse & J.E. Kinsella. 1993. Inhibition of human LDL oxidation by resveratrol. *Lancet.* **341:** 1103–1104.
4. Huang C., W.Y. Ma, A. Goranson, *et al.* 1999. Resveratrol suppresses cell transformation and induces apoptosis through a p53-dependent pathway. *Carcinogenesis* **20:** 237–242.
5. Potter G.A., L.H. Patterson, E. Wanogho, *et al.* 2002. The cancer preventative agent resveratrol is converted to the anticancer agent piceatannol by the cytochrome P450 enzyme CYP1B1. *Br. J. Cancer.* **86:** 774–778.
6. Szekeres T., K. Gharehbaghi, M. Fritzer, *et al.* 1994. Biochemical and antitumor activity of trimidox, a new inhibitor of ribonucleotide reductase. *Cancer Chemother. Pharmacol.* **34:** 63–66.
7. Elford H.L., M. Freese, E. Passamani, *et al.* 1970. Ribonucleotide reductase and cellproliferation. I. Variations of ribonucleotide reductase activity with tumor growth rate in a series of rat hepatomas. *J. Biol. Chem.* **245:** 5228–5233.
8. Elford, HL, G.L. Wampler & B. van't Riet. 1979. New ribonucleotide reductase inhibitors with antineoplastic activity. *Cancer Res.* **39:** 844–851.
9. Fontecave, M., M. Lepoivre, E. Elleingand, *et al.* 1998. Resveratrol, a remarkable inhibitor of ribonucleotide reductase. *FEBS Lett.* **421:** 277–279.
10. Boocock, D.J., G.E. Faust, K.R. Patel, *et al.* 2007. Phase I dose escalation pharmacokinetic study in healthy volunteers of resveratrol, a potential cancer chemopreventive agent. *Cancer Epidemiol. Biomarkers Prev.* **16:** 1246–1252.
11. Burkon, A. & V. Somoza. 2008. Quantification of free and protein-bound trans-resveratrol metabolites and identification of trans-resveratrol-C/O-conjugated diglucuronides—two novel resveratrol metabolites in human plasma. *Mol. Nutr. Food Res.* **52:** 549–557.
12. Maier-Salamon, A., B. Hagenauer, M. Wirth, *et al.* 2006. Increased transport of resveratrol across monolayers of the human intestinal Caco-2 cells is mediated by inhibition and saturation of metabolites. *Pharm. Res.* **23:** 2107–21015.
13. Maier-Salamon, A., B. Hagenauer, G. Reznicek, *et al.* 2008. Metabolism and disposition of resveratrol in the isolated perfused rat liver: role of Mrp2 in the biliary excretion of glucuronides. *J. Pharm. Sci.* **97:** 1615–1628.
14. Vitrac, X. A., Desmoulière, B. Brouillaud, *et al.* 2003. Distribution of [^{14}C]-trans-resveratrol, a cancer chemopreventive polyphenol, in mouse tissues after oral administration. *Life Sci.* **72:** 2219–2233.
15. Hoshino, J. , E.J. Park,T.P Kondratyuk, *et al.* 2010. Selective synthesis and biological evaluation of sulfate-conjugated resveratrol metabolites. *J. Med. Chem.* **53:** 5033–5043.
16. Calamini, B., K. Ratia, M.G. Malkowski, *et al.* 2010. Pleiotropic mechanisms facilitated by resveratrol and its metabolites. *Biochem. J.* **429:** 273–282.
17. Mikulski, D. & M. Molski. 2010. Quantitative structure-antioxidant activity relationship of trans-resveratrol oligomers, trans-4,4'-dihydroxystilbene dimer, trans-resveratrol-3-O-glucuronide, glucosides: trans-piceid, cis-piceid, trans-astringin and trans-resveratrol-4'-O-beta-D-glucopyranoside. *Eur. J. Med. Chem.* **45:** 2366–2380.
18. Murias, M., N. Handler, T. Erker, *et al.* 2004. Resveratrol analogues as selective cyclooxygenase-2 inhibitors: synthesis and structure-activity relationship. *Bioorg. Med. Chem.* **12:** 5571–5578.
19. Murias, M., W. Jäger, N. Handler, *et al.* 2005. Antioxidant, prooxidant and cytotoxic activity of hydroxylated resveratrol analogues: structure-activity relationship. *Biochem. Pharmacol.* **69:** 903–912.
20. Miksits, M., K. Wlcek, M. Svoboda, *et al.* 2009. Antitumor activity of resveratrol and its sulfated metabolites against human breast cancer cells. *Planta Med.* **75:** 1227–1230.
21. Bader, Y., S. Madlener, S. Strasser, *et al.* 2008. Stilbene analogues affect cell cycle progression and apoptosis independently of each other in an MCF-7 array of clones with distinct genetic and chemoresistant backgrounds. *Oncol. Rep.* **19:** 801–810.
22. Saiko, P., M. Pemberger, Z. Horvath, *et al.* 2008. Novel resveratrol analogs induce apoptosis and cause cell cycle arrest in HT29 human colon cancer cells: inhibition of ribonucleotide reductase activity. *Oncol. Rep.* **19:** 1621–1626.
23. Fritzer-Szekeres, M., I. Savinc, Z. Horvath, *et al.* 2008. Biochemical effects of piceatannol in human HL-60 promyelocytic leukemia cells—synergism with Ara-C. *Int. J. Oncol.* **33:** 887–8892.
24. Horvath, Z., S. Marihart-Fazekas, P. Saiko, *et al.* 2007. Novel resveratrol derivatives induce apoptosis and cause cell cycle arrest in prostate cancer cell lines. *Anticancer Res.* **27:** 3459–3464.
25. Vo, T.P., S. Madlener, Z. Bago-Horvath, *et al.* 2009. Pro- and anti-carcinogenic mechanisms of piceatannol are activated dose-dependently in MCF-7 breast cancer cells. *Carcinogenesis* **22:** 845–852.
26. Murias, M., M.W. Luczak, A. Niepsuj, *et al.* 2008. Cytotoxic activity of 3,3',4,4',5,5'-hexahydroxystilbene against breast

cancer cells is mediated by induction of p53 and down-regulation of mitochondrial superoxide dismutase. *Toxicol. In Vitro.* **22:** 1361–1370.
27. Horvath, Z., P. Saiko, C. Illmer, *et al.* 2005. Synergistic action of resveratrol, an ingredient of wine, with Ara-C and tiazofurin in HL-60 human promyelocytic leukemia cells. *Exp. Hematol.* **33:** 329–335.
28. Horvath, Z., M. Murias, P. Saiko, *et al.* 2006. Cytotoxic and biochemical effects of 3,3',4,4',5,5'-hexahydroxystilbene, a novel resveratrol analogue in HL-60 human promyelocytic leukemia cells. *Exp. Hematol.* **34:** 1377–1384.
29. Wachek, V., Z. Horvath, S. Strommer, *et al.* 2005. Resveratrol analogue M8 chemosensitizes malignant melanoma to dacarbacin *in vivo. Clin. Cancer Res.* Part 2(Suppl S 11): 8971S–8971S.
30. Paulitschke, V., N. Schicher, T. Szekeres, *et al.* 2010. 3,3',4,4',5,5'-hexahydroxystilbene impairs melanoma progression in a metastatic mouse model. *J. Invest. Dermatol.* **130:** 1668–1679.

Ann. N.Y. Acad. Sci. ISSN 0077-8923

ANNALS OF THE NEW YORK ACADEMY OF SCIENCES
Issue: *Resveratrol and Health*

The beneficial effect of resveratrol on severe acute pancreatitis

Qingyong Ma, Min Zhang, Zheng Wang, Zhenhua Ma, and Huanchen Sha
Department of Hepatobiliary and Pancreas Surgery, First Affiliated Hospital of Xi'an Jiaotong University, Xi'an, China

Address for correspondence: Qingyong Ma, Department of Hepatobiliary and Pancreas Surgery, First Affiliated Hospital of Xi'an Jiaotong University, 277 West Yanta Road, Xi'an, Shaanxi, 710061, China. qyma56@mail.xjtu.edu.cn

Acute pancreatitis is a common kind of acute abdominal disease. The management of severe acute pancreatitis (SAP) is a challenge because of its high morbidity, which is due to systemic inflammatory response syndrome and multiorgan dysfunction syndrome. Therefore, it is important to explore therapies to control the disease's progression. A series of *in vivo* and *in vitro* experiments has demonstrated that resveratrol—an extract from Chinese herbs, grapes, and many plants—exhibits a wide range of biological and pharmacological activities, including anti-inflammatory, antioxidation, and chemopreventive effects, as well as the inhibition of platelet aggregation, which could benefit the treatment of SAP. Here, we examine the possible mechanism of resveratrol in treating the progression of SAP. Resveratrol could inhibit the production and progression of SAP through down-regulating pro-inflammatory cytokines, improving microcirculation, modulating cell apoptosis, and blocking calcium overload. We propose that resveratrol has a potentially therapeutic effect on the progression of SAP.

Keywords: resveratrol; severe acute pancreatitis; inflammatory mediators; microcirculation; apoptosis

Introduction

Acute pancreatitis is a sudden inflammation of the pancreas. Depending on the severity of the disease, it has major complications and high mortality despite treatment. In about 15–20% of patients with acute pancreatitis, severe damage to the pancreas may lead to a life-threatening illness that is often associated with prolonged hospitalization, multiple surgical procedures, and death. Severe acute pancreatitis (SAP) usually develops when parts of the pancreas become necrotic from the acute inflammation. Patients with SAP can experience shock, acute respiratory failure, renal failure, multiorgan dysfunction syndrome (MODS), and even death.[1] Since the middle 1990s, when treating acute pancreatitis, most specialists consider the most common treatment the best option and use different therapeutic regimens according to the particular patient's needs and the stage of the disease.[2] The management of SAP is a challenge because of its high morbidity, which is due to systemic inflammatory response syndrome (SIRS) and multiple organ failure. Since it is an inflammatory disease, some compounds with anti-inflammatory properties have received much attention in recent studies.

Resveratrol (*trans*-3, 4′, 5-trihydroxystilbene) is a plant-derived polyphenolic phytoalexin that is in high concentrations in the skin and seeds of grapes.[3] A series of experiments has demonstrated that resveratrol exhibits a wide range of biological and pharmacological activities, including anti-inflammatory, antioxidation, and chemopreventive effects, and inhibits platelet aggregation.[4–11] Due to its wide pharmacological activity and ease of extraction, resveratrol has received an increasing amount of attention in treating many diseases. In previous studies, it has been shown that resveratrol exerts therapeutic effects in inflammatory diseases through inhibiting the expression of pro-inflammatory factors, reversing the microcirculatory disturbance, and promoting antioxidant effects.[12–14] Recently, the drug has shown potential therapeutic effects in cases of SAP.

SAP is a life-threatening cause of acute abdomen and requires urgent treatment. Studies have found that pancreatic microcirculation disturbance, leukocyte over activation, the expression of

doi: 10.1111/j.1749-6632.2010.05847.x

inflammatory mediators, apoptosis of pancreatic acinar cells, and calcium overload play important roles in disease progression.[15–17] Resveratrol possesses various biological properties including anti-inflammatory, antioxidative, and anti-coagulative characteristics. The aim of this study was an incorporation of all of the relevant findings of resveratrol on SAP.

Inhibitory effect of resveratrol on the production and release of inflammatory mediators

A series of studies have shown that resveratrol has favorable anti-inflammatory effects in both acute and chronic inflammation.[18–20] Intra-articular injection of resveratrol may protect cartilage against the development of experimental inflammatory arthritis induced by intra-articular injection of lipopolysaccharide.[21] As we know, activated nuclear factor kappa B (NF-κB) and activator protein-1 (AP-1), which are both nuclear transcription factors with multiple functions, can induce the transcription of cytokines, adhesion molecules and chemokines, which play key roles in inflammation. Resveratrol can block the activation of NF–κB and AP-1, which induce the transcription of tumor necrosis factor-α (TNF-α), PMA, and H_2O_2, and this anti-inflammatory effect or resveratrol may thus be ascribed to the inhibition of activation of NF-κB and AP-1 and the associated kinases.[22] It has also been confirmed that resveratrol can inhibit the activation of NF-κB and then reduce the expression of inducible nitric oxide synthase (iNOS) in macrophages induced by lipopolysaccharide (LPS).[23,24] In another study, it has been found that resveratrol inhibited the production and release of interleukin-8 (IL-8) induced by PMA and the binding of AP-1 with DNA, but there was little influence on the activation of NF-κB.[25] In addition, resveratrol could (1) inhibit IL-8 secretion by blocking the activation of mitogen-activated protein kinase (MAPK) phosphorylation and NF-κB and (2) have significant antibiotic activity against many kinds of bacteria and fungi, such as *Staphylococcus aureus*, *Enterococcus faecalis*, *Trichophyton tonsurans*, *Trichophyton rubrum*, and *Epidermophyton floccosum*.[26,27]

Suppressing the production and release of inflammatory mediators plays an important role in the treatment of SAP. In recent years, studies *in vivo* experiment and *in vitro* suggested that TNF-α plays an initial role in the occurrence and development of histological damage in the pancreas and in extra-pancreatic organs in acute pancreatitis.[28] TNF-α, caused by various factors in an inflammatory reaction, can induce the production and release of many kinds of cytokines (IL-1, IL-6, IL-8) that can damage tissue cells directly and indirectly.[29] Moreover, they can decompose arachidonic acid into several kinds of inflammatory mediators, such as platelet-activating factor, leukotrienes, and thromboxane A2, which aggravates the inflammatory reaction and microcirculatory disturbance.[30] In addition, the cascade chain reaction of inflammatory mediators is the main reason that mild acute pancreatitis progresses to SIRS and multi-organ system failure (MOSF).[31] Current therapies, which include inhibiting the production of inflammatory mediators and antagonizing the functions of these molecules, may improve the prognosis of SAP.

As reported in other studies, it has been shown that resveratrol can decrease the serum levels of TNF-α and IL-6 in SAP modeled rats, and it can reduce mortality, which indicates that resveratrol has therapeutic effect in SAP.[32] The mechanism is possibly related to suppression of the production, activation, and release of the inflammatory mediators. In addition, some data have demonstrated that the levels of NF-κB in rats with SAP increased after modeling, while resveratrol inhibited the activity of NF-κB and decreased the expression of TNF-α and IL-8 significantly, which means that resveratrol could modulate the expression of cytokines by inhibiting the activity of NF-κB and the inflammatory response in SAP.[33] Studies concerning the activation of macrophages have found that NF-κB in macrophages was activated in the early stage of SAP. Macrophages then produce and release significant amounts of TNF-α, IL-1, and NO. However, the activation of NF-κB, as well as production and release of TNF-α, IL-1, and NO, in macrophages decreased significantly after the rats were treated with resveratrol. At the same time, the levels of TNF-α, IL-1, and NO in serum decreased, which showed a significant therapeutic effect in rats with SAP.[34]

In short, the reduction of the expression of inflammatory mediators in the blood can relieve the damage in pancreatic tissue and suppress SIRS, which is the initiating factor of extra-pancreatic organ damage and MOSF. Resveratrol can decrease the production and release of various inflammatory mediators, such as TNF-α, IL-1, IL-6, IL-8, and NO, by inhibiting the activation of NF-κB, which can directly or indirectly relieve pancreatic damage. Meanwhile, the reduction of inflammatory mediators in the blood and in extra-pancreatic organ tissues can interrupt the progression of SIRS caused by SAP.

Protective effect of resveratrol on microcirculation

Stable microcirculation can decrease the incidence of inflammatory disease. Resveratrol could suppress leukocyte adhesion and destruction of vascular endothelial barriers in rats with mesentery ischemia/reperfusion injury.[35] A hypercoagulable state exists in every stage of SAP, which is attributed to the increase of platelet adhesion and accumulation, and resveratrol was verified to inhibit platelet adhesion *in vitro* and *in vivo*. Examining the platelet aggregation ratio using Born's method, it has been proven that resveratrol can inhibit the function of collagen, thromboxane, and adenosine diphosphate (ADP).[36] In addition, it was found that the application of resveratrol in washed platelet inhibited platelet adhesion induced by LPS or LPS combined with thromboxane, and this effect reached the limit of detection when the dose of resveratrol was 100 mg/mL.[37] Hence, resveratrol was able to reduce platelet adhesion and decrease the activity of platelets in the inflammatory reaction, as was described in another report.[38] Since tissue factor is a kind of receptor in the cellular membrane, the combination between $VII_{(a)}$ and tissue factor can lead to a coagulation cascade reaction. Some work has verified that resveratrol suppressed the expression of tissue factor in endothelial cells and monocytes.[39]

Microcirculatory disturbance in SAP involves a series of changes including vasoconstriction, ischemia, increased vascular permeability, impairment of nutritive tissue perfusion, ischemia/reperfusion, leukocyte adhesion, hemorheological changes, and impaired lymphatic drainage.[40] Improvement of microcirculatory disturbance is key in the treatment of SAP. Hemorheologic changes in rats with SAP are significant, due to the activation of the pancreatic enzymes, which causes increased permeability in capillary vessels and a number of plasma-like liquids exude from the blood circulation to the abdominal cavity and tissue space.[41,42] As a result, pachydermia with the high HCT and blood viscosity has been reported. Tissue factors such as ADP released from the fragmentized erythrocyte can activate the coagulation system and lead to disseminated intra-vascular coagulation, which consumes abundant fibrinogen. Congestion of the capillary vessel and microthrombus formation owing to the reduction of the blood flow volume and hypoperfusion of microcirculation may be one of the primary causes of progressive pancreatic necrosis and MOSF. Thus, the therapeutic tactics of improving the microcirculatory disturbance of the pancreas and extra-pancreatic organs should be considered.

Studies on SAP have shown that resveratrol might stabilize erythrocytes, thus decreasing thrombus formation and the expression of intercellular adhesion molecule-1 (ICAM-1) and platelet/endothelial cell adhesion molecule-1 (PECAM-1) genes, which could mediate leukocyte adhesion and migration.[43] Resveratrol also suppressed leukocyte adhesion to endothelium and reduced infiltration of leukocytes into inflammatory sites. At the same time, plasmatic expressions levels of plasma renin activity, angiotensinII (AngII), endothelin (ET), and NO in the resveratrol group were lower than those in the SAP group, in accordance with pathologic changes in pancreatic tissue.[44] Therefore, resveratrol could improve microcirculatory disturbance and relieve pathologic pancreatic injury caused by SAP via inhibition of the renin-angiotensin system (RAS) system, reduction of ET and NO, and regulation of an unbalanced ET/NO status.

Protective effect of resveratrol on oxidative stress

Oxidative stress mechanisms are integral in the pathogenesis of various diseases, including SAP. Reactive oxygen species (ROS) can cause tissue injury and activate genes that are responsible for the initiation of the inflammatory process. Resveratrol could protect intestinal tissue against I/R injury with its potent antioxidant properties and have a beneficial effect in inhibiting the oxidative damage in the liver induced by chronic ethanol administration.[45,46] Moreover, resveratrol could alleviate the oxidative damage to mouse hepatocytes caused by H_2O_2 to some extent and increase the activity of superoxide dismutase (SOD) and GSH-PX, decrease the consumption of GSH, reduce the level of malondialdehyde (MDA), and prevent damage to hepatocytes induced by the oxidant.[47] These results show that resveratrol can protect the organs in MODS through antagonizing oxidant molecules. Resveratrol could down-regulate the level of MDA and up-regulate the expression of SOD, catalase, and glutathione peroxidase in a rat liver ischemia/reperfusion injury model, which suggested that resveratrol may alleviate liver ischemia/reperfusion injury through the benefit of antioxidant activity.[48] Leonard *et al.*[49] demonstrated that resveratrol is a high-efficiency scavenger that can clear hydroxyl, superoxide, and metal inductive radicals. In addition, it can also protect against lipid peroxidation in the membrane and the DNS damage derived from ROS. The potential mechanism is resveratrol-mediated inhibition of the NF-κB pathway and subsequent suppression of ROS products.

In 1988, Wisner *et al.* found that oxygen free radicals played an important role in caerulein-induced acute pancreatitis in rats.[50] In humans, acute pancreatitis has been associated with glutathione deficiency and involvement of oxygen free radicals. Under physiological conditions, the production and elimination of oxygen free radicals is balanced. When the scavenging system of oxygen free radicals becomes weak, it can lead to pancreatic damage due to ROS. Oxygen free radicals can cause damage to macromolecules, such as proteins, nucleic acids, lipids and polysaccharides, thereby increasing the vascular permeability of the pancreas and resulting in pancreatic edema, hemorrhage, degeneration, and necrosis. Moreover, oxygen free radicals can reduce the stability of the membrane, which leads to the release and activation of pancreatic enzymes in the lysosomes of pancreatic acinar cells. Phospholipase A can be activated by oxygen free radicals, breaking down lecithins in the membrane of the pancreatic cells, which causes further damage to pancreatic tissue. Meanwhile, many oxygen free radicals are released in SAP, which is an important factor of MOSF.[51–53]

The levels of SOD and MDA, an antioxidant enzyme and a lipid peroxidation product, can reflect the level of free radicals in tissues. Some studies have demonstrated that resveratrol exhibits a protective effect against lipid peroxidation in cellular membranes caused by oxygen free radicals and results in a benefit effect on pathologic injury to the pancreas in the early stages of SAP.[54] In addition, some work has shown that lipid peroxidation occurred in the pancreas, small intestine, liver, lung, and kidney tissues, which causes damage to organs in the early stage of SAP. Resveratrol can increase SOD activity in organs and reduce the activity of MDA, thereby reducing the production of oxygen free radicals and reliving damage of multiple organs in the early stage of SAP.[55] In one study, resveratrol was shown to exert a protective effect against lipid peroxidation in the cell membrane and eventually to prevent DNA damage caused due to circulating ROS during SAP induction. It has been observed that SOD levels decrease and MDA levels increase significantly in the early phases of SAP. In addition, the levels of PECAM-1, serum amylase, and pancreatic histological scores increase gradually with the progression of SAP.[56] In short, resveratrol is an effective scavenger of hydroxyl superoxide and meta-induced radicals in SAP.

Roles of resveratrol on cell apoptosis caused by SAP

More than half of SAP-induced mortality is due to multiple organ failure, including liver and renal failure. This kind of organ failure can be attributed to apoptosis. Apoptosis is known to occur via two established pathways: the death receptor and mitochondrial pathway. SAP-induced apoptosis was shown to be of mitochondrial origin. This was proven by the release of cytochrome-C in early SAP.[57] The release of mitochondrial cytochrome-C is inhibited by Bcl-2, an anti-apoptotic member of the Bcl-2 family. Bcl-2 is known to clear ROS and to prevent the release of Ca^{2+} from the endoplasmic reticulum.

The mechanism and treatment of apoptosis have become a well-studied issue in the field of MODS in SAP. Cell apoptosis is a basic physiological mechanism to maintain homeostasis. The maintenance of physiological functions in organs depends on the balance between proliferation and cell apoptosis, and excessive apoptosis may cause organ dysfunction. One study has shown that cell apoptosis is prevalent in extra-pancreatic organs in SAP, and that the extent of apoptosis and organ damage is correlated.[58] It was demonstrated that apoptosis control genes, Bax, caspase-3, Fas, and FasL, were up-regulated significantly in the lung, liver and small intestine after SAP modeling in rats, and histologic damage to organs could be observed.[59,60] The extent of histologic damage in organs was correlated with the expression level of apoptosis control genes. In addition, mitochondria play an important role in the occurrence of cell apoptosis. Cytochrome C, a precursor of apoptosis, is released into the cytoplasm through a permeability transition pore in the mitochondrial membrane, subsequently activating caspase-3 and leading to cellular apoptosis. Bcl-2, an apoptosis inhibitory protein, is able to prevent directly and indirectly the opening of permeability transition pore induced by Bax.[61]

It has been demonstrated that resveratrol is able to relieve injury in extra-pancreatic organs in SAP—such as lungs, liver, and intestines—through up-regulating the expression of Bcl-2 and down-regulating the expression of Bax, caspase-3, and cytochrome c levels.[59,62,63]

Effect of resveratrol on calcium overload

Under physiological conditions, calcium remains stable, whereas in pathological conditions, some factors can influence intracellular calcium regulation and cause an increase in intracellular calcium content resulting in calcium overload. In SAP, in addition to microcirculatory disturbance, increases in oxygen free radicals and production of inflammatory cytokines, the concentration of intracellular Ca^{2+} can also increase. Studies indicate that calcium overload is not only one of the etiological factors of SAP, it also aggravates the disease state by affecting other organs.[64] The integrity of the cellular membrane and endocytoplasmic/sarcoplasmic reticulum membrane is important in intracellular calcium regulation; the Ca^{2+}–Mg^{2+}–ATPase activity of the cellular membrane and the Ca^{2+}–ATPase activity of the endocytoplasmic/sarcoplasmic reticulum membrane are also involved in this regulation.[65]

The regulation of the concentration of intracellular calcium depends on the following factors: the structural integrity of the plasma membrane and the endoplasmic reticulum/sarcoplasmic reticulum, the activity of Ca^{2+}–Mg^{2+}–ATPase in plasma membrane, and the activity of the Na^{+}–Ca^{2+} exchanger and Ca^{2+}–ATPase in the ER/SR membrane. In SAP, the activity of Ca^{2+}–Mg^{2+}–ATPase and Ca^{2+}–ATPase decreases, while the activity of PLA_2 increases, which can become a dangerous cycle, leading to anomalies in the regulation and concentration of intracellular calcium and which can result in serious injury to pancreatic cells.[66–69] In summary, intracellular calcium overload can induce severe pancreatic and lung tissue injury in SAP, and resveratrol can beneficially reduce the severity of acute pancreatitis by restoring the intracellular calcium regulatory mechanisms and by reducing calcium overload.[69]

Conclusions

Acute pancreatitis is a common kind of acute abdominal disease. There are two levels that depend on the severity of the disease: mild acute pancreatitis and SAP. Acute pancreatitis takes a mild course in about 80% of patients, who also report slight organ dysfunction. An increasing amount of data has confirmed that resveratrol is beneficial in the treatment of SAP. First, resveratrol could relieve the pathologic injury of the pancreas and extra-pancreatic organs (lung, liver, intestine) induced by SAP and increase the survivability of SAP patients. Second, resveratrol could inhibit the activation of NF-κB and modulate the production of inflammatory cytokines, eventually ameliorating the pancreatic injury and progression of systemic inflammatory response syndrome. Third, resveratrol could improve microvascular permeability, reducing leukocyte migration into tissues and inhibiting the inflammation reaction through modulating the levels of ICAM-1, 6-Keto-PGF1α and TBX2. In addition, studies have indicated that resveratrol could improve the hemodynamic abnormalities. Moreover, resveratrol could induce cellular apoptosis in the pancreas and extra-pancreatic organs through modulating the levels of Bax, caspase-3 and Bcl-2, which could relieve organ pathologic injury. Finally, resveratrol could modulate the calcium concentrations of cells and relieve the tissue injury induced by calcium overload. Therefore, resveratrol inhibits the production and progression of SAP through reducing inflammatory cytokines, thus improving microcirculation, modulating cell apoptosis, and blocking calcium overload. Further studies in this area would provide wider insights into the successful management of SAP.

Conflicts of interest

The authors declare no conflicts of interest.

References

1. Algül, H. & R.M. Schmid. 2009. Acute pancreatitis: etiology, diagnosis and therapy. *Med. Monatsschr. Pharm.* **32:** 242–247.

2. Bradley, E.L., 3rd & N.D. Dexter. 2010. Management of severe acute pancreatitis: a surgical odyssey. *Ann. Surg.* **251:** 6–17.
3. Soleas, G.J., E.P. Diamandis & D.M. Goldberg. 1997. Wine as a biological fluid: history, production, and role in disease prevention. *J. Clin. Lab. Anal.* **11:** 287–313.
4. Nakamura, M., H. Saito, M. Ikeda, *et al.* 2010. An antioxidant resveratrol significantly enhanced replication of hepatitis C virus. *World J. Gastroenterol.* **16:** 184–192.
5. Shah, V.O., J.E. Ferguson, L.A. Hunsaker, *et al.* 2010. Natural products inhibit LPS-induced activation of pro-inflammatory cytokines in peripheral blood mononuclear cells. *Nat. Prod. Res.* **24:** 1177–1188.
6. Gullett, N.P., A.R. Ruhul Amin, S. Bayraktar, *et al.* 2010. Cancer prevention with natural compounds. *Semin. Oncol.* **37:** 258–281.
7. Lu, X., L. Ma, L. Ruan, *et al.* 2010. Resveratrol differentially modulates inflammatory responses of microglia and astrocytes. *J. Neuroinflammation* **7:** 46.
8. Je, H.D., M.H. Lee, J.H. Jeong, *et al.* 2010. Protective effect of resveratrol on agonist-dependent regulation of vascular contractility via inhibition of rho-kinase activity. *Pharmacology* **86:** 37–43.
9. Fan, E., L. Zhang, S. Jiang, *et al.* 2008. Beneficial effects of resveratrol on atherosclerosis. *J. Med. Food* **11:** 610–614.
10. Sinha, K., G. Chaudhary & Y.K. Gupta. 2002. Protective effect of resveratrol against oxidative stress in middle cerebral artery occlusion model of stroke in rats. *Life Sci.* **71:** 655–665.
11. Bloomfield Rubins, H., J. Davenport, V. Babikian, *et al.* 2001. Reduction in stroke with gemfibrozil in men with coronary heart disease and low HDL cholesterol: the veterans affairs HDL intervention trial (VA-HIT). *Circulation* **103:** 2828–2833.
12. Shin, J.A., H. Lee, Y.K. Lim, *et al.* 2010. Therapeutic effects of resveratrol during acute periods following experimental ischemic stroke. *J. Neuroimmunol.* **227:** 93–100.
13. Jha, R.K., Q. Ma, H. Sha, *et al.* 2009. Acute pancreatitis: a literature review. *Med. Sci. Monit.* **15:** RA147–RA156.
14. Gonzales, A.M. & R.A. Orlando. 2008. Curcumin and resveratrol inhibit nuclear factor-kappaB-mediated cytokine expression in adipocytes. *Nutr. Metab. (Lond).* **5:** 17.
15. Rinderknecht, H. 1988. Fatal pancreatitis, a consequence of excessive leukocytestimulation? *Int. J. Pancreatol.* **3:** 105–112.
16. Zhang, X.P., Z.J. Li & J. Zhang. 2009. Inflammatory mediators and microcirculatory disturbance in acute pancreatitis. *Hepatobiliary Pancreat. Dis. Int.* **8:** 351–357.
17. Pandol, S.J. 2006. Acute pancreatitis. *Curr. Opin. Gastroenterol.* **22:** 481–486.
18. Das, S. & D.K. Das. 2007. Anti-inflammatory responses of resveratrol. *Inflamm Allergy Drug Targets.* **6:** 168–173.
19. Zhang, F., J. Liu & J.S. Shi. 2010. Anti-inflammatory activities of resveratrol in the brain: role of resveratrol in microglial activation. *Eur. J. Pharmacol.* **636:** 1–7.
20. de la Lastra, C.A. & I. Villegas. 2005. Resveratrol as an anti-inflammatory and anti-aging agent: mechanisms and clinical implications. *Mol. Nutr. Food Res.* **49:** 405–430.
21. Elmali, N., O. Baysal, A. Harma, *et al.* 2007. Effects of resveratrol in inflammatory arthritis. *Inflammation* **30:** 1–6.
22. Manna, S.K., A. Mukhopadhyay & B.B. Aggarwal. 2000. Resveratrol suppresses TNF-induced activation of nuclear transcription factors NF-kappa B, activator protein-1, and apoptosis: potential role of reactive oxygen intermediates and lipid peroxidation. *J. Immunol.* **164:** 6509–6519.
23. Ma, Z.H., Q.Y. Ma, H.C. Sha, *et al.* 2006. Effect of resveratrol on lipopolysaccharide-induced activation of rat peritoneal macrophages. *Nan Fang Yi Ke Da Xue Xue Bao.* **26:** 1363–1365.
24. Bi, X.L., J.Y. Yang, Y.X. Dong, *et al.* 2005. Resveratrol inhibits nitric oxide and TNF-alpha production by lipopolysaccharide-activated microglia. *Int. Immunopharmacol.* **5:** 185–193.
25. Shen, F., S.J. Chen, X.J. Dong, *et al.* 2003. Suppression of IL-8 gene transcription by resveratrol in phorbol ester treated human monocytic cells. *J. Asian Nat. Prod. Res.* **5:** 151–157.
26. Oh, Y.C., O.H. Kang, J.G. Choi, *et al.* 2009. Anti-inflammatory effect of resveratrol by inhibition of IL-8 production in LPS-induced THP-1 cells. *Am. J. Chin. Med.* **37:** 1203–1214.
27. Shen, F., S.J. Chen, X.J. Dong, *et al.* 2003. Suppression of IL-8 gene transcription by resveratrol in phorbol ester treated human monocytic cells. *J. Asian Nat. Prod. Res.* **5:** 151–157.
28. Malleo, G., E. Mazzon, A.K. Siriwardena, *et al.* 2007. Role of tumor necrosis factor-alpha in acute pancreatitis: from biological basis to clinical evidence. *Shock* **28:** 130–140.
29. Bhatia M. 2005. Inflammatory response on the pancreatic acinar cell injury. *Scand. J. Surg.* **94:** 97–102.
30. Sendur, R. & W.W. Pawlik. 1996. Vascular factors in the mechanism of acute pancreatitis. *Przegl Lek.* **53:** 41–45.
31. Hirota, M., F. Nozawa, A. Okabe, *et al.* 2000. Relationship between plasma cytokine concentration and multiple organ failure in patients with acute pancreatitis. *Pancreas* **21:** 141–146.
32. Huang, J.Y., Q.Y. Ma, Q. Sun, *et al.* 2005. Experimental study of resveratrol on severe acute pancreatitis. *J. Xi'an Jiaotong Univ.(Med. Sci.)* **26:** 163–165.
33. Meng, Y., Q.Y. Ma, X.P. Kou, *et al.* 2005. Effect of resveratrol on activation of nuclear factor kappa-B and inflammatory factors in rat model of acute pancreatitis. *World J. Gastroenterol.* **11:** 525–528.
34. Ma, Z.H. & Q.Y. Ma. 2005. Effect of resveratrol on peritoneal macrophages in rats with severe acute pancretitis. *Inflamm. Res.* **54:** 522–527.
35. Shigematsu, S., S. Ishida, M. Hara, *et al.* 2003. Resveratrol, a red wine constituent polyphenol, prevents superoxide-dependent inflammatory responses induced by ischemia/reperfusion, platelet-activating factor, or oxidants. *Free Radic. Biol. Med.* **34:** 810–817.
36. Wang, Z., Y. Huang, J. Zou, *et al.* 2002. Effects of red wine and wine polyphenol resveratrol on platelet aggregation in vivo and in vitro. *Int. J. Mol. Med.* **9:** 77–79.
37. Olas, B. & B. Wachowicz. 2001. Biological activity of resveratrol. *Postepy Hig Med. Dosw* **55:** 71–79.
38. Suttnar, J., L. Másová, T. Scheiner, *et al.* 2002. Role of free radicals in blood platelet activation. *Cas Lek Cesk.* **141**(Suppl): 47–49.

39. Pendurthi, U.R., F. Meng, N. Mackman, *et al.* 2002. Mechanism of resveratrol-mediated suppression of tissue factor gene expression. *Thromb. Haemost.* **87:** 155–162.
40. Zhou, Z.G. & Y.D. Chen. 2002. Influencing factors of pancreatic microcirculatory impairment in acute panceatitis. *World J. Gastroenterol.* **8:** 406–412.
41. Yan, L., J. Wei, H. Wu, *et al.* 1990. The role of hemorheologic changes in the pathogenesis of acute hemorrhagic necrotizing pancreatitis. *Hua Xi Yi Ke Da Xue Xue Bao.* **21:** 25–29.
42. Yan, L., Z. Lei, X. Cui, *et al.* 1993. The role of hemorheologic disturbance in experimental acute pancreatitis. *Hua Xi Yi Ke Da Xue Xue Bao.* **24:** 71–74.
43. Meng, Y., M. Zhang, J. Xu, *et al.* 2005. Effect of resveratrol on microcirculation disorder and lung injury following severe acute pancreatitis in rats. *World J. Gastroenterol.* **11:** 433–435.
44. Ma, Z.H., Q.Y. Zhang, Q. Li, *et al.* 2009. The effect of resveratrol to microcirculatory dysfunction on rats with SAP. *Chin. J. Exp. Surg.* **26:** 1225
45. Ozkan, O.V., M.F. Yuzbasioglu, H. Ciralik, *et al.* 2009. Resveratrol, a natural antioxidant, attenuates intestinal ischemia/reperfusion injury in rats. *Tohoku J. Exp. Med.* **218:** 251–258.
46. Kasdallah-Grissa, A., B. Mornagui, E. Aouani, *et al.* 2007. Resveratrol, a red wine polyphenol, attenuates ethanol-induced oxidative stress in rat liver. *Life Sci.* **80:** 1033–1039.
47. Mo, Z.X. & H.X. Shao. 2000. Protective effects of polydatin on mouse hepatocyte injury induced by hydrogen peroxide in vitro. *Chin. Pharmacol. Bull.* **16:** 519–521.
48. Gedik, E., S. Girgin, H. Ozturk, *et al.* 2008. Resveratrol attenuates oxidative stress and histological alterations induced by liver ischemia/reperfusion in rats. *World J. Gastroenterol.* **14:** 7101–7106.
49. Leonard, S.S., C. Xia, B.H. Jiang, *et al.* 2003. Resveratrol scavenges reactive oxygen species and effects radical-induced cellular responses. *Biochem. Biophys. Res. Commun.* **309:** 1017–1026.
50. Wisner, J., D. Green, L. Ferrell, *et al.* 1988. Evidence for a role of oxygen derived free radicals in the pathogenesis of caerulein induced acute pancreatitis in rats. *Gut* **29:** 1516–1523.
51. Leung, P.S. & Y.C. Chan. 2009. Role of oxidative stress in pancreatic inflammation. *Antioxid Redox Signal* **11:** 135–165.
52. Modzelewski, B. & A. Janiak. 2005. Lipid peroxidation product as prognostic factors in acute necrotizing pancreatitis. *Pol. Merkur. Lekarski.* **19:** 511–513.
53. Chvanov, M., O.H. Petersen & A. Tepikin. 2005. Free radicals and the pancreatic acinar cells: role in physiology and pathology. *Philos. Trans. R. Soc. Lond. B Biol. Sci.* **360:** 2273–2284.
54. Li, Z.D., Q.Y. Ma & C.A. Wang. 2006. Effect of resveratrol on pancreatic oxygen free radicals in rats with severe acute pancreatitis. *World J. Gastroenterol.* **12:** 137–140.
55. Qin, Y., Q.Y. Ma, X.Y. Dang, *et al.* 2007. Effect of resveratrol on oxygen free radicals of multiple organs in rats with severe acute pancreatitis of early stage. *J. Xian Jiaotong Univ. (Med. Sci.)* **28:** 572–574.
56. Wang, Z., O.Y. Ma, L. Ren, *et al.* 2006. The study in resveratrol function to acute lung injury sourced from severe acute pancreatitis. *Sichuan Da Xue Xue Bao Yi Xue Ban.* **37:** 904–907.
57. Mutinga, M., A. Rosenbluth, S.M. Tenner, *et al.* 2000. Does mortality occur early or late in acute pancreatitis? *Int. J. Pancreatol.* **28:** 91–95.
58. Takeyama, Y. 2005. Significance of apoptotic cell death in systemic complications with severe acute pancreatitis. *J. Gastroenterol.* **40:** 1–10.
59. Sha, H., Q. Ma, R.K. Jha, *et al.* 2009. Resveratrol ameliorates lung injury via inhibition of apoptosis in rats with severe acute pancreatitis. *Exp. Lung Res.* **35:** 344–358.
60. Ni, Y., J.S. Wu, P.P. Fang, *et al.* 2008. Mechanism of liver injury in severe acute pancreatitis rats and role of melatonin. *Zhonghua Yi Xue Za Zhi* **88:** 2867–2871.
61. Xia, T., C. Jiang, L. Li, *et al.* 2002. A study on permeability transition pore opening and cytochrome c release from mitochondria, induced by caspase-3 in vitro. *FEBS Lett.* **510:** 62–66.
62. Sha, H., Q. Ma, R.K. Jha, *et al.* 2008. Resveratrol ameliorates hepatic injury via the mitochondrial pathway in rats with severe acute pancreatitis. *Eur. J. Phamacology.* **601:** 136–142.
63. Jha, R.K., M.Q. Yong, S.H. Chen. 2008. The protective effect of resveratrol on the intestinal mucosal barrier in rats with severe acute pancreatitis. *Med. Sci. Monit.* **14:** BR14–BR19.
64. Li, Y.Y. & H. Zhang. 2001. Pancreatic acinous cell calcium overload and acute pancreatitis. *Chin. J. Surg. Integr. Tradit. West. Med.***7:** 123–125.
65. Pariente, J.A., A.I. Lajas, M.J. Pozo, *et al.* 1999. Oxidizing effects of vanadate on calcium mobilization and amylase release in rat pancreatic acinar cells. *Biochem. Pharmacol.* **58:** 77–84.
66. Pu, Q., L. Yan & J. Shen. 1999. Effects of calcium overload in the conversion of acute edematous pancreatitis to necrotizing pancreatitis in rats. *Zhonghua Yi Xue Za Zhi* **79:** 143–145.
67. Redondo Valdeolmillos, M., M.L. del Olmo Martínez, A. Almaraz Gómez, *et al.* 1999. The effects of an oral calcium overload on the rat exocrine pancreas. *Gastroenterol. Hepatol.* **22:** 211–217.
68. Criddle, D.N., J.V. Gerasimenko, H.K. Baumgartner, *et al.* 2007. Calcium signalling and pancreatic cell death: apoptosis or necrosis? *Cell Death Differ.* **14:** 1285–1294.
69. Wang, L.C., Q.Y. Ma, X.L. Chen, *et al.* 2008. Effects of resveratrol on calcium regulation in rats with severe acute pancreatitis. *Eur. J. pharm.* **500:** 271–276.

Ann. N.Y. Acad. Sci. ISSN 0077-8923

ANNALS OF THE NEW YORK ACADEMY OF SCIENCES
Issue: *Resveratrol and Health*

Neuroprotective properties of resveratrol and derivatives

Tristan Richard, Alison D. Pawlus, Marie-Laure Iglésias, Eric Pedrot, Pierre Waffo-Teguo, Jean-Michel Mérillon, and Jean-Pierre Monti
Université de Bordeaux, Villenave d'Ornon, France

Address for correspondence: Jean-Pierre Monti, GESVAB (EA 3675), Université de Bordeaux, ISVV Bordeaux-Aquitaine, 71 Avenue Edouard Bourleaux, 33883 Villenave d'Ornon Cedex, France. jean-pierre.monti@u-bordeaux2.fr

Stilbenoid compounds consist of a family of resveratrol derivatives. They have demonstrated promising activities *in vitro* and *in vivo* that indicate they may be useful in the prevention of a wide range of pathologies, such as cardiovascular diseases and cancers, as well have anti-aging effects. More recently stilbenoid compounds have shown promise in the treatment and prevention of neurodegenerative disorders, such as Huntington's, Parkinson's, and Alzheimer's diseases. This paper primarily focuses on the impact of stilbenoids in Alzheimer's disease and more specifically on the inhibition of β-amyloid peptide aggregation.

Keywords: stilbenoid; Alzheimer's disease; β-amyloid peptide; inhibition of aggregation

Introduction

Natural products are an abundant source of polyphenolic compounds, including a less common group, the stilbenoids. These constitute a large class of resveratrol derivatives—including monomers, dimers, and oligomers—resulting from different oxidative condensations of the individual monomers. Stilbenoids are naturally occurring in several plant families, such as Cyperaceae, Dipterocarpaceae, Gnetaceae, and Vitaceae.[1,2] Grapes, which include the wine grape, *Vitis vinifera* L., are viewed as the most important dietary sources of these substances.[3,4] Numerous labs have been involved in the chemical characterization of stilbenes, primarily those found in grapes, red wine, and vine stems.

The impetus for the numerous research studies involving the chemical characterization of stilbenoids is their many promising biological activities, particularly those of resveratrol. The biological activities of resveratrol, and several of its derivatives, include the prevention or direct interference of numerous degenerative processes, such as neurodegenerative diseases.[5]

Due to the increase in the aging population, neurodegenerative diseases continue to involve a greater percentage of the population. These disorders, which result from the deterioration of neurons, are classified into two pathological classes: movement disorders such as Parkinson's disease (PD), and cognitive deterioration and dementia, such as Alzheimer's disease (AD).

AD is the most common type of neurodegenerative disorder, accounting for 65% of all dementias, with a prevalence estimated to be between 1 and 5% among people age 65 years and older.[6] Histopathological evaluations reveal that one of the major characteristics of AD is the excessive accumulation of two types of proteins: tau proteins and β-amyloid peptide (βA).[7] βA originates from the proteolytic cleavages of the transmembrane amyloid precursor protein (APP).[8] βA accumulation leads to the formation and deposit of amyloid plaques and neurofibrilary tangles, which promote inflammation and activate neurotoxic pathways, leading to dysfunction and death of brain cells.[9]

Recently, numerous studies have shown that a wide range of polyphenols have neuroprotective effects both *in vitro* and *in vivo*.[10–20] Among these polyphenols, resveratrol and several of its derivatives have demonstrated some of the most promising anti-neurodegenerative activities.[12–19] In this paper, we present investigations on the role of stilbenoids in neurodegenerative diseases, with an emphasis on

doi: 10.1111/j.1749-6632.2010.05865.x

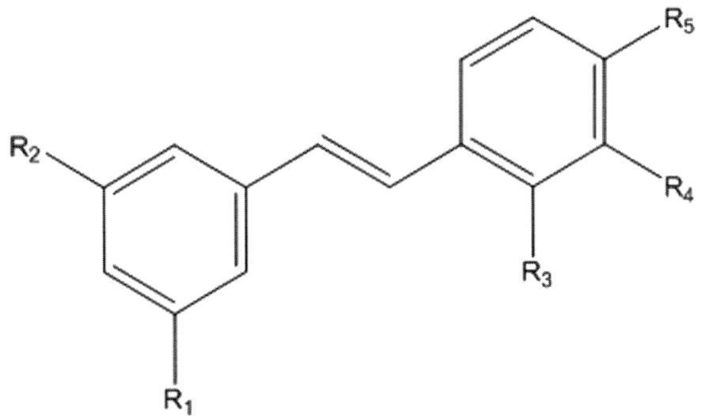

Compound Name	R^1	R^2	R^3	R^4	R^5
Resveratrol	OH	OH	H	H	OH
Piceid	OGlc	OH	H	H	OH
Piceatannol	OH	OH	H	OH	OH
Resveratroloside	OH	OH	H	H	OGlc
Pterostilbene	OMe	OMe	H	H	OH
Oxyresveratrol	OH	OH	OH	H	OH
Gob C	OH	OH	H	H	OGlc
Gob D	OH	OH	H	OMe	OGlc

Figure 1. Structures of stilbene monomer derivatives.

the studies demonstrating their ability to inhibit βA fibril formation in AD.

Resveratrol derivatives

Stilbenoids are defined by their structural skeleton, which comprises two aromatic rings joined by an ethylene bridge (C_6–C_2–C_6). From this relatively simple structure, a large array of compounds has been synthesized by plants. The stilbene monomers are primarily modified via the number and the position of hydroxyl groups, substitutions with methyl and methoxy groups, sugars, and differences due to the *cis* and *trans* configurations (Fig. 1). In addition to stilbene monomers, many dimers and oligomers have also been identified. These compounds are the result of the oxidative condensations of the resveratrol monomer. The dimers are divided into two major groups: group A, with compounds having five-membered oxygen heterocyclic ring, such as the viniferins (Fig. 2), and group B, without the oxygen containing ring, such as pallidol (Fig. 3). Finally, stilbene trimers and, more recently, tetramers, such as hopeaphenol, are known. Due to the multiple variations that can be achieved, it is reasonable to assume there are considerably more stilbene oligomers to be found in plants that exist as minor constituents (Fig. 4).[21,22]

Stilbenoids and neurodegenerative diseases

While numerous stilbenes exit, the majority of studies have involved resveratrol, which has shown to have antiaging effects in several organisms, including yeasts, nematodes, mice, and rats. Resveratrol has also been shown to have specific activities that can delay or alter the progression of neurological disorders, such as brain ischemia, Huntington's disease (HD), PD, and AD.[23]

In a study involving rat hippocampal neurons, voltage-activated potassium currents were inhibited by resveratrol, suggesting that it may be useful for treating ischemic brain injury.[24] Resveratrol also demonstrated protective effects in a PD model, using midbrain dopaminergic neurons, against several type of insults that have shown a correlation to PD pathogenesis. These included the cytotoxic effects induced by 1-methyl-4-phenyl pyrimidium, sodium azide, thrombin, and DNA damage.[25] In two HD models, resveratrol rescued mutant polyglutamine-specific cell death in neuronal cells derived from HdhQ111 knock-in mice and in transgenic *Caenorhabditis elegans*.[26] In a brain ischemia model, administration of resveratrol to gerbils during the early stage of cerebral ischemia protected against neuronal death in the

	R^1	R^2	R^3	R^4
Gnetin C	OH	H	H	OH
Gneafricanin A	OH	OMe	OMe	OH
Gnemenoside A	OGlc	H	H	OGlc

	R^1	R^2
Scirpusin A	OH	OH
ε-Viniferin	OH	H
ε-Viniferin glucoside	OGlc	H

Figure 2. Structures of stilbene dimer derivatives (group A).

hippocampal CA1 area and concomitantly inhibited glial cell activation.[27] Moreover, in this *in vivo* model, results showed that resveratrol, after formation of glucuronide conjugates, enters the bloodstream and could cross the blood–brain barrier. Required concentration of resveratrol to have neuroprotective actions were in range of 10–100 μM.[27] Different mechanisms, such as antioxidant and regulation of gene transcription, have been suggested to be involved in the protective actions of resveratrol.

Stilbenoids and Alzheimer's disease

The majority of neurodegenerative studies is currently focused on AD, most likely due to the fact that AD is the most common type of neurodegenerative disorder. So far, the effects of resveratrol and several others stilbenes in AD models suggest that stilbenes may be very effective modulators of AD development and progression.[28]

For example, in an APP695-transfected mouse neuroblastoma N2a cells, resveratrol could reduce the secretion of βA; in this study, the treatment of cells with selective proteasome inhibitors significantly blocked the resveratrol-induced decrease of βA levels.[14] These results suggest the antiamyloid activity of resveratrol is proteasome dependant. In another study using hippocampal primary neurons, resveratrol significantly attenuated the βA-induced

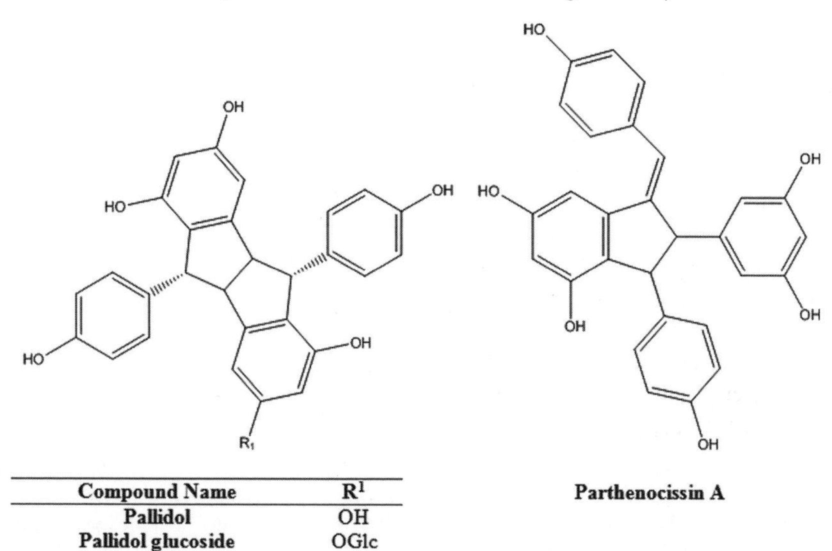

Compound Name	R^1
Pallidol	OH
Pallidol glucoside	OGlc

Figure 3. Structures of stilbene dimer derivatives (group B).

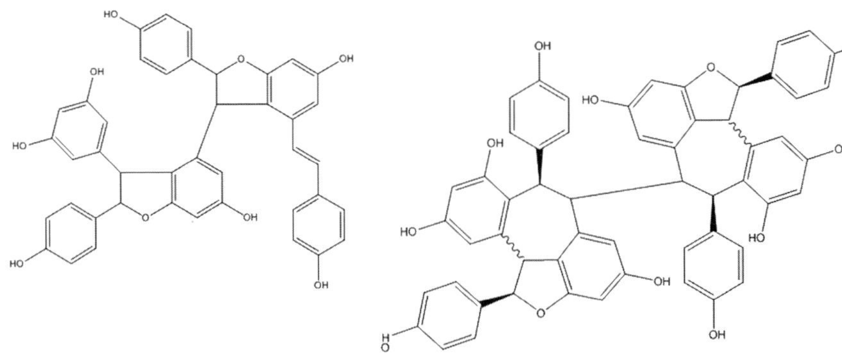

Miyabenol C **Hopeaphenol** **Isohopeaphenol**

Figure 4. Structures of some stilbenoid oligomers.

cell death in a concentration dependant manner.[15] This study suggested that the protein kinase C pathway was involved in the neuroprotective action of resveratrol.[15] Jeon *et al.* reported that resveratrol, oxyresveratrol, and scirpusin A have potent inhibitory activities on β-secretase with IC_{50} values at 15, 7.6, and 10 μM, respectively.[16] Among the secretases, β-secretase is one of the most attractive targets for the inhibition of amyloid production. Finally, in PC12 neural cells, resveratrol protected against the βA peptide-induced toxicity by influencing apoptotic signaling pathways, reducing changes in mitochondrial membrane potential and inhibiting the accumulation of intracellular reactive oxygen intermediates.[13]

Recently, a therapeutic approach that interferes directly with the neurodegenerative process in AD, especially the accumulation and the aggregation of βA, is considered one of the most promising targets to alter the progression of the disease, rather than merely treating the symptoms.[29] Several studies suggest that polyphenols could prevent AD or delay its onset by directly inhibiting the formation of βA fibril deposits in the brain.[17–20] Indeed, pathologic and biochemical studies suggest that βA fibrils are reactive oxygen species generators, whereas monomeric βA acts as natural antioxidant that prevents neuronal cell death caused by oxidative stress.[30–32] Using *in vitro* assays, we recently reported that resveratrol derivatives inhibit the aggregation of the peptide and destabilize the preformed oligomers and fibrils.[17–19] Initial screening of 9 monomers, 6 dimers, and 3 stilbenoid oligomers for inhibition was performed at a concentration of 10 μM and their inhibiting effect was compared to that of curcumin as the positive control.[20] Five stilbenoids exhibited peptide aggregation inhibition activity equal to or better than that of curcumin. These were further tested to determine their IC_{50} values. The IC_{50} values of all compounds are summarized in Table 1. Among all the

Table 1. Inhibition of βA fibril formation[a]

Compound	Inhibition %	IC_{50} (μM)
Curcumin	**45 ± 9**	**10 ± 2**
Stilbene Monomers		
Moracin M	9 ± 7	–
Resveratrol	**63 ± 6**	**6 ± 2**
Piceid	**62 ± 6**	**6 ± 2**
Piceatannol	25 ± 9	–
Pterostilbene	35 ± 7	–
Oxyresveratrol	32 ± 7	–
Gob C	11 ± 6	–
Gob D	28 ± 5	–
Stilbene Dimers		
Gnetin C	39 ± 5	–
Gneafricanin A	40 ± 6	–
Gnemenoside A	**46 ± 7**	**10 ± 2**
Scirpusin A	**80 ± 9**	**0.7 ± 0.3**
ε-Viniferin	25 ± 9	–
ε-Viniferin glucoside	**93 ± 3**	**0.2 ± 0.3**
Stilbene Oligomers		
Myabenol C	15 ± 5	–
Hopeaphenol	13 ± 6	–
Isohopeaphenol	21 ± 9	–

[a]Bold values indicate the molecules exhibiting inhibitory activity equal to, or greater than, the postive control, curcumin.

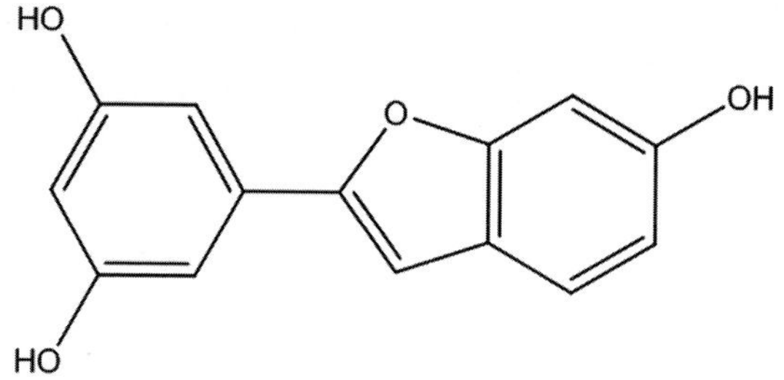

Figure 5. Structures of moracin M.

compounds tested, resveratrol, piceid, and the dimers, scirpusin A and ε-viniferin glucoside, exhibited an efficient inhibition of βA aggregation.

Resveratrol and its glucoside both inhibited the aggregation of the peptide, more so than the other stilbene monomers tested. Examination of inhibitory data for stilbene monomers suggests several potential structure-activity relationships. Additional substitutions on the aromatic rings reduced the protective activity against aggregation. Methoxy or glucosyl derivatives and derivatives with an additional hydroxyl group were weaker inhibitors, suggesting that hydrogen bonds might play an important role in the binding process. Finally, the benzofuranyl group in moracin M led to a decrease in activity, potentially due to the greater rigidity of its structure (Fig. 5).

Of the six stilbene dimers, three had an inhibitory activity equal to or greater than curcumin, which was found to be 10 ± 2 μM. Two dimers had considerable inhibitory activity: scirpusin A (0.7 ± 0.3 μM) and ε-viniferin glucoside (0.2 ± 0.3 μM). These two compounds differed only by two substituents. Scirpusin A had an additional hydroxyl group and ε-viniferin glucoside has a glucose unit (Fig. 2). It is very difficult to draw any conclusions on the structure-activity relationship using this limited data set. Nevertheless, their strong inhibitory activity *in vitro* warrants further investigation into their potential as therapeutic agents in AD treatment and prevention.[18]

Unlike the stilbene dimers, the stilbene oligomers (trimers and tetramers), were weak inhibitors. These results suggest that spatial constraints are critical in the binding process. However, other oligomers need to be tested to confirm that bulk compounds are not active since results indicated that the inhibitory effect depends not only on the specific ring substituents, but may also depend on the overall 3-D structure.[18]

Conclusion

A limited number of therapeutic options are available to treat neurodegenerative diseases; however, these are primarily treating symptoms and not interfering with the disease process. This paper demonstrates that resveratrol derivatives are highly promising molecules that could be used to both prevent and/or treat neurodegenerative diseases such as AD. Indeed, resveratrol derivatives may effectively modulate multiple mechanisms of the neurodegenerative disease pathology. For example, in AD, resveratrol inhibits activity on β-secretase, the generation of the reactive oxygen intermediates, and the aggregation of βA peptide. Moreover, other resveratrol derivatives, such as the dimers scirpusin A and ε-viniferin glucoside, could be potent preventive and therapeutic agents. However, the efficacy and utility of resveratrol derivatives in treating neurodegenerative pathology also depends on their bioavailability and activity *in vivo*. Future research is warranted to address these issues.

Conflicts of interest

The authors declare no conflicts of interest.

References

1. Hart, J.H. 1981. Role of phytostilbenes in decay and disease resistance. *Rev. Phytopathol.* **19:** 437–458.
2. Sotheeswaran, S. & V. Pasapathy. 1993. Distribution of resveratrol oligomers in plants. *Phytochemistry* **32:** 1083–1092.
3. Mattivi, F., F. Reniero & S. Korhammer. 1995. Isolation, characterization, and evolution in Red Wine Vinification of resveratrol monomers. *J. Agric. Food Chem.* **43:** 1820–1823.
4. Goldberg, D.M., E. Ng, A. Karumanchiri, *et al.* 1996. Resveratrol glucosides are important components of commercial wines. *Am. J. Enol. Vitic.* **47:** 415–420.
5. Saiko, P., A. Szakmary, W. Jaeger, *et al.* 2008. Resveratrol and its analogs: defense against cancer, coronary disease and neurodegenerative maladies or just a fad? *Mut. Res.* **658:** 68–94.
6. Ritchie, K. & S. Lovestone. 2002. The dementias. *The Lancet* **360:** 1759–1766.
7. Sisodia, S.S. & D.L. Price. 1995. Role of the β-amyloid protein in Alzheimer's disease. *FASEB J.* **9:** 366–370.
8. Selkoe, D.J. 2001. Alzheimer's disease: genes, proteins, and therapy. *Physiol. Rev.* **81:** 741–766.
9. Pereira, C., P. Agostinho, P.I. Moreira, *et al.* 2005. Alzheimer's disease-associated neurotoxic mechanisms and neuroprotective strategies. *Curr. Drug Targets CNS Neurol. Disord.* **4:** 383–403.

10. Inanami, O., Y. Watanabe, B. Syuto, *et al.* 1998. Oral administration of (-)-catechin protects against ischemia-reperfusion-induced neuronal death in the gerbil. *Free Radic. Res.* **29:** 359–365.
11. Choi, Y.T., C.H. Jung, S.R. Lee, *et al.* 2001. The green tea polyphenol (-)-epigallocatechin gallate attenuates β-amyloid-induced neurotoxicity in cultured hippocampal neurons. *Life Sci.* **70:** 603–614.
12. Bastianetto, S., W.H. Zheng & R. Quirion. 2000. Neuroprotective abilities of resveratrol and other red wine constituents against nitric oxide-related toxicity in cultured hippocampal neurons. *Br. J. Pharmacol.* **131:** 711–720.
13. Jang, J.H. & Y.J. Surh. 2003. Protective effect of resveratrol on β-amyloid-induced oxidative PC12 cell death. *Free Radical Bio. Med.* **34:** 1100–1110.
14. Marambaud, P., P.H. Zhao & P. Davies. 2005. Resveratrol promotes clearance of Alzheimer's disease amyloid-β peptides. *J. Biol. Chem.* **280:** 37377–37382.
15. Han, Y.S., W.H. Zheng, S. Bastianetto, *et al.* 2004. Neuroprotective effects of resveratrol against β-amyloid-induced neurotoxicity in rat hippocampal neurons: Involvement of protein kinase C. *Br. J. Pharmacol.* **141:** 997–1005.
16. Jeon, S.Y., S.H. Kwon, Y.H. Seong, *et al.* 2007. β-secretase (BACE1)-inhibiting stilbenoids from Smilax Rhizoma. *Phytomedecine* **14:** 403–408.
17. Rivière, C., T. Richard, L. Quentin, *et al.* 2007. Inhibitory activity of stilbenes on Alzheimer's β-amyloid fibrils *in vitro*. *Bioorg. Med. Chem.* **15:** 1160–1167.
18. Rivière, C., Y. Papastamoulis, P. Y. Fortin, *et al.* 2010. New stilbene dimers against amyloid fibril formation. *Bioorg. Med. Chem. Lett.* **20:** 3441–3443.
19. Rivière, C., J.C. Delaunay, F. Immel, *et al.* 2009. The polyphenol piceid destabilizes preformed amyloid fibrils and oligomers in vitro: hypothesis on possible molecular mechanisms. *Neurochem. Res.* **34:** 1120–1128.
20. Ono, K., Y. Yoshiike, A. Takashima, *et al.* 2003. Potent anti-amyloidogenic and fibril-destabilizing effects of polyphenols *in vitro*: implications for the prevention and therapeutics of Alzheimer's disease. *J. Neurochem.* **87:** 172–181.
21. Chong, J., A. Poutaraud & P. Hugueney. 2009. Metabolism and roles of stilbenes in plants. *Plant Sci.* **177:** 143–155.
22. Xiao, K., H.J. Zhang, L.J. Xuan, *et al.* 2008. Stilbenoids: chemistry and bioactivities. In: *Studies in Natural Products Chemistry*. F.R.S. Atta-ur-Rahman, Ed.: 453–646. Elsevier, Amsterdam.
23. Ramassamy, C. 2006. Emerging role of polyphenolic compounds in the treatment of neurodegenerative diseases: a review of their intracellular targets. European Journal of Pharmacology. *Eur. J. Pharmacol.* **545:** 51–64.
24. Gao, Z.B. & G.Y. Hu. 2005. Trans-resveratrol, a red wine ingredient, inhibits voltage-activated potassium currents in rat hippocampal neurons. *Brain Res.* **1056:** 68–75.
25. Okawa, M., H. Katsuki, E. Kurimoto, *et al.* 2007. Resveratrol protects dopaminergic neurons in midbrain slice culture from multiple insults. *Biochem. Pharmacol.* **73:** 550–560.
26. Parker, J.A., M. Arango, S. Abderahmane, *et al.* 2005. Resveratrol rescues mutant polyglutamine cytotoxicity in nematode and mammalian neurons. *Nat. Genet.* **37:** 349–350.
27. Wang, Q., J. Xu, G.E. Rottinghaus, *et al.* 2002. Resveratrol protects against global cerebral ischemic injury in gerbils. *Brain Res.* **958:** 439–447.
28. Anekonda, T.S. 2006. Resveratrol–A boon for treating Alzheimer's disease? *Brain Res.Rev.* **52:** 316–326.
29. Golde, T.E. 2006. Disease modifying therapy for AD? *J. Neurochem.* **99:** 689–707.
30. Gervais, F.G., D. Xu, G.S. Robertson, *et al.* 1999. Involvement of caspases in proteolytic cleavage of Alzheimer's amyloid-β precursor protein and amyloidogenic Aβ peptide formation. *Cell* **97:** 395–406.
31. Tabner, B.J., S. Turnbull, O.M.A. El-Agnaf, *et al.* 2002. Formation of hydrogen peroxide and hydroxyl radicals from Aβ and α-synuclein as a possible mechanism of cell death in Alzheimer's disease and Parkinson's disease. *Free Radical Bio. Med.* **32:** 1076–1083.
32. Zou, K., J.S. Gong, K. Yanagisawa, *et al.* 2002. A Novel Function of Monomeric Amyloid β-protein serving as an antioxidant molecule against metal-induced oxidative damage. *J. Neurosci.* **22:** 4833–4841.

Ann. N.Y. Acad. Sci. ISSN 0077-8923

ANNALS OF THE NEW YORK ACADEMY OF SCIENCES
Issue: *Resveratrol and Health*

MicroRNA signatures of resveratrol in the ischemic heart

Partha Mukhopadhyay,[1] Pal Pacher,[1] and Dipak K. Das[2]

[1]Laboratory of Physiologic Studies, NIAAA, National Institute of Health, Bethesda, Maryland. [2]Cardiovascular Research Center, University of Connecticut School of Medicine, Farmington, Connecticut

Address for correspondence: Partha Mukhopadhyay, Ph.D., Laboratory of Physiologic Studies, NIAAA, National Institutes of Health, 5625 Fishers Lane, Room 2N-17 MSC 9413, Rockville, MD 20852-1758. mpartha@mail.nih.gov

Until the middle of the last decade, few people had heard of microRNAs (miRNAs), 21- to 23-nucleotide conserved RNAs. MicroRNAs represent a new paradigm because they regulate most physiological processes and thus have immense potential for medical advancement. Resveratrol, a red wine-derived polyphenolic compound, has been shown to have significant effects in various disease models, such as cardioprotection in ischemic heart, diabetes, and chemoprevention of cancers. The targets of resveratrol include various pathways and molecules, such as sirtuins, FOXOs, and autophagy. The successful application of resveratrol lies in understanding its mechanisms of action through direct and indirect interactions with pathways, including miRNAs. For example, a unique miRNA footprint is present in the heart treated with resveratrol. Targets of those miRNAs have potential implications for physiological and pathophysiological processes in health and disease.

Keywords: miRNA expression profile; heart; translational repression or activation; stability of microRNA

Introduction

The rapid pace of outstanding findings in RNA interference (RNAi) research has led to an expanding array of tools to understand the basic processes of life and disease. After the completion of the Human Genome Project, a relatively small number (~5%) of protein-coding genes have been found relative to genome size.[1] The rest are non-coding sequences from which a variety of small RNAs are produced: miRNAs (20–22 nt), trans-acting endogenous siRNA (small interfering RNA), and piRNA (piwi interacting RNA, 16–29 nt with repeat sequence). Among them, miRNAs are a key group identified by bioinformatics, genetics, and the molecular biology approach of cloning and characterization.[2–4] Genes for miRNAs are essential components of the genetic program of all species, many of which are also evolutionarily conserved.[5]

RNA that regulates target molecules or pathways has been observed as of a complex regulatory system in bacteria.[6] The first report of RNA silencing was found in the plant system.[7] The fundamental study on miRNAs was done in *Caenorhabditis elegans*, where the gene locus lin-4 was found to be a regulator of developmental gene expression. Later, the conserved 21-nt RNA let-7, a miRNA from that locus, was found to suppress the expression of a target transcript.[8]

Resveratrol, a constituent of red wine and many plant roots used for Asian medicine, is known for its role of chemoprevention in cancers (reviewed in Refs. 9 and 10). In many tissue injury models, resveratrol also protects against such as neuronal damage.[11,12] Several pieces of evidence in the literature suggests that resveratrol plays a key role in cardioprotection (reviewed in Refs. 13 and 14). The molecular mechanism of cardioprotection is still under investigation; and the role of miRNA has not been looked at so far.

MicroRNA: mechanism of action

MicroRNAs (miRNAs) are the mature form of processed pre-miRNAs. Pre-miRNAs are processed by Drosha from longer polyadenylated transcripts, known as p-miRNA, in the nucleus and are exported to the cytoplasm by exportin 5.[15] Further maturation of pre-miRNA to miRNA occurs in both the nucleus and the cytoplasm through Dicer and

doi: 10.1111/j.1749-6632.2010.05866.x

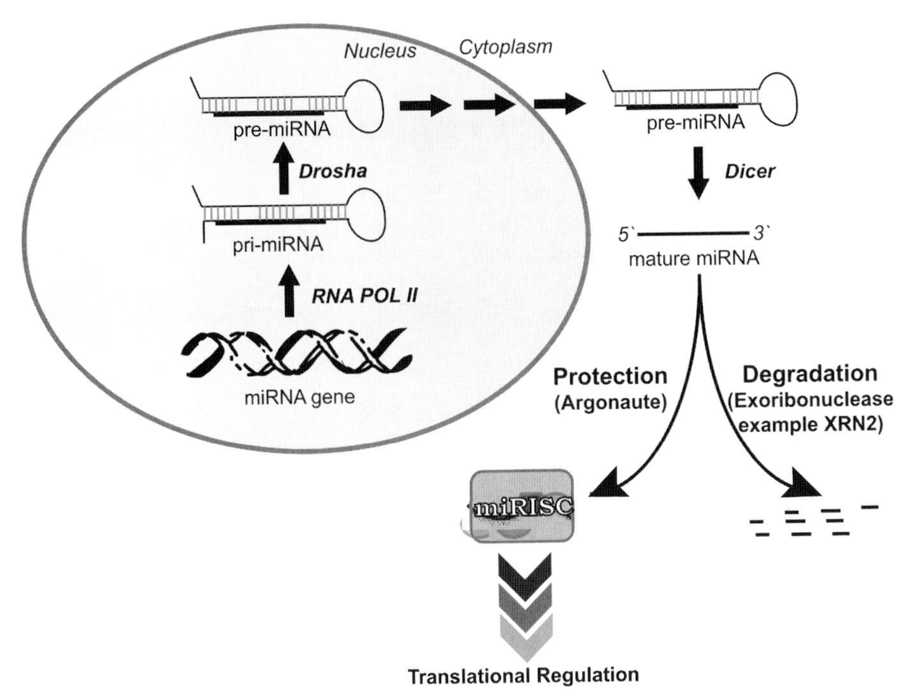

Figure 1. MicroRNA biogenesis and stability. After synthesis by RNA polymerase II (RNA POL II), primary transcripts of miRNA (pri miRNA) are recognized by Drosha and Pasha, which excise the hairpin-generating precursor miRNA (pre-miRNA). These are transported to cytoplasm by exportin 5 and further processed by Dicer into mature ~23nt miRNA. Mature miRNA associated with argonaute and other factors leads to the targeted translational regulation. Release from argonaute or absence of protection machinery leaves miRNA prone to degradation by exoribonucleases, such as XRN-2 or SDN.

other protein complexes (Fig. 1). miRNAs target their mRNA by complimentary base pairing sequences located mainly at 3′UTR (untranslated region), though miRNAs can also target 5′UTR or coding regions of mRNA.[16,17] In addition to sequence specific targeting of mRNA, miRNA functions as a ribonucleoprotein complex (miRNPs) known as miRISCs (miRNA-induced silencing complex). Key components of miRISCs include AGO (argonaute) and GW182 (glycine-tryptophan repeat-containing protein family), which have conserved domains such as PAZ, PIWI, and MID. These proteins are found enriched in processing bodies (commonly known as P or GW bodies) for degradation of mRNA.[18]

Although mature miRNAs are generally thought to be stable due to their small size, they are prone to degradation by both 5′ to 3′ and 3′ to 5′ exoribonucleases present in cells, such as XRN2.[19,20] miRNA stability is also determined by its sequence complexity.[21] The stability of miRNAs depends on binding to miRISC proteins, such as argonaute (Fig. 1).

miRNAs are well known for their role as inhibitors of protein synthesis. The inhibition mechanism of protein synthesis has been hypothesized differently as (1) deadenylation of mRNA followed by degradation, (2) blocking of translational initiation, and (3) blocking at post-initiation stage of translation either by elongation block, ribosome drop-off, or proteolysis of nascent polypeptide. Details of the regulation have been reviewed previously[22,23] and are summarized in Figure 2. Some specific initial studies are described briefly in the following paragraphs.

As previously mentioned, miRNA-mediated deadenylation and degradation of mRNA were initially reported in *C. elegans* let-7 miRNA targeting lin-41 mRNA and in Zebrafish miR-430.[24,25] Various mechanisms of deadenylation and degradation of mRNA have been proposed after their initial discovery, as they appeared in different organisms. mRNA degradation is often mediated by removal of the poly(a) tail by 3′–5′ ribo-nucleases (RNases) such as PARN, CCR4p, POP (Pop2p), and

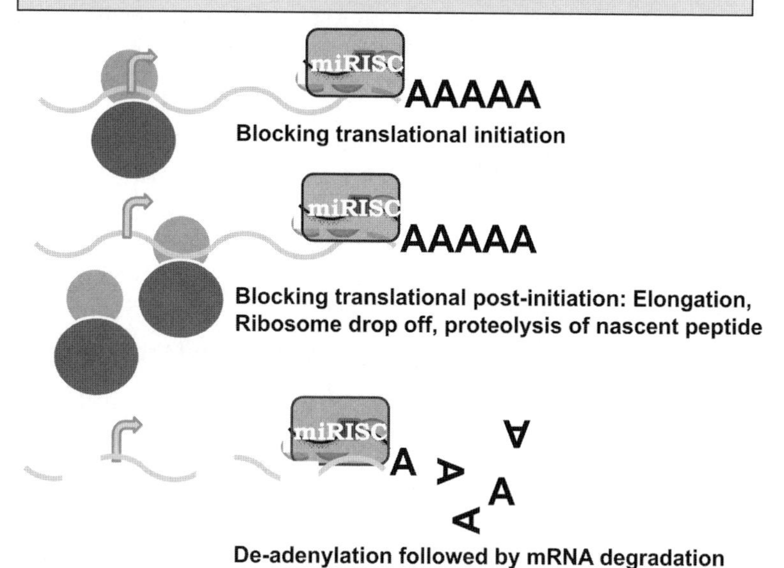

Figure 2. Translational regulation by miRNA. miRNA modulates translation either by repression or activation, although the mechanism is different. miRNA-repressed translation has three mechanisms, as described in different steps of the translational process.

PAN (Pan2p).[26–29] mRNA stability is often under the control of *cis*-acting elements within the 3′-untranslated regions (UTR), which recruit enzymes/factors followed by recruitment of deadenylation enzymes.

Blocking of translational initiation was first observed in HeLa cells using reporter mRNAs whose 3′-UTRs were targeted by endogenous (let-7) miRNA.[30] Similar observations have been reported by inhibiting eukaryotic initiation factors, such as 4E/cap, and poly(A) tail function using artificial miRNA (CXCR4).[31] Much evidence supports the idea that miRNAs destabilize target mRNAs through deadenylation and subsequent decapping, followed by 5′–3′ exonucleolytic digestion.

Lin-4, the original miRNA in *C. elegans*, was initially shown to cause inhibition of translation of lin-14, which is important for postembryonic development: either no reduction in mRNA levels or a shift in polysomes was observed, which lead to the conclusion that miRNAs inhibit mRNA translation at the elongation step.[8]

Recently, miRNAs were also shown to activate protein synthesis.[17,32,33] AU-rich elements (AREs) and microRNA target sites are conserved sequences in 3′-UTRs of mRNA. During the cell cycle arrest, the AREs in tumor necrosis factor-α mRNA are transformed into a translation activation signal, recruiting many factors associated with micro-ribonucleoproteins (miRNPs) (Fig. 2). Vasudevan *et al.* have shown that human microRNA miR369-3 directs the association of these proteins with the AREs to activate translation.[34]

MicroRNA in cardiovascular research

The development of various cardiovascular disease models is a complex process involving different cell types including fibroblasts, cardiomyocytes, endothelial cells, smooth muscle cells, and many others. The patterns of miRNA expression are different in those cell types and thus can be explained based on different models.

In cardiac fibrosis, cardiac fibroblasts are central in the development of many diseases, such as cardiomyopathy, hypertension, myocardial infraction, and chronic Doxorubicin-induced cardiomyopathy, and regulate the cardiac extracellular matrix components.[35–37] Initial studies demonstrated the dysfunction of miRNA metabolism in the heart using a conditional deletion mutant (of dicer), which leads to hypertrophy and ventricular fibrosis.[38] In an acute myocardial infarction model, dysregulation of a specific miRNA (miR-29) was observed, and targeting miR-29 led to reduced collagen expression in fibroblasts.[39] In addition to miR-29 dysregulation, increased expression of miR-21, miR-214, and miR-224, and reduced expression of miR-29b and miR-149, were found in myocardial infarction based on microarray analyses followed by real-time PCR.[39] Similar studies with microarray and Northern blot analyses led to the discovery of miR-21 overexpression in failing heart; miR-21 was observed to regulate ERK-MAP kinase pathways.[40] Later, miR-21 was also found to regulate MMP2 in fibroblast in myocardial infraction models.[41] One of the key players in fibrosis, CTGF, was regulated at the posttranslational level by miR-133 and miR-30.[42]

In ischemic heart disease, miR-1 has been shown to be up-regulated in human studies. Overexpression studies in rats correlate miR-1 expression with arrhythmogenesis, cardiac conduction disturbance, and membrane potential abnormality.[43] Another miRNA (miR-133), encoded by the same loci of miR-1, induces myoblast proliferation *in vitro* and proliferates skeletal as well as cardiac muscle after overexpression in Xenopus embryos.[44] Thus, miRNA seems to play an important role in the development of the pathophysiological condition of heart disease as well as therapeutics.

Recently, comparative profile studies by microarray analyses between healthy patients and patients with coronary artery disease (CAD) led to the discovery of many circulating miRNA in blood.[45,46] Some miRNAs such as miR-126, miR-17, miR-92a, and mir155 are reduced in CAD patients, whereas miR-133 and miR-208 are increased in blood.[45] These studies may be helpful in the future to development miRNA biomarkers in cardiac disease.

Resveratrol and the French paradox

In their seminal article in 1992 about the French paradox, Renaud and de Lorgeril presented evidence that dietary fat and blood cholesterol may not be the determining factors for mortality and morbidity due to heart disease.[47] The mortality due to coronary heart disease is only 78 per 100,000 population in Toulouse, France, and 105 in Lille, France, compared to 182 in Stanford, California, United States, 348 in Belfast, Ireland, and 380 in Glasgow, Scotland. Despite the same saturated fat intake of 15% of the total calories and similar serum cholesterol in other parts of Europe (Belfast 232, Glasgow 244, compared to Toulouse 230 and Lille 252) or even lower in the United States (Stanford 209), the French had the lowest mortality due to myocardial infarction.[47] The authors noted that the countries that had the lowest mortality due to heart disease had one thing in common: the population of these countries had regular wine consumption, suggesting that wine provides cardioprotection. They described this phenomenon as the French paradox. Comparing the population of several countries, the authors concluded that wine consumption inhibits platelet aggregation, which in turn lowers mortality due to ischemic heart disease.[48]

Resveratrol was not well known to people until 1991, when the phenomenon of the French paradox was attributed to red wine, which is routinely consumed by the French with their meals. Subsequent studies determined that wine, especially red wine, is rich in certain flavonoids and antioxidants that can neutralize damaging free radicals continuously generated by the human body.[49] Within a short period of time, one of the polyphenolic compounds present in red wine, resveratrol, was identified; later, the cardioprotective property of red wine was attributed to resveratrol.[49]

Resveratrol is a trihydroxystilbene present in grape vines and skins that protects grapes from fungal infection.[50] Grape vines produce increasing amounts of resveratrol when they are subjected to stresses in their environment, such as dehydration, nutrient deprivation, and attacks by pathogenic organisms. Defense-promoting molecules such as resveratrol are called phytoalexins, a Greek word meaning "plant-protector", and resveratrol fulfils this definition. Besides wine, resveratrol is also present in peanuts and certain berries. The main source of resveratrol is the dried roots of *Polygonum cuspidatum*, which are used in traditional Japanese and Chinese medicine, Kojo-kon, to treat fungal diseases of heart, liver, and blood vessels.[51] As mentioned, resveratrol became the central issue of the French paradox because it is found in high amounts in wine. Resveratrol was found to possess heart health, antiobesity, and antiaging properties.[52–54] However, whether resveratrol alone is responsible for the French paradox is still under considerable debate.

Striking similarities exist between the cardioprotective properties of wine and resveratrol. Both can reduce blood pressure,[52] increase HDL cholesterol and decrease LDL cholesterol,[55,56] possess antiplatelet and anti-thrombin activities (aspirin-like effects),[57,58] exhibit insulin-like effects and lower blood sugar,[59,60] reduce obesity,[61,62] and activate longevity genes.[63,64] Although overwhelming evidence exists supporting the effects of resveratrol on cardiovascular health,[33–63,64] the concentration of resveratrol necessary to achieve the positive effects is under debate. The amount of wine necessary for maintaining a healthy heart is about two glasses of wine a day,[65] which may contain about 3–6 mg of resveratrol, depending on the source of the wine. On average, American red wine contains only about 2–3 mg of resveratrol in two glasses while wines from Spain, Italy, and Germany can contain double this amount.[66–68] Yet, published articles have stated that one needs hundreds of bottles of red wine to obtain the amount of resveratrol necessary to maintain a healthy heart,[69] and higher amounts of resveratrol could even be cytotoxic, as resveratrol appears to behave like a hormetin.[70,71] Similar to alcohol and wine, resveratrol health effects exhibit an inverted U-shaped or a J-shaped curve, indicating that like high doses of alcohol and wine, resveratrol might also exert harmful effects in the heart.[70,71] Further research is necessary to resolve the problem of the correct amount of resveratrol necessary to achieve heart health.

One of the important findings in recent years is that resveratrol induces autophagy.[71] This supports previous reports that resveratrol combats heart disease by preconditioning, that is, by adapting the heart to stress.[72] At lower doses (0.1–1 μM in H9c2 cardiac myoblast cells and 2.5 mg/kg/day in rats) resveratrol induces autophagy in the ischemic mycardium, as evidenced by the formation of autophagosomes and its markers LC3-II and beclin-1.[73] Autophagy is attenuated with the higher doses of resveratrol, which induce apoptotic cell death. The induction of autophagy with low doses of resveratrol is correlated with enhanced cell survival and decreased apoptosis.[71] The activation of mammalian target of rapamycin (mTOR) is differentially regulated with low doses of resveratrol: the phosphorylation of mTOR at serine 2448 is inhibited whereas the phosphorylation of mTOR at serine 2481 is enhanced (the latter is attenuated with a higher dose of resveratrol).[71] Interestingly, even though resveratrol attenuates the activation of mTOR complex 1, low-dose resveratrol significantly induces the expression of rictor, a component of mTOR complex 2, leading to the activation of its downstream survival kinase Akt (Ser 473).[71] Resveratrol-induced rictor eventually binds to mTOR and rictor-specific siRNA and thereby attenuates resveratrol-induced autophagy. Thus, this suggests that at a lower dose resveratrol-mediated cell survival, at least in part, is mediated through the induction of autophagy.[71]

The fact that resveratrol renders numerous health benefits from chemoprevention to cardioprotection makes it a target for stem cell therapy. Recent studies have indicated that resveratrol can prolong stem cell survival by altering the intracellular redox environment of the heart, thus resolving one of the major problems associated with cell therapy. Resveratrol was introduced in two ways. The first was by feeding the animals with resveratrol for up to three weeks; this changed the intracellular redox environment and was followed by cell therapy.[74] The second was by pretreating the stem cells with low concentration of resveratrol before cell therapy.[74] In both protocols, resveratrol prolonged the life of stem cells in

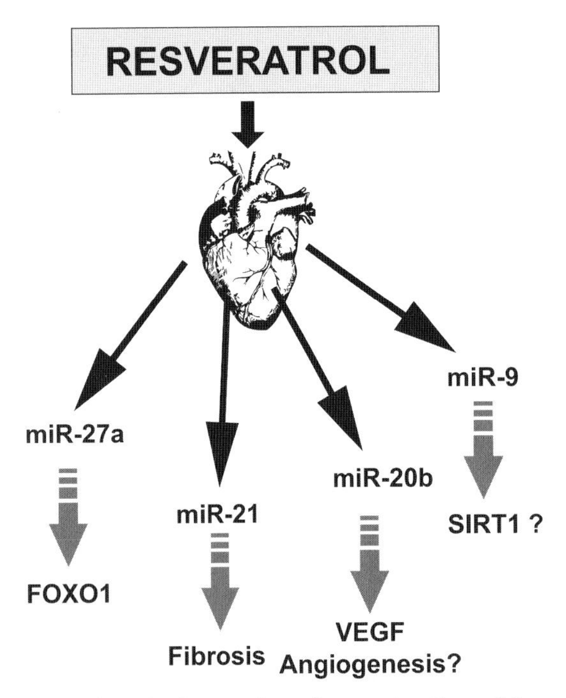

Figure 3. Role of miRNA in cardioprotection. Preconditioning of the heart by resveratrol, involves differential changes as significant miRNAs are up or down when treated with resveratrol for three weeks in rats. Few miRNAs were shown as key regulators in cardioprotection by various pathways.

the infracted heart, as evidenced by active proliferation and differentiation suggesting that resveratrol can regenerate the infracted myocardium.

MicroRNA expression profile in response to resveratrol

Differential expression of over 50 miRNAs was observed in ischemic reperfused (IR) myocardium, some of which were previously implicated in cardiac remodeling (Mukhopadhyay and Das, unpublished data). The target genes for the differentially expressed miRNA include genes of various molecular functions such as metal ion modulation and transcription factors, which may play key roles in reducing ischemic reperfusion injury.[78] IR samples pretreated with resveratrol, or its commercial formulation Longevinex®, had attenuated the up- or downregulation in more than 50% of differentially expressed miRNAs compared to IR exposed hearts. Resveratrol modulation of miRNAs in IR includes miR-21, miR-20b, mir-27a, and miR-9, just to mention a few.[78] miR-21 was shown to regulate the ERK-MAP kinase signaling pathway in cardiac fibroblasts, regulating cardiac structure and function,[40] and VEGF was reported to be modulated by miR-20b through HIF1α, whereas FOXO1 is regulated by miR-27a in cancer cells.[75,76] SIRT1 was observed to be regulated by miR-9 in stem cells.[77] Thus, resveratrol treatment leads to a unique signature of miRNA expression. Some of the miRNA were demonstrated to be involved in various cardiac remodeling, such as fibrosis. Few differentially expressed miRNAs target key transcription factors, such as FOXO1 and SIRT1, based on bioinformatics analyses and observations found in different cell types (Fig. 3). Future studies will be based on the mechanistic action and stability of miRNA as described before. Further detailed *in vivo* and *in vitro* studies, such as targeting those miRNA followed by loss/gain of function, will enable us to explore the complex mechanism underlying the cardioprotection of resveratrol. Targeting miRNA responsible for cardiac damage may lead to identifying novel therapeutic targets.

Acknowledgments

This study was supported in part by NIH HL 34360, HL 22559, and HL 33889 (DKD) and the Intramural Research Program of the NIAAA/NIH (PP).

Conflicts of interest

The authors declare no conflicts of interest.

References

1. Venter, J.C. *et al.* 2001. The sequence of the human genome. *Science* **291:** 1304–1351.
2. Siomi, H. & M.C. Siomi. 2009. On the road to reading the RNA-interference code. *Nature* **457:** 396–404.
3. Katiyar-Agarwal, S. *et al.* 2006. A pathogen-inducible endogenous siRNA in plant immunity. *Proc. Natl. Acad. Sci. USA* **103:** 18002–18007.
4. Liu, Q. & Z. Paroo. 2010. Biochemical principles of small RNA pathways. *Annu. Rev. Biochem.* **79:** 295–319.
5. Ambros, V. 2004. The functions of animal microRNAs. *Nature* **431:** 350–355.
6. Beisel, C.L. & G. Storz. 2010. Base pairing small RNAs and their roles in global regulatory networks. *FEMS Microbiol. Rev.* **34:** 866–882.
7. Hamilton, A.J. & D.C. Baulcombe. 1999. A species of small antisense RNA in posttranscriptional gene silencing in plants. *Science* **286:** 950–952.
8. Olsen, P.H. & V. Ambros. 1999. The lin-4 regulatory RNA controls developmental timing in *Caenorhabditis elegans* by blocking LIN-14 protein synthesis after the initiation of translation. *Dev. Biol.* **216:** 671–680.
9. Delmas, D. *et al.* 2006. Resveratrol as a chemopreventive agent: a promising molecule for fighting cancer. *Curr. Drug Targets.* **7:** 423–442.

10. Goswami, S.K. & D.K. Das. 2009. Resveratrol and chemoprevention. *Cancer Lett*. **284:** 1–6.
11. de Almeida, L.M. *et al.* 2008. Resveratrol protects against oxidative injury induced by H2O2 in acute hippocampal slice preparations from Wistar rats. *Arch. Biochem. Biophys.* **480:** 27–32.
12. Quincozes-Santos, A. *et al.* 2009. The janus face of resveratrol in astroglial cells. *Neurotox Res.* **16:** 30–41.
13. Das, D.K. & N. Maulik. 2006. Resveratrol in cardioprotection: a therapeutic promise of alternative medicine. *Mol. Interv*. **6:** 36–47.
14. Das, S. & D.K. Das. 2007. Resveratrol: a therapeutic promise for cardiovascular diseases. *Recent Pat. Cardiovasc. Drug Discov*. **2:** 133–138.
15. Hammond, S.M. 2005. Dicing and slicing: the core machinery of the RNA interference pathway. *FEBS Lett*. **579:** 5822–5829.
16. Rigoutsos, I. 2009. New tricks for animal microRNAs: targeting of amino acid coding regions at conserved and nonconserved sites. *Cancer Res.* **69:** 3245–3248.
17. Orom, U.A., F.C. Nielsen & A.H. Lund. 2008. MicroRNA-10a binds the 5′UTR of ribosomal protein mRNAs and enhances their translation. *Mol. Cell* **30:** 460–471.
18. Tolia, N.H. & L. Joshua-Tor. 2007. Slicer and the argonautes. *Nat. Chem. Biol.* **3:** 36–43.
19. Kai, Z.S. & A.E. Pasquinelli. 2010. MicroRNA assassins: factors that regulate the disappearance of miRNAs. *Nat. Struct. Mol. Biol.* **17:** 5–10.
20. Krol, J., I. Loedige & W. Filipowicz. 2010. The widespread regulation of microRNA biogenesis, function and decay. *Nat. Rev. Genet*. **11:** 597–610.
21. Bail, S. *et al.* 2010. Differential regulation of microRNA stability. *RNA* **16:** 1032–1039.
22. Chekulaeva, M. & W. Filipowicz. 2009. Mechanisms of miRNA-mediated post-transcriptional regulation in animal cells. *Curr. Opin. Cell Biol.* **21:** 452–460.
23. Filipowicz, W., S.N. Bhattacharyya & N. Sonenberg. 2008. Mechanisms of post-transcriptional regulation by microRNAs: are the answers in sight? *Nat. Rev. Genet*. **9:** 102–114.
24. Bagga, S. *et al.* 2005. Regulation by let-7 and lin-4 miRNAs results in target mRNA degradation. *Cell* **122:** 553–563.
25. Giraldez, A.J. *et al.* 2006. Zebrafish MiR-430 promotes deadenylation and clearance of maternal mRNAs. *Science* **312:** 75–79.
26. Korner, C.G. & E. Wahle. 1997. Poly(A) tail shortening by a mammalian poly(A)-specific 3′-exoribonuclease. *J. Biol. Chem.* **272:** 10448–10456.
27. Brown, C.E. & A.B. Sachs. 1998. Poly(A) tail length control in Saccharomyces cerevisiae occurs by message-specific deadenylation. *Mol. Cell Biol.* **18:** 6548–6559.
28. Tollervey, D. 1996. Genetic and biochemical analyses of yeast RNase MRP. *Mol. Biol. Rep.* **22:** 75–79.
29. Collart, M.A. 2003. Global control of gene expression in yeast by the Ccr4-Not complex. *Gene* **313:** 1–16.
30. Pillai, R.S. *et al.* 2005. Inhibition of translational initiation by Let-7 MicroRNA in human cells. *Science* **309:** 1573–1576.
31. Humphreys, D.T. *et al.* 2005. MicroRNAs control translation initiation by inhibiting eukaryotic initiation factor 4E/cap and poly(A) tail function. *Proc. Natl. Acad. Sci. USA*. **102:** 16961–16966.
32. Buchan, J.R. & R. Parker. 2007. Molecular biology. The two faces of miRNA. *Science* **318:** 1877–1878.
33. Vasudevan, S. & J.A. Steitz. 2007. AU-rich-element-mediated upregulation of translation by FXR1 and Argonaute 2. *Cell* **128:** 1105–1118.
34. Vasudevan, S., Y. Tong & J.A. Steitz. 2007. Switching from repression to activation: microRNAs can up-regulate translation. *Science* **318:** 1931–1934.
35. Mukhopadhyay, P. *et al.* 2010. CB1 cannabinoid receptors promote oxidative stress and cell death in murine models of doxorubicin-induced cardiomyopathy and in human cardiomyocytes. *Cardiovasc. Res.* **85:** 773–784.
36. Raman, S.V. 2010. The hypertensive heart. An integrated understanding informed by imaging. *J. Am. Coll. Cardiol.* **55:** 91–96.
37. Jellis, C. *et al.* 2010. Assessment of nonischemic myocardial fibrosis. *J. Am. Coll. Cardiol.* **56:** 89–97.
38. da Costa Martins, P.A. *et al.* 2008. Conditional dicer gene deletion in the postnatal myocardium provokes spontaneous cardiac remodeling. *Circulation*. **118:** 1567–1576.
39. van Rooij, E. *et al.* 2008. Dysregulation of microRNAs after myocardial infarction reveals a role of miR-29 in cardiac fibrosis. *Proc. Natl. Acad. Sci. USA* **105:** 13027–13032.
40. Thum, T. *et al.* 2008. MicroRNA-21 contributes to myocardial disease by stimulating MAP kinase signalling in fibroblasts. *Nature* **456:** 980–984.
41. Roy, S. *et al.* 2009. MicroRNA expression in response to murine myocardial infarction: miR-21 regulates fibroblast metalloprotease-2 via phosphatase and tensin homologue. *Cardiovasc. Res.* **82:** 21–29.
42. Duisters, R.F. *et al.* 2009. miR-133 and miR-30 regulate connective tissue growth factor: implications for a role of microRNAs in myocardial matrix remodeling. *Circ. Res.* **104:** 170–178, 176p following 178.
43. Yang, B. *et al.* 2007. The muscle-specific microRNA miR-1 regulates cardiac arrhythmogenic potential by targeting GJA1 and KCNJ2. *Nat. Med.* **13:** 486–491.
44. Chen, J.F. *et al.* 2006. The role of microRNA-1 and microRNA-133 in skeletal muscle proliferation and differentiation. *Nat. Genet*. **38:** 228–233.
45. Fichtlscherer, S. *et al.* 2010. Circulating MicroRNAs in patients with coronary artery disease. *Circ. Res.* **107:** 810–817.
46. Kumarswamy, R., S.D. Anker & T. Thum. 2010. MicroRNAs as circulating biomarkers for heart failure: questions about MiR-423–5p. *Circ Res.* **106:** e8; author reply e9.
47. Renaud, S. & M. de Lorgeril. 1992. Wine, alcohol, platelets, and the French paradox for coronary heart disease. *Lancet* **339:** 1523–1526.
48. Renaud, S. & M. de Lorgeril. 1993. The French paradox: dietary factors and cigarette smoking-related health risks. *Ann N.Y. Acad. Sci.* **686:** 299–309.
49. Troup, G.J. *et al.* 1994. Free radicals in red wine, but not in white? *Free Radic. Res.* **20:** 63–68.
50. Farina, A., C. Ferranti & C. Marra. 2006. An improved synthesis of resveratrol. *Nat. Prod. Res.* **20:** 247–252.

51. Wenzel, E. & V. Somoza. 2005. Metabolism and bioavailability of trans-resveratrol. *Mol. Nutr. Food Res.* **49:** 472–481.
52. Wallerath, T. *et al.* 2002. Resveratrol, a polyphenolic phytoalexin present in red wine, enhances expression and activity of endothelial nitric oxide synthase. *Circulation* **106:** 1652–1658.
53. Das, D.K. *et al.* 1999. Cardioprotection of red wine: role of polyphenolic antioxidants. *Drugs Exp. Clin. Res.* **25:** 115–120.
54. de la Lastra, C.A. & I. Villegas. 2005. Resveratrol as an anti-inflammatory and anti-aging agent: mechanisms and clinical implications. *Mol. Nutr. Food Res.* **49:** 405–430.
55. Fan, E. *et al.* 2008. Beneficial effects of resveratrol on atherosclerosis. *J. Med. Food* **11:** 610–614.
56. Frankel, E.N., A.L. Waterhouse & J.E. Kinsella. 1993. Inhibition of human LDL oxidation by resveratrol. *Lancet* **341:** 1103–1104.
57. Bertelli, A.A. *et al.* 1995. Antiplatelet activity of synthetic and natural resveratrol in red wine. *Int. J. Tissue React.* **17:** 1–3.
58. Stef, G. *et al.* 2006. Resveratrol inhibits aggregation of platelets from high-risk cardiac patients with aspirin resistance. *J. Cardiovasc. Pharmacol.* **48:** 1–5.
59. Szkudelska, K. & T. Szkudelski. 2010. Resveratrol, obesity and diabetes. *Eur. J. Pharmacol.* **635:** 1–8.
60. Chen, W.P. *et al.* 2007. Resveratrol enhances insulin secretion by blocking K(ATP) and K(V) channels of beta cells. *Eur. J. Pharmacol.* **568:** 269–277.
61. Bertelli, A.A. & D.K. Das. 2009. Grapes, wines, resveratrol, and heart health. *J. Cardiovasc. Pharmacol.* **54:** 468–476.
62. Pfluger, P.T. *et al.* 2008. Sirt1 protects against high-fat diet-induced metabolic damage. *Proc. Natl. Acad. Sci. USA* **105:** 9793–9798.
63. Das, D.K., S. Mukherjee & D. Ray. 2010. Resveratrol and red wine, healthy heart and longevity. *Heart Fail Rev.* **15:** 467–477.
64. Mukherjee, S. *et al.* 2009. Expression of the longevity proteins by both red and white wines and their cardioprotective components, resveratrol, tyrosol, and hydroxytyrosol. *Free Radic. Biol. Med.* **46:** 573–578.
65. Goldberg, I.J. *et al.* 2001. AHA Science Advisory: Wine and your heart: a science advisory for healthcare professionals from the Nutrition Committee, Council on Epidemiology and Prevention, and Council on Cardiovascular Nursing of the American Heart Association. *Circulation* **103:** 472–475.
66. Stervbo, U., O. Vang & C. Bonnesen. 2007. A review of the content of the putative chemopreventive phytoalexin resveratrol in red wine. *Food Chem.* **101:** 449–457.
67. Lamuelaraventos, R.M. *et al.* 1995. Resveratrol and piceid levels in wine production and in finished wines. *Abstr. Pap. Am. Chem. S* **210:** 10-Agfd.
68. Goldberg, D.M. *et al.* 1995. A global survey of *trans*-Resveratrol concentrations in commercial wines. *Am. J. Enol. Viticult.* **46:** 159–165.
69. E.W. 2006. Wine ingredient resveratrol as anti-aging pill? Not just yet. USA TODAY. 11/29/2006.
70. Mukherjee, S, L.I., Das DK. Hormetic response of resveratrol against cardioprotection. *J. Exp. Clin. Cardiology*. In press.
71. Gurusamy, N. *et al.* 2010. Cardioprotection by resveratrol: a novel mechanism via autophagy involving the mTORC2 pathway. *Cardiovasc. Res.* **86:** 103–112.
72. Das, S. *et al.* 2005. Pharmacological preconditioning with resveratrol: role of CREB-dependent Bcl-2 signaling via adenosine A3 receptor activation. *Am. J. Physiol. Heart Circ. Physiol.* **288:** H328–H335.
73. Lekli, I., Ray D., S. Mukherjee, *et al.* 2010. Coordinated autophagy with resveratrol and gamma tocotrienol confers synergetic cardioprotection. *J. Cell Mol. Med.* **20:** 1–13.
74. Gurusamy, N., Ray D., I. Lekli, *et al.* 2010. Regeneration of infracted myocardium by nutritionally modified cardiac stem cells. *J. Cell Mol. Med.* doi: 10.1111/j.1582 4934.2010.01140.x.
75. Cascio, S. *et al.* 2010. miR-20b modulates VEGF expression by targeting HIF-1 alpha and STAT3 in MCF-7 breast cancer cells. *J. Cell Physiol.* **224:** 242–249.
76. Guttilla, I.K. & B.A. White. 2009. Coordinate regulation of FOXO1 by miR-27a, miR-96, and miR-182 in breast cancer cells. *J. Biol. Chem.* **284:** 23204–23216.
77. Saunders, L.R. *et al.* 2010. miRNAs regulate SIRT1 expression during mouse embryonic stem cell differentiation and in adult mouse tissues. *Aging*.
78. Mukhopadhyay, P., Mukherjee, S. *et al.* 2010. *PLOSone.* **2:** 415–431.

Ann. N.Y. Acad. Sci. ISSN 0077-8923

ANNALS OF THE NEW YORK ACADEMY OF SCIENCES
Issue: *Resveratrol and Health*

Anti-inflammatory effects of resveratrol: possible role in prevention of age-related cardiovascular disease

Anna Csiszar

University of Oklahoma Health Sciences Center, Reynolds Oklahoma Center on Aging, The Donald W. Reynolds Department of Geriatric Medicine and Department of Physiology, Oklahoma City, Oklahoma

Address for correspondence: Anna Csiszar, M.D., Ph.D., Reynolds Oklahoma Center on Aging, Department of Geriatric Medicine, University of Oklahoma HSC, 975 N. E. 10th Street – BRC 1315, Oklahoma City, OK 73104. anna-csiszar@ouhsc.edu

Cardiovascular diseases are the most common cause of death among the elderly in the Western world. Resveratrol (3,5,4′-trihydroxystilbene) is a plant-derived polyphenol that was shown to exert diverse anti-aging activity mimicking some of the molecular and functional effects of caloric restriction. This mini-review focuses on the molecular and cellular mechanisms activated by resveratrol in the vascular system, and explores the links between its anti-oxidative and anti-inflammatory effects, which could be exploited for the prevention or amelioration of vascular aging in the elderly.

Keywords: resveratrol; aging; vascular dysfunction

Introduction

Age-specific mortality rates from heart disease and stroke increase exponentially with age, which imposes a huge financial burden on the health care systems in the Western world. There is thus an urgent need for effective therapeutic strategies that have the potential to promote cardiovascular health in the elderly and to prevent or delay the development of atherosclerotic vascular diseases. During the past decade, dietary supplementation with the plant-derived polyphenol resveratrol (3,5,4′-trihydroxystilbene) has emerged as a promising approach to counteract aging-induced pro-atherogenic phenotypic changes in the vasculature. The first population-based studies demonstrated that Mediterranean diets, which are rich in resveratrol, are associated with significantly reduced risk of cardiovascular disease in humans.[1,2] Subsequently resveratrol was shown to exert significant anti-aging action in invertebrates,[3,4] mimicking many aspects of caloric restriction.[5–8] Importantly, resveratrol supplementation was also shown to exert anti-inflammatory and anti-oxidant effects in various mammalian models of aging and cardiovascular diseases.[5,7] In this review, the potential mechanisms underlying the vasoprotective effects of resveratrol are considered and its use as a dietary supplement to promote cardiovascular health in the elderly is discussed.

Role of oxidative stress and inflammation in cardiovascular aging

Our current understanding of the pathogenesis of age-associated cardiovascular diseases is that age-related oxidative stress may promote vascular inflammation even in the absence of traditional risk factors associated with atherogenesis (e.g., hypertension or metabolic diseases; reviewed recently elsewhere[9]) (Fig. 1). It is well-established that age-associated low-grade inflammation accelerates the incidence of coronary artery disease and stroke in the elderly.[9] There is abundant experimental data suggesting that increased activity of NAD(P)H oxidases and mitochondrial overproduction of reactive oxygen species underlie age-related oxidative stress in the vasculature, promoting inflammation and endothelial damage.[10–14] Nitric oxide (NO) is a critical factor for the health and function of endothelial cells. Consequences of increased oxidative stress in aging include functional inactivation of NO by high concentrations of O_2^-, which results in significant vasomotor dysfunction (recently reviewed

doi: 10.1111/j.1749-6632.2010.05848.x

elsewhere),[15] increased apoptosis of endothelial cells,[16,17] microvascular rarefaction, and impaired mitochondrial biogenesis.[18–20] The key role of endothelium-derived NO in protecting the cardiovascular system during aging is underscored by the findings that endothelial nitric oxide synthase (eNOS) knockout mice exhibit a premature cardiac aging phenotype that is associated with early mortality.[21] The existing data also point to an important cross-talk between reactive oxygen species (ROS) production and inflammatory processes in the pathogenesis of cardiovascular aging.[9] On the one hand, ROS per se can function as a signaling molecule that activates pathways that regulate inflammatory processes,[22] including endothelial activation and secretion of inflammatory mediators. Specifically, mitochondria-derived H_2O_2 is thought to contribute to the activation of nuclear factor kappa beta (NF-κB) in the vasculature, resulting in a pro-inflammatory shift in endothelial gene expression profile.[22] Increased NF-κB binding in aging is likely responsible for the increased expression of inducible nitric oxide synthase (iNOS),[10,22,23] which is a major source of vascular peroxynitrite production. On the other hand, inflammation itself promotes cellular oxidative stress (e.g., by TNF-α-mediated activation of NAD(P)H oxidases).[24] In that regard it is important to note that both in laboratory rodents and humans there is an age-related up-regulation of TNF-α and that disruption of TNF-α signaling confers vasoprotection in aging.[24]

Anti-oxidative and anti-inflammatory effects of resveratrol in aging

Previous studies have established that resveratrol can exert significant cardiovascular protective effects in various models of myocardial injury,[25–27] hypertension,[26,28] and type 2 diabetes.[7,29–31] Recent studies provide clear evidence that resveratrol treatment can also confer vasoprotection in aged mice and rats,[7,22] which attenuates ROS production, improves endothelial function, inhibits inflammatory processes, and decreases the rate of endothelial apoptosis. The mechanisms underlying the cardiovascular protective action of resveratrol are likely multifaceted (Fig. 1). Resveratrol was shown to upregulate eNOS and increase NO bioavailability.[29,32,33] Resveratrol can also induce major cellular anti-oxidant enzymes (e.g. glutathione peroxidase, heme oxygenase, superoxide dismutase) in cardiac and vascular cells,[34–37] which results in a marked attenuation of oxidative stress. Resveratrol both downregulates vascular and cardiac expression of TNF-α and inhibits NADPH oxidases in the vasculature.[29,31,38] It is significant that resveratrol was also shown to inhibit mitochondrial production of reactive oxygen species in the vasculature.[30] In addition, resveratrol, both *in vivo* and at nutritionally relevant concentrations *in vitro*, was shown to inhibit inflammatory processes, including NF-κB activation, inflammatory gene expression and attenuation of monocyte adhesiveness to endothelial cells,[7,25,39–47] all of which may contribute to its cardioprotective effects in aging. Recent studies showed that resveratrol, via an eNOS-dependent pathway, induces mitochondrial biogenesis both in cultured endothelial cells and in endothelia of mice with accelerated vascular aging.[20] Further studies are needed to determine whether the aforementioned vasoprotective effects of resveratrol are manifested in the cardiovascular system of elderly humans as well. In that regard, studies on vessels isolated from non-human primates treated with resveratrol will also be highly informative.

The molecular targets of resveratrol, which mediate its diverse cellular effects, are the subject of ongoing investigations. On the basis of the structural similarity of resveratrol to diethylstilbestrol, resveratrol was characterized as a phytoestrogen.[48] Given the cardioprotective benefits attributed to estrogens at the time, this idea lead to a number of follow-up studies suggesting that some of the cardiovascular effects of high doses of resveratrol may indeed be modulated by activation of the estrogen receptor.[49] Yet, more recently the cardioprotective effects of estrogen replacement have become subjects of debate, and there are also a number of studies that suggest that the estrogen receptor is not the main cellular target of resveratrol in the vasculature. Since the original observation of Sinclair and co-workers,[4] a large body of evidence has been published linking the cellular action of resveratrol to regulation of a pathway dependent on SIRT1, a mammalian homolog of the *Saccharomyces cerevisae* silent information regulator 2 (Sir2) protein.[3,5,50–55] There is an ongoing debate whether resveratrol is a direct activator of SIRT1, which catalyzes NAD^+-dependent protein deacetylation and is a critical regulator of transcription, genome stability, apoptosis, and metabolism. Although recent

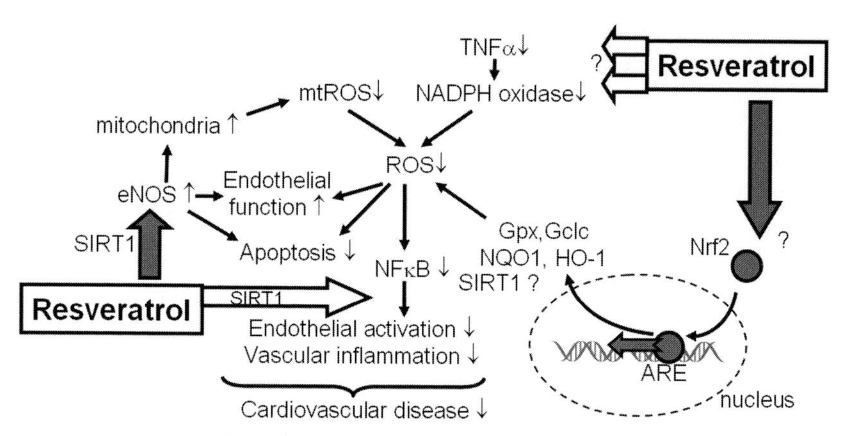

Figure 1. Proposed scheme for the mechanisms by which resveratrol confers anti-oxidative and anti-inflammatory vasoprotective effects in aging. During aging, increased NADPH oxidase- and mitochondria-derived ROS production enhances NF-κB activation, which promotes inflammatory cytokine and chemokine expression, endothelial activation, and leukocyte adhesion. Increased oxidative stress also impairs endothelial vasomotor function and promotes endothelial apoptosis, which together with chronic low-grade vascular inflammation significantly increases the risk for the development of vascular diseases in the elderly. The model predicts that resveratrol, via up-regulating Nrf2-driven antioxidant enzymes and eNOS, down-regulating TNF-α-activated NADPH oxidases, exerting mitochondrial protective effects, and inhibiting NF-κB, significantly attenuates vascular oxidative stress and inflammation in aging. Empty block arrows: inhibition; filled block arrows: induction/activation.

studies suggest that in cell-free assays resveratrol may not activate SIRT1 directly,[56] there is strong evidence that resveratrol and its metabolites, both *in vivo* and *ex vivo*, can promote SIRT1-dependent cellular responses, as demonstrated by resveratrol-induced decreases in acetylation of various known SIRT1 targets.[8] Furthermore, resveratrol has been shown to upregulate protein expression of SIRT1 in multiple cell types, including endothelial cells.[20] In addition, overexpression of SIRT1 in endothelial cells (similar to many other cell types) can mimic many of the effects of resveratrol,[57] whereas depletion of SIRT1 tends to attenuate resveratrol-induced cellular effects.[20,30,43,57] There is also solid evidence that inhibition of NF-κB by resveratrol is mediated via SIRT1.[58] SIRT1 is also needed for resveratrol-mediated induction of mitochondrial biogenesis and attenuation of mitochondrial oxidative stress in cardiovascular cells.[20,30] In light of recent controversies regarding the interaction of resveratrol and SIRT1, further studies are needed to elucidate the role of SIRT1-regulated pathways in the vasoprotective action of resveratrol in aging and the cellular mechanisms responsible for resveratrol-induced up-regulation of SIRT1 in cardiovascular tissues. Future studies should also address the interaction of SIRT1-dependent pathways with alternative cellular targets of resveratrol (e.g. NQO2,[59,60] cyclooxygenase, etc.) in vascular endothelial and smooth muscle cells.

Nrf2 activation: a new target for resveratrol

There is increasing evidence that activation of NF-E2-related factor 2 (Nrf2) is a key mechanism by which resveratrol confers its cytoprotective effects in the cardiovascular system.[27] Nrf2 is a basic leucine zipper transcription factor that regulates the coordinated expression of key antioxidant mechanisms in the cell by binding to the antioxidant response elements in the promoter regions of target genes. The first evidence that resveratrol can activate Nrf2 came from studies on cultured PC12 cells[61] and human lung epithelial cells.[62] Subsequent studies demonstrated that in cultured endothelial cells resveratrol also significantly increases transcriptional activity of Nrf2, which is associated with up-regulation of several Nrf2 target genes.[37] Many of these Nrf2 targets (e.g., NAD(P)H:quinone oxidoreductase 1, heme oxygenase-1) have been shown to promote endothelial health under conditions of increased oxidative stress.[34] Because Nrf2-driven pathways can be activated by concentrations of resveratrol readily achievable *in vivo*, future studies should elucidate whether Nrf2 activation contributes to the vasoprotective effects of resveratrol in aging. Recent evidence obtained using high fat diet-fed $Nrf2^{-/-}$

mice lend support to the hypothesis that Nrf2 activation plays an important role in the vasoprotective action of resveratrol.[37] Further, induction of the Nrf2 target HO-1 has also been implicated in the cardioprotective effects of resveratrol under conditions of experimentally-induced ischemia.[36] At present it is not well understood how pathways governed by Nrf2 and SIRT1 cross-talk. Further studies are warranted to test the possibility that SIRT1 acts as a permissive factor, modulating Nrf2-driven responses in the vasculature. Future studies also should test the possibility that Nrf2-dependent pathways regulate SIRT1 expression at the level of transcription.

Perspectives

Although significant progress has been achieved in elucidating the cellular mechanisms activated by resveratrol, the specific roles for pathways regulating mitochondrial function, cellular antioxidant defenses, and mechanisms involved in macromolecular repair need to be elucidated further. There is reasonable consensus that oxidative stress plays a central role in the development of atherosclerosis and that redox-sensitive molecular pathways (e.g., NF-κB) promote vascular inflammation in aging. Yet, recent large randomized clinical trials have shown no significant benefit when anti-oxidants targeted to the cell membranes (such as vitamin E) were given to patients with a high-risk coronary arterial disease profile. At present, it is unknown whether administration of mitochondria-targeted antioxidants would affect progression of cardiovascular diseases in elderly patients. In various experimental settings, including studies in aged laboratory rodents, resveratrol was shown to attenuate free radical production in multiple cellular compartments (i.e., both in the mitochondria and the cytosol). Thus, further studies on the effects of resveratrol on aging-induced oxidative stress and inflammation and their role in cardiovascular pathology are warranted. Importantly, these studies should determine whether anti-oxidative and anti-inflammatory effects of resveratrol are manifested in primates. Finally, research efforts should persist to fully elucidate the effects of resveratrol on microvascular alterations in aging. Future studies should extend the results of earlier investigations[63–65] assessing whether treatment with resveratrol can delay/prevent the age-associated decline in cerebral regional blood flow, the reduction in capillary and arteriolar density and angiogenesis, and the decline in spatial learning and memory in rodent and primate species and determine the roles of SIRT1 and Nrf2 in the microvascular protective effects of resveratrol treatment.

Acknowledgements

This work was supported by grants from the American Federation for Aging Research (to AC), the University of Oklahoma College of Medicine Alumni Association (to AC), and the NIH (AT006526, HL077256, and AG031085).

Conflicts of interest

The authors declare no conflicts of interest.

References

1. Keys, A., A. Menotti, M.J. Karvonen, *et al.* 1986. The diet and 15-year death rate in the seven countries study. *Am. J. Epidemiol.* **124:** 903–915.
2. de Lorgeril, M., P. Salen, J.L. Martin, *et al.* 1999. Mediterranean diet, traditional risk factors, and the rate of cardiovascular complications after myocardial infarction: final report of the Lyon Diet Heart Study. *Circulation* **99:** 779–785.
3. Wood, J.G., B. Rogina, S. Lavu, *et al.* 2004. Sirtuin activators mimic caloric restriction and delay ageing in metazoans. *Nature* **430:** 686–689.
4. Howitz, K.T., K.J. Bitterman, H.Y. Cohen, *et al.* 2003. Small molecule activators of sirtuins extend Saccharomyces cerevisiae lifespan. *Nature* **425:** 191–196.
5. Baur, J.A., K.J. Pearson, N.L. Price, *et al.* 2006. Resveratrol improves health and survival of mice on a high-calorie diet. *Nature* **444:** 337–342.
6. Kim, D., M.D. Nguyen, M.M. Dobbin, *et al.* 2007. SIRT1 deacetylase protects against neurodegeneration in models for Alzheimer's disease and amyotrophic lateral sclerosis. *EMBO J.* **26:** 3169–3179.
7. Pearson, K.J., J.A. Baur, K.N. Lewis, *et al.* 2008. Resveratrol delays age-related deterioration and mimics transcriptional aspects of dietary restriction without extending life Span. *Cell Metab.* **8:** 157–168.
8. Lagouge, M., C. Argmann, Z. Gerhart-Hines, *et al.* 2006. Resveratrol improves mitochondrial function and protects against metabolic disease by activating SIRT1 and PGC-1alpha. *Cell* **127:** 1109–1122.
9. Csiszar, A., M. Wang, E.G. Lakatta & Z. I. Ungvari. 2008. Inflammation and endothelial dysfunction during aging: role of NF-{kappa}B. *J. Appl. Physiol.* **105:** 1333–1341.
10. Csiszar, A., Z. Ungvari, J.G. Edwards, *et al.* 2002. Aging-induced phenotypic changes and oxidative stress impair coronary arteriolar function. *Circ. Res.* **90:** 1159–1166.
11. Van Der Loo, B., R. Labugger, J.N. Skepper, *et al.* 2000. Enhanced peroxynitrite formation is associated with vascular aging. *J. Exp. Med.* **192:** 1731–1744.
12. Adler, A., E. Messina, B. Sherman, *et al.* 2003. NAD(P)H oxidase-generated superoxide anion accounts for reduced

control of myocardial O2 consumption by NO in old Fischer 344 rats. *Am. J. Physiol. Heart Circ. Physiol.* **285:** H1015–H1022.
13. Donato, A.J., I. Eskurza, A.E. Silver, *et al.* 2007. Direct evidence of endothelial oxidative stress with aging in humans: relation to impaired endothelium-dependent dilation and upregulation of nuclear factor-kappaB. *Circ. Res.* **100:** 1659–1666.
14. Jacobson, A., C. Yan, Q. Gao, *et al.* 2007. Aging enhances pressure-induced arterial superoxide formation. *Am. J. Physiol. Heart Circ. Physiol.* **293:** H1344–H1350.
15. Ungvari, Z., R. Buffenstein, S.N. Austad, *et al.* 2008. Oxidative stress in vascular senescence: lessons from successfully aging species. *Front Biosci.* **13:** 5056–5070.
16. Hoffmann, J., J. Haendeler, A. Aicher, *et al.* 2001. Aging enhances the sensitivity of endothelial cells toward apoptotic stimuli: important role of nitric oxide. *Circ. Res.* **89:** 709–715.
17. Csiszar, A., Z. Ungvari, A. Koller, *et al.* 2004. Proinflammatory phenotype of coronary arteries promotes endothelial apoptosis in aging. *Physiol. Genomics* **17:** 21–30.
18. Ungvari, Z.I., N. Labinskyy, S.A. Gupte, *et al.* 2008. Dysregulation of mitochondrial biogenesis in vascular endothelial and smooth muscle cells of aged rats. *Am. J. Physiol. Heart Circ. Physiol.* **294:** H2121–H2128.
19. Addabbo, F., B. Ratliff, H.C. Park, *et al.* 2009. The Krebs cycle and mitochondrial mass are early victims of endothelial dysfunction: proteomic approach. *Am. J. Pathol.* **174:** 34–43.
20. Csiszar, A., N. Labinskyy, J.T. Pinto, *et al.* 2009. Resveratrol induces mitochondrial biogenesis in endothelial cells. *Am. J. Physiol. Heart Circ. Physiol.* **297:** H13–H20.
21. Li, W., S. Mital, C. Ojaimi, *et al.* 2004. Premature death and age-related cardiac dysfunction in male eNOS-knockout mice. *J. Mol. Cell Cardiol.* **37:** 671–680.
22. Ungvari, Z.I., Z. Orosz, N. Labinskyy, *et al.* 2007. Increased mitochondrial H2O2 production promotes endothelial NF-kB activation in aged rat arteries. *Am. J. Physiol. Heart Circ. Physiol.* **293:** H37–H47.
23. Cernadas, M.R., L. Sanchez de Miguel, M. Garcia-Duran, *et al.* 1998. Expression of constitutive and inducible nitric oxide synthases in the vascular wall of young and aging rats. *Circ. Res.* **83:** 279–286.
24. Csiszar, A., N. Labinskyy, K. Smith, *et al.* 2007. Vasculoprotective effects of anti-TNFalfa treatment in aging. *Am. J. Pathol.* **170:** 388–698.
25. Hattori, R., H. Otani, N. Maulik & D.K. Das. 2002. Pharmacological preconditioning with resveratrol: role of nitric oxide. *Am. J. Physiol. Heart Circ. Physiol.* **282:** H1988–H1995.
26. Juric, D., P. Wojciechowski, D.K. Das & T. Netticadan. 2007. Prevention of concentric hypertrophy and diastolic impairment in aortic-banded rats treated with resveratrol. *Am. J. Physiol.* **292:** H2138–H2143.
27. Gurusamy, N., D. Ray, I. Lekli & D.K. Das. 2010. Red wine antioxidant resveratrol-modified cardiac stem cells regenerate infarcted myocardium. *J Cell Mol. Med.* **14:** 2235–9.
28. Csiszar, A., N. Labinskyy, S. Olson, *et al.* 2009. Resveratrol prevents monocrotaline-induced pulmonary hypertension in rats. *Hypertension* **54:** 668–675.
29. Zhang, H., J. Zhang, Z. Ungvari & C. Zhang. 2009. Resveratrol improves endothelial function: role of TNF{alpha} and vascular oxidative stress. *Arterioscler Thromb. Vasc. Biol.* **29:** 1164–1171.
30. Ungvari, Z., N. Labinskyy, P. Mukhopadhyay, *et al.* 2009. Resveratrol attenuates mitochondrial oxidative stress in coronary arterial endothelial cells. *Am. J. Physiol. Heart Circ. Physiol.* **297:** H1876–H1881.
31. Zhang, H., B. Morgan, B.J. Potter, *et al.* 2010. Resveratrol improves left ventricular diastolic relaxation in type 2 diabetes by inhibiting oxidative/nitrative stress. *Am. J. Physiol. Heart Circ. Physiol* **299:** H985–94.
32. Bradamante, S., L. Barenghi, F. Piccinini, *et al.* 2003. Resveratrol provides late-phase cardioprotection by means of a nitric oxide- and adenosine-mediated mechanism. *Eur. J. Pharmacol.* **465:** 115–123.
33. Taubert, D. & R. Berkels. 2003. Upregulation and activation of eNOS by resveratrol. *Circulation* **107:** e78–e79; author reply e78–e79.
34. Ungvari, Z., Z. Orosz, A. Rivera, *et al.* 2007. Resveratrol increases vascular oxidative stress resistance. *Am. J. Physiol.* **292:** H2417–H2424.
35. Thirunavukkarasu, M., S.V. Penumathsa, S. Koneru, *et al.* 2007. Resveratrol alleviates cardiac dysfunction in streptozotocin-induced diabetes: role of nitric oxide, thioredoxin, and heme oxygenase. *Free Radic. Biol. Med.* **43:** 720–729.
36. Das, S., C.G. Fraga & D.K. Das. 2006. Cardioprotective effect of resveratrol via HO-1 expression involves p38 map kinase and PI-3-kinase signaling, but does not involve NFkappaB. *Free Radic. Res.* **40:** 1066–1075.
37. Ungvari, Z., Z. Bagi, A. Feher, *et al.* 2010. Resveratrol confers endothelial protection via activation of the antioxidant transcription factor Nrf2. *Am. J. Physiol. Heart Circ. Physiol.* **299:** H18–H24.
38. Chow, S.E., Y.C. Hshu, J.S. Wang & J.K. Chen. 2007. Resveratrol attenuates oxLDL-stimulated NADPH oxidase activity and protects endothelial cells from oxidative functional damages. *J. Appl. Physiol.* **102:** 1520–1527.
39. Wung, B.S., M.C. Hsu, C.C. Wu & C.W. Hsieh. 2005. Resveratrol suppresses IL-6-induced ICAM-1 gene expression in endothelial cells: effects on the inhibition of STAT3 phosphorylation. *Life Sci.* **78:** 389–397.
40. Shigematsu, S., S. Ishida, M. Hara, *et al.* 2003. Resveratrol, a red wine constituent polyphenol, prevents superoxide-dependent inflammatory responses induced by ischemia/reperfusion, platelet-activating factor, or oxidants. *Free Radic. Biol. Med.* **34:** 810–817.
41. Csiszar, A., K. Smith, N. Labinskyy, *et al.* 2006. Resveratrol attenuates TNF-{alpha}-induced activation of coronary arterial endothelial cells: role of NF-{kappa}B inhibition. *Am. J. Physiol.* **291:** H1694–H1699.
42. Ungvari, Z., Z. Orosz, N. Labinskyy, *et al.* 2007. Increased mitochondrial H2O2 production promotes endothelial NF-kappaB activation in aged rat arteries. *Am. J. Physiol. Heart Circ. Physiol.* **293:** H37–H47.
43. Csiszar, A., N. Labinskyy, A. Podlutsky, *et al.* 2008. Vasoprotective effects of resveratrol and SIRT1: attenuation of cigarette smoke-induced oxidative stress and proinflammatory phenotypic alterations. *Am. J. Physiol. Heart Circ. Physiol.* **294:** H2721–H2735.

44. Tsai, S.H., S.Y. Lin-Shiau & J.K. Lin. 1999. Suppression of nitric oxide synthase and the down-regulation of the activation of NFkappaB in macrophages by resveratrol. *Br. J. Pharmacol.* **126:** 673–680.
45. Manna, S.K., A. Mukhopadhyay & B.B. Aggarwal. 2000. Resveratrol suppresses TNF-induced activation of nuclear transcription factors NF-kappa B, activator protein-1, and apoptosis: potential role of reactive oxygen intermediates and lipid peroxidation. *J. Immunol.* **164:** 6509–6519.
46. Ferrero, M.E., A.E. Bertelli, A. Fulgenzi, *et al.* 1998. Activity in vitro of resveratrol on granulocyte and monocyte adhesion to endothelium. *Am. J. Clin. Nutr.* **68:** 1208–1214.
47. Moon, S.O., W. Kim, M.J. Sung, *et al.* 2006. Resveratrol suppresses tumor necrosis factor-alpha-induced fractalkine expression in endothelial cells. *Mol. Pharmacol.* **70:** 112–119.
48. Gehm, B.D., J.M. McAndrews, P.Y. Chien & J.L. Jameson. 1997. Resveratrol, a polyphenolic compound found in grapes and wine, is an agonist for the estrogen receptor. *Proc. Natl. Acad. Sci. USA* **94:**14138–14143.
49. Klinge, C.M., N.S. Wickramasinghe, M.M. Ivanova & S.M. Dougherty. 2008. Resveratrol stimulates nitric oxide production by increasing estrogen receptor alpha-Src-caveolin-1 interaction and phosphorylation in human umbilical vein endothelial cells. *FASEB J.* **22:** 2185–2197.
50. Danz, E.D., J. Skramsted, N. Henry, *et al.* 2009. Resveratrol prevents doxorubicin cardiotoxicity through mitochondrial stabilization and the Sirt1 pathway. *Free Radic. Biol. Med.* **46:** 1589–1597.
51. Miyazaki, R., T. Ichiki, T. Hashimoto, *et al.* 2008. SIRT1, a longevity gene, downregulates angiotensin II type 1 receptor expression in vascular smooth muscle cells. *Arterioscler. Thromb. Vasc. Biol.* **28:** 1263–1269.
52. Gracia-Sancho, J., G. Villarreal Jr., Y. Zhang & G. Garcia-Cardena. 2009. Activation of SIRT1 by resveratrol induces KLF2 expression conferring an endothelial vasoprotective phenotype. *Cardiovasc. Res.* **85:** 514–519.
53. Baur, J.A. & D.A. Sinclair. 2006. Therapeutic potential of resveratrol: the in vivo evidence. *Nat. Rev. Drug Discov.* **5:** 493–506.
54. Milne, J.C., P.D. Lambert, S. Schenk, *et al.* 2007. Small molecule activators of SIRT1 as therapeutics for the treatment of type 2 diabetes. *Nature* **450:** 712–716.
55. Dasgupta, B. & J. Milbrandt. 2007. Resveratrol stimulates AMP kinase activity in neurons. *Proc. Natl. Acad. Sci. USA* **104:** 7217–7222.
56. Pacholec, M., B.A. Chrunyk, D. Cunningham, *et al.* 2010. SRT1720, SRT2183, SRT1460, and resveratrol are not direct activators of SIRT1. *J. Biol. Chem.* **285:** 8340–8351.
57. Csiszar, A., N. Labinskyy, R. Jimenez, *et al.* 2009. Anti-oxidative and anti-inflammatory vasoprotective effects of caloric restriction in aging: role of circulating factors and SIRT1. *Mech. Ageing Dev.* **130:** 518–527.
58. Yeung, F., J.E. Hoberg, C.S. Ramsey, *et al.* 2004. Modulation of NF-kappaB-dependent transcription and cell survival by the SIRT1 deacetylase. *Embo J.* **23:** 2369–2380.
59. Hsieh, T.C., Z. Wang, C.V. Hamby & J.M. Wu. 2005. Inhibition of melanoma cell proliferation by resveratrol is correlated with upregulation of quinone reductase 2 and p53. *Biochem. Biophys. Res. Commun.* **334:** 223–230.
60. Wang, Z., T.C. Hsieh, Z. Zhang, *et al.* 2004. Identification and purification of resveratrol targeting proteins using immobilized resveratrol affinity chromatography. *Biochem. Biophys. Res. Commun.* **323:** 743–749.
61. Chen, C.Y., J.H. Jang, M.H. Li & Y.J. Surh. 2005. Resveratrol upregulates heme oxygenase-1 expression via activation of NF-E2-related factor 2 in PC12 cells. *Biochem. Biophys. Res. Commun.* **331:** 993–1000.
62. Kode, A., S. Rajendrasozhan, S. Caito, *et al.* 2008. Resveratrol induces glutathione synthesis by activation of Nrf2 and protects against cigarette smoke-mediated oxidative stress in human lung epithelial cells. *Am. J. Physiol. Lung Cell Mol. Physiol.* **294:** L478–L488.
63. Abraham, J. & R.W. Johnson. 2009. Consuming a diet supplemented with resveratrol reduced infection-related neuroinflammation and deficits in working memory in aged mice. *Rejuvenation Res.* **12:** 445–453.
64. Ranney, A. & M.S. Petro. 2009. Resveratrol protects spatial learning in middle-aged C57BL/6 mice from effects of ethanol. *Behav. Pharmacol.* **20:** 330–336.
65. Oomen, C.A., E. Farkas, V. Roman, *et al.* 2009. Resveratrol preserves cerebrovascular density and cognitive function in aging mice. *Front Aging Neurosci.* **1:** 4.

Ann. N.Y. Acad. Sci. ISSN 0077-8923

ANNALS OF THE NEW YORK ACADEMY OF SCIENCES
Issue: *Resveratrol and Health*

The phenomenon of resveratrol: redefining the virtues of promiscuity

John M. Pezzuto
College of Pharmacy, University of Hawaii at Hilo, Hilo, Hawaii

Address for correspondence: John M. Pezzuto, College of Pharmacy, University of Hawaii at Hilo, 34 Rainbow Drive, Hilo, HI 96720. pezzuto@hawaii.edu

Cancer chemoprevention entails the ingestion of dietary or pharmaceutical agents that can prevent, delay, or reverse the process of carcinogenesis. With support provided by the National Cancer Institute, we have been actively engaged in the systematic discovery and characterization of natural chemopreventive agents. The typical approach involves identifying active crude substances such as extracts derived from terrestrial plants or marine organisms, utilizing *in vitro* bioassay systems, followed by the isolation of pure active components. As part of this project, an extract obtained from a nonedible Peruvian legume, *Cassia quinquangulata* Rich. (Leguminosae), was evaluated and found to be active as an inhibitor of cyclooxygenase. The active component was identified as resveratrol. Surprisingly broad spectrum activity was observed, indicative of potential to inhibit carcinogenesis at the stages of initiation, promotion, and progression. This discovery has led to many additional research efforts. There are now around 3,500 papers concerning some aspect of resveratrol action, yet the molecule is unusually promiscuous and specific mechanisms remain elusive. Considering the structural simplicity of this stilbene, the intensity of interest is phenomenal.

Keywords: resveratrol; discovery; pleiotropic mechanisms; metabolites; chemoprevention

Introduction

Cancer may be prevented or delayed through the use of naturally occurring chemicals. Unlike the treatment of existing tumors, cancer chemoprevention targets the process of carcinogenesis itself, and the desired outcome is the prevention or delay of the disease. Ideally, cancer chemoprevention would work as well as vaccines for the prevention of other human ailments. Although this has yet to be accomplished, proof-of-principal has been established, and some pharmaceutical agents are available for the prevention of breast (tamoxifen relatives such as raloxifene; aromatase inhibitors) and prostate (finasteride) cancer. In addition, therapeutic options are available for preventing familial colon polyps, the precursor of colon cancer (e.g., Celebrex®).

However, the pipeline of new cancer chemopreventive drugs is thin. The current armamentarium of agents has resulted largely from epidemiological observations, off-shoots of cancer therapeutic agents, or agents that were used for other therapeutic indications. FDA approval for cancer prevention is cumbersome and pharmaceutical companies are not heavily committed to this area despite the potential for long-term economic gain that can result from developing successful chemopreventive drugs (e.g., the case of statins). As emphasized by a poignant remark in an article by Gina Kolata ("Medicines to Deter Some Cancers Are Not Taken," *NY Times*, Nov. 12, 2009), "You have to think that in boardrooms they are saying, Man, did we learn a lesson. We will stay as far away as possible from cancer prevention." Indeed, R&D programs are rarely focused on cancer chemopreventive agents.

Following an approach called "activity-guided fractionation," which utilizes a battery of state-of-the-art *in vitro* assays, we monitored the natural product purification process to isolate the most active agents in their pure form.[1] Although edible and nonedible terrestrial plants have yielded valuable leads, a new and exciting area of research involves exploring the biodiversity provided by

doi: 10.1111/j.1749-6632.2010.05849.x

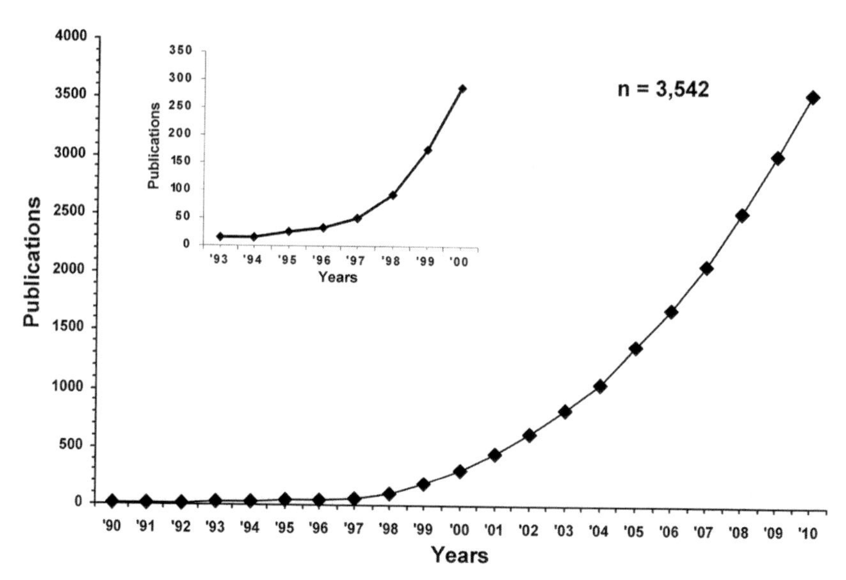

Figure 1. Graph depicting the number of papers concerning resveratrol during the period of 1990–2010. Numbers were derived from a search of PubMed.gov ($n = 3542$). Inset: Number of publications concerning resveratrol during the period of 1993–2000. The trend illustrates the impact of "Cancer chemopreventive activity of resveratrol, a natural product derived from grapes," which appeared in *Science* on January 10, 1997.

microbes of the marine environment.[2] Irrespective of the starting material, once purified, the structures of the molecules are determined using advanced NMR, mass spec, and X-ray crystallographic methods. This work is augmented by sophisticated new techniques involving LS/MS/MS and ultrafiltration-mass spectrometry and, most recently, macromolecular X-ray crystallographic assisted deconvolution of natural product mixtures. Compounds selected for further investigation are subjected to mechanistic investigation in order to make priority decisions about next steps. Using cell culture models, in conjunction with LC/MS analysis, the potential of oral absorption and the extent of metabolism is investigated. Depending on the structure, methods of synthesis or derivatization may be devised increase potency and to decrease metabolic instability, cytotoxicity, etc. All of these factors are critical for making decisions about next steps.

Based on the overall chemical and biological profile, a decision is made in regard to moving on to the next stage of early preclinical development. The first study generally requires 20–50 mg of material, which is generated by isolation or synthesis. An animal study is performed to investigate oral absorption and dose-tolerance. Following oral administration, tissues and serum are examined for parent compound and metabolites, and tissue biomarkers, if available. Based on these data, a decision can be made about proceeding to a longer-term animal study to assess chemopreventive efficacy. These tests require a larger quantity of material (perhaps 20–50 g).

Therefore, with a concerted effort involving a range of expertise, new natural product chemopreventive agents with clinical potential are being uncovered using a systematic approach of drug discovery. Some recent discoveries include isolates that interact with RxR or Keap1, and inhibition of quinone reductase 2 or NF-κB. A fascinating array of chemical structures have been uncovered, some of which show promise for clinical trials. Of course, synthetic organic chemistry also plays an important role in the program, and recently, after testing thousands of inactive extracts and compounds for interaction with the retinoid X receptor (RXR), the synthetic substance, 3-amino-6-(3-aminopropyl)-5,6-dihydro-5,11-dioxo-11*H*-indeno[1,2-*c*]isoquinoline dihydrochloride,[3] was finally discovered as a superior lead.

On the other end of the spectrum in terms of chemical complexity, one of our most notable discoveries is the structurally simplistic stilbene known as resveratrol.[4] This common constituent of grapes and grape products was originally reported by us to mediate cancer chemopreventive activity. Stimulated by this report, resveratrol is now the subject of nearly 3,500 manuscripts (Fig. 1). Some reflections on this discovery and the ramifications thereof are currently presented.

Discovery of resveratrol as a cancer chemopreventive agent

Fellow scientists and peer-reviewers commonly criticize the discipline of natural product drug discovery. In fact, natural products have proven to be valuable drugs, molecular probes, chemical prototypes, scaffolds for combinatorial chemistry, and tools for defining novel mechanisms. Chemical diversity is fascinating and unpredictable by any means other than discovery. Each case is unique, and the experience, intuition, and insight of each individual investigator comes into play. It is this innate ability, and not serendipity, that led to the discovery of renowned drugs such as taxol (Monroe E. Wall, personal communication) and vincristine (Gordon Svoboda, personal communication).

As part of our project designed for the discovery of unique cancer chemopreventive agents, an extract obtained from a nonedible Peruvian legume, *Cassia quinquangulata* Rich. (Leguminosae), was evaluated with a panel of bioassays. This was acquisition number 46 in a sample collection now in the thousands; there was no particular reason for the selection other than knowing the power of nature and the proven success of natural product drug discovery. One of the biological targets introduced for this project was cyclooxygenase. Anti-inflammatory agents are known to target cyclooxygenase and known to serve as chemopreventive agents. The *Cassia* extract was found to mediate significant inhibitory activity; the active component was identified as resveratrol (3,4′,5-trihydroxystilbene) (Fig. 2). From the vantage point of a phytochemist, resveratrol did not inspire enthusiasm. The compound was known (i.e., lacked novelty), and there were no challenging aspects of structure-elucidation or stereochemistry. By conventional thinking, this was an all-round loser.

As an inhibitor of cylclooxygenase, however, remarkable activity was observed, and this inspired further evaluation of biological potential. Activity was tested with a panel of *in vitro* bioassay systems. As a harbinger of things to come, a surprisingly broad spectrum activity was observed. Within the panel of assays, the only test that failed was inhibition of cylcooxygenase 2, which led the editors of *Science* to quip "at least there is one activity it does not mediate." Of course, lack of activity with cyclooxygenase 2 was a technical limitation of the assay system, and many studies have subsequently demonstrated such a response (cf. Ref. 5). Next, in order to demonstrate the physiological relevance of the *in vitro* responses, greater quantities of resveratrol were required. Resveratrol was not, however, readily available. Industrial contacts were made seeking help, but this did not lead to cooperation.

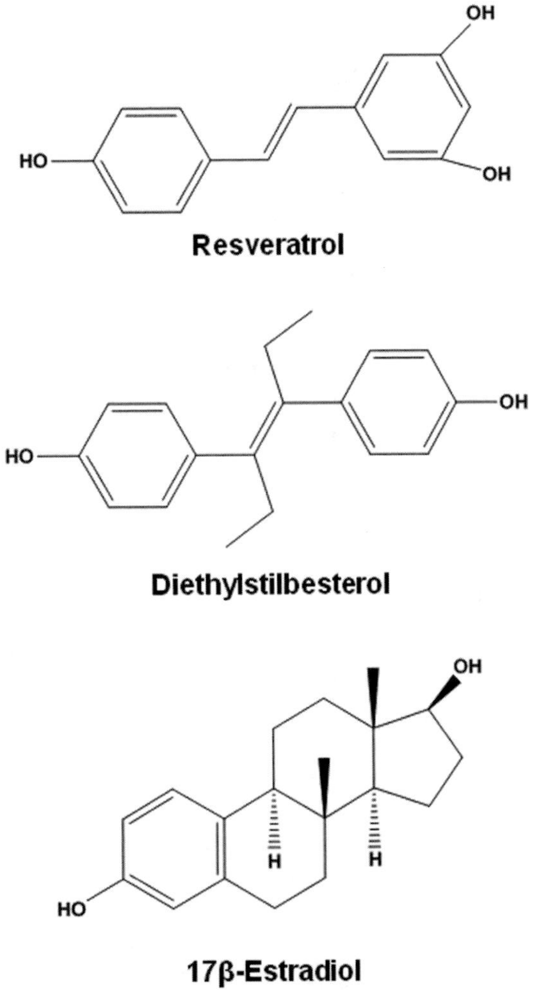

Figure 2. Structures of resveratrol, diethystilbesterol, and estrogen.

Figure 3. Cartoon that was created following the publication of "Cancer chemopreventive activity of resveratrol, a natural product derived from grapes," which appeared in *Science* on January 10, 1997. *Chip Bok Editorial Cartoon* ©1997 Chip Bok. All rights reserved. Used with the permission of Chip Bok and Creators Syndicate.

Finally, as a leap of faith, a decision was made to make a major capital investment and purchase the material from a commercial source. Following publication of the work, Sigma Chemical Co. sent a letter asking how they could help to further advance the discovery, but a request for supplying more material went unanswered. In any case, using material purchased from Sigma, impressive anti-inflammatory activity was demonstrated with a rat model, and significant antitumor activity was demonstrated with the two-stage mouse skin model. We surmised these data were indicative of potential to inhibit carcinogenesis at the stages of initiation, promotion and progression. Previous reports found unacceptable by *Science* appeared in *Nature Medicine*;[6,7] with good fortune, our paper was accepted and published by *Science*.[4]

Ramifications of the discovery

Our report in *Science* let to extensive media coverage. Although resveratrol is ubiquitous in nature, it is found in only a few edible substances, most notably the grape. The grape, of course, is used for the production of wine, and resveratrol is found predominantly in red wines due to the processing methodology. This is the reason resveratrol received such intense and immediate attention: the notion of red wine preventing cancer. The stage had already been set for notoriety as a result of the French Paradox—our discovery extended the vision to cancer prevention. Moreover, a molecule had been identified that might provide a plausible explanation.

In addition to television, radio, newspaper, and magazine coverage, cartoons emerged (Fig. 3). The compound became so well-known it was mentioned on an episode of the television series *CSI* and used as part of the evidence to apprehend a criminal. In addition to the large body of primary scientific papers, symposia have been conducted,[8] reviews and monographs have been prepared,[9–15] companies have been created (Royalmount Pharma; Sirtris Pharmaceuticals Inc.), and many commercial products are available. Of particular note, our work was brought to the attention of the California Table Grape Commission. Over the past 10 years, this organization has demonstrated an unyielding and laudable commitment to the pure, basic, and applied science of the grape, has evidenced by providing financial support for many research projects.[16]

As might be expected, at least in retrospect, a discovery of such magnitude leads to controversy as well as benefits. The first bit of controversy appeared almost immediately, on the front page of the *NY Times* (Jane E. Brody, "Study Hints That Grapes and Red Wine May Inhibit Cancer," January 10, 1997). David M. Goldberg was quoted as saying,

"It doesn't matter how potent a compound is in the test tube. If it doesn't get into the bloodstream, it won't have any effect." At that time, little was know about the absorption and metabolism of resveratrol. The intent of this statement was clear, however, implying that the discovery may have been interesting but meaningless. Of course, some evidence was available from the outset to prove this criticism was unfounded: our original report demonstrated an anti-inflammatory response in rats and the resveratrol was administered orally. Nonetheless, to provide additional self-assurance, I personally consumed a sample of pure resveratrol and discovered urinary metabolites by LC/MS analysis the following day. It has always been certain that resveratrol is absorbed following oral administration. It has been known at least since 1997 that metabolites may have greater importance in human beings ($n = 1$). A great deal of work has demonstrated these points in the ensuing years.

The next bit of notable controversy resulted from a paper published in *Proc. Natl. Acad. Sci. USA* that characterized resveratrol as a super estrogen agonist.[17] Employing a cell-based reporter assay, resveratrol was shown to mediate a genetic response much greater than estrogen. While the authors concluded this might imply a beneficial effect for cardiovascular health, the observation led to concern regarding the potential of promoting cancer or untoward genetic defects. Visual inspection of the structure of resveratrol, in comparison with diethylstilbesterol (DES) (Fig. 3), illustrates some obvious similarities. In our studies that followed, however, utilizing the same cell-based reporter assay system, we were not able to reproduce the response. Moreover, utilizing the ovaryectomized rat as a model, we were completely unable to demonstrate an estrogenic response (Table 1).[11] In sum, the potential of resveratrol to serve as an estrogen agonist or antagonist warrants further investigation. Nonetheless, existing data does not strongly support such activity in mammals.

The promiscuous nature of resveratrol

A combination of publishing in *Science* as well as the ensuing media coverage acutely brought the discovery of resveratrol to the attention of the scientific community. There was an immediate response in terms of publication rate (Fig. 1, inset) and, as mentioned above, there are now around 3,500 papers concerning some aspect of resveratrol action. In addition to cancer chemopreventive potential, wherein hundreds of reports have appeared,[15] the potential of resveratrol to modulate numerous other disease states has been investigated.[14]

Of the multitude of pharmaceutical agents that have been discovered throughout history, only a few seem to function by a clearly-defined mechanism of action. For example, penicillin inhibits the formation of peptidoglycan cross-links in bacterial cell walls, 5-fluorouracil serves as a pyrimidine analogue, taxol stabilizes tubulin, tamoxifen is a SERM, and methotrexate interferes with one carbon metabolism. For the majority of drugs, however, mechanistic definition is extremely complex. This is especially apparent with resveratrol.

As a small sampling, modes of action identified for resveratrol include induction of phase II drug-metabolizing enzymes, inhibition of cyclooxygenase (COX), and cellular differentiation.[4] Resveratrol inhibits cytochromes P450, cell invasion, transformation, and angiogenesis.[18] Resveratrol has been shown to up-regulate antioxidant enzymes such as glutathione peroxidase, catalase, and quinone reductase. It inhibits lipid peroxidation, ornithine decarboxylase (ODC), protein kinases, and cellular, proliferation.[19] Resveratrol effectively induces apoptosis modulated through multiple pathways including up-regulation of p53, activation of caspases, decreases in Bcl-2 and Bcl-x^L, increases in Bax, inhibition of D-type cyclins, and interference with NF-κB and AP-1 mediated cascades.[20]

A multitude of *in vitro* and *in vivo* studies implicate resveratrol in a large web of anticancer pathways (cf. Ref. 21). Resveratrol treatment resulted in growth arrest at G1 and G1/S phases of the cell cycle by inducing the expression of p21 and p27. It reduced inflammation via inhibition of prostaglandin production and COX-2 activity. Resveratrol has been shown to regulate cathepsin D, inhibit hypoxia-induced protein, and downregulate telomerase. Resveratrol pretreatment suppressed activation of ERK2, JNK and p38 in association with inhibition of protein kinase C (PKC) and protein tyrosine kinase.[22] Resveratrol blocked activation of NF-κB through suppression of IκB activation, inhibited activation of MEK, and abrogated TNF-induced caspase activation. And the list goes on and on.

Table 1. Effect of resveratrol on ovariectomized rats. Female Sprague–Dawley rats were ovariectomized and treated as indicated below. As indicated by uterus weight, in the absence of estrogen, resveratrol demonstrated no significant estrogenic activity and, in the presence of estrogen, resveratrol demonstrated no significant antiestrogenic activity. Similarly, resveratrol had no significant effect on vaginal cell morphology, in the presence or absence of estrogen. Modified from Bhat and Pezzuto.[11]

Group	Uterus weight (g)	% Cells from vaginal smears in different stages of estrous cycle[a]
Control	0.09 ± 0.003	L = 96 N = 4 C = 0
Estradiol (50 μg/kg body weight, s.c.)	0.3 ± 0.01	L = 1 N = 10 C = 89
Resveratrol (3000 mg/kg diet; 30 days)	0.087 ± 0.007	L = 97 N = 2 C = 1
Resveratrol + estradiol	0.34 ± 0.004	L = 4 N = 12 C = 84
Resveratrol (Intact rats)	0.43 ± 0.014	Normal $3^1/_2$ day cycle

[a]L, Leucocytes; N, Nucleated; C, Cornified.

Our supposition, while not very intellectually satisfying, fundamentally hypothesizes that a critical, straightforward pathway leading to the chemoprevention of cancer will never be known. It may be necessary to finally accept a superb therapeutic response as being empirical in nature and due to a fortuitous sequence of events leading to a good outcome. Of utmost importance is the ability to facilitate a predicable clinical response in the absence of toxicity.[23] It is clear that many contemporary basic and clinical scientists, as well as health authorities and regulatory agencies, will not find the proposition of empiricism to be sufficiently gratifying, but we should be willing to forge ahead on a semi-empirical basis in our fight against dreadful disease states.

The complexities of promiscuity

It is clear that resveratrol does not function by virtue of interacting with a single well-defined molecular target. Indeed, the virtue of resveratrol is most likely related to the ability of functioning through pleiotropic mechanisms. Many other drugs probably function in such a manner, be it explicitly recognized or not. To complicate the matter further, it is known that resveratrol is rapidly absorbed, but serum levels are low (e.g., 0.1–0.5 μmol/L). On the other hand, metabolites such as resveratrol-3- and 4′-sulfate, and resveratrol-3-glucuronide are generated and may be present at higher serum concentrations (e.g., 2–10 μmol/L). Accordingly, it is obligatory to investigate the biological activity of resveratrol metabolites since any response may in fact be attributed to these species. We have recently performed some studies along these lines,[24] investigating some targets such as cylcooxygenase[25] and the target with great avidity for unmetabolized resveratrol, quinone reductase 2.[26] Thus, although a myriad of studies have been performed with resveratrol as the test substance, the corresponding activity of metabolites may be the more relevant question. Considering the enormous number of published papers, this is an alarming concept.

The agility provided by promiscuity versus a single target supposition is well illustrated by SIRT1. Activation of SIRT1, an enzyme that deacetylates proteins as the expense of NAD, has been associated with mimicking caloric restriction and promoting life extension. This enzyme was touted as a target for resveratrol and used as a justification for resveratrol-mediated longevity.[27] Thinking along these lines apparently led to the creation of a company named Sirtrus Pharmaceuticals and the subsequent sale to GlaxoSmithKline for the astounding sum of $720 million (Andrew Pollack, "Glaxo Says Compound in Wine May Fight Aging," *New York Times*, April 23, 2008). Shortly thereafter, it was unequivocally demonstrated that SIRT1 is not a real target for resveratrol;[28] rather, the response was an artifact generated by interaction of high concentration of resveratrol (or other compounds) with the fluorophore-containing peptide substrates. In addition, it has been shown recently that resveratrol does not extend the lifespan of genetically heterogeneous mice.[29] Nonetheless, given the virtues of promiscuity, numerous other explanations have become readily available to rationalize the perceived benefits.

The journey continues

From a realistic point of view, it is not likely that dietary resveratrol has much influence on human health, at least as a single entity. Concentrations found in the diet are too low and, beyond this, resveratrol is only one of over 1,600 components found in the most prevalent dietary source, the grape.[16] Accordingly, when considering grapes and health, a holistic view appears to be more meaningful, taking into account all chemical components, metabolism, biological potential, biodistribution, absorption, processing, etc. In order to fathom such a massive amount of information, we recently proposed the following:[30] by utilizing a next generation intelligent system, attempts can be made to leverage complexity, by bringing together all known information in conjunction with expert judgment, processing this information through a computational "engine" or engines to provide suggested solutions (or implicit functional relationships).

As a single drug entity, given as a dietary supplement, for example, resveratrol may have the potential of mediating some type of curative or preventative effects in human beings. Heretofore, any such medicinal value has not been formally proven. Obviously, properly designed and conducted clinical trials are required to establish clinical responses that can be utilized on a widespread basis following a standard protocol. Whether investigations of this type will be undertaken with resveratrol is not clear. At least one clinical trial has been terminated (GlaxoSmithKline ClinicalTrials.gov Identifier: NCT00920556), and history suggests future investments might focus on a patented derivative rather than the parent compound itself. Nonetheless, this simple, promiscuous molecule has stimulated a great deal of basic and applied research, and captured the attention of the public. No doubt, the public has a greater awareness of the virtues of disease prevention and the relationship between diet and health. Although any true benefit for human health remains to be seen, some have been able to exploit the molecule for great financial gain. The story is ongoing. The journey has been interesting and somewhat surprising. In the end, the journey may prove more important than the destination itself, whatever that may be.

Acknowledgments

Work conducted in the laboratory of the author is supported by program project P01 CA48112 awarded by the National Cancer Institute.

Conflicts of interest

The author declares no conflicts of interest.

References

1. Pezzuto, J.M., J.W. Kosmeder, E.-J. Park, *et al.* 2005. Characterization of natural product chemopreventive agents. In *Cancer Chemoprevention, Volume 2: Strategies for Cancer Chemoprevention.* G.J. Kelloff, E.T. Hawk & C.C. Sigman, Eds.: 1–37 Humana Press Inc., Totowa, New Jersey.
2. Fenical, W. & P.R. Jensen. 2006. Developing a new resource for drug discovery: marine actinomycete bacteria. *Nat. Chem. Biol.* **2:** 666–673.
3. Park, E.-J., T.P. Kondratyuk, A. Morrell, *et al.* 2010. Induction of retinoid X receptor activity and consequent up-regulation of p21$^{WAF1/CIP1}$ by indenoisoquinolines in MCF7 cells. *Cancer Prevent. Res.* In press.
4. Jang, M., L. Cai, G.O. Udeani, *et al.* 1997. Cancer chemopreventive activity of resveratrol, a natural product derived from grapes. *Science* **275:** 218–220.
5. Subbaramaiah, K., W.J. Chung, P. Michaluart, *et al.* 1998. Resveratrol inhibits cyclooxygenase-2 transcription and activity in phorbol ester-treated human mammary epithelial cells. *J. Biol. Chem.* **273:** 21875–21882.
6. Gerhäuser, C., W. Mar, S.-K. Lee, *et al.* 1995. Rotenoids mediate potent chemopreventive activity through transcriptional regulation of ornithine decarboxylase. *Nature Med.* **1:** 260–266.
7. Pisha, E., H. Chai, I.-S. Lee, *et al.* 1995. Discovery of betulinic acid as a selective inhibitor of human melanoma that functions by induction of apoptosis. *Nat. Med.* **1:** 1046–1051.
8. Pezzuto, J.M. & V. Steele (Eds.). 1998. *Proceedings of a Conference Exploring the Power of Phytochemicals: Research*

Advances on Grape Compounds. Swets and Zeitlinger. Lisse, The Netherlands, pp. 80 (A supplement of *Pharm. Biol.*)
9. Bhat, K.P.L., J.W. Kosmeder & J.M. Pezzuto. 2001. Biological effects of resveratrol. *Antioxid. Redox. Signal.* **3:** 1041–1064.
10. Jang, M. & J.M. Pezzuto. 1999. Cancer chemopreventive activity of resveratrol. *Drugs Exp. Clin. Res.* **25:** 65–77.
11. Bhat, K.P. & J.M. Pezzuto. 2002. Cancer chemopreventive activity of resveratrol. *Ann. N.Y. Acad. Sci.* **957:** 210–229.
12. Pezzuto, J.M., T. Kondratyuk & E. Shalaev. 2006. Cancer chemoprevention by wine polyphenols and resveratrol. In *Carcinogenic and Anticarcinogenic Food Components*. W. Baer-Dubowska, A. Bartoszek & D. Malejka-Giganti, Eds.: 239–282. CRC Press. Boca Raton, Florida.
13. Bagchi, D. 2000. *Resveratrol and Human Health*. McGraw Hill. Columbus, Ohio.
14. Aggarwal, B. & S. Shishodia. 2006. *Resveratrol in Health and Disease*. Taylor & Francis. Boca Raton, Florida.
15. Pezzuto, J.M. 2008. Resveratrol as an inhibitor of carcinogenesis. *Pharm. Biol.* **46:** 443–573.
16. Pezzuto, J.M. 2008. Grapes and human health: a perspective. *J. Agric. Food Chem.* **56:** 6777–6784.
17. Gehm, B.D., J.M. McAndrews, P.-Y. Chien & J.L. Jameson. 1997. Resveratrol, a polyphenolic compound found in grapes and wine, is an agonist for the estrogen receptor. *Proc. Natl. Acad. Sci. USA* **94:** 14138–14143.
18. Brakenhielm, E., R. Cao & Y. Cao. 2001. Suppression of angiogenesis, tumor growth, and wound healing by resveratrol, a natural compound in red wine and grapes. *FASEB J.* **15:** 1798–1800.
19. Agarwal, C., Y. Sharma, & R. Agarwal. 2000. Anticarcinogenic effect of a polyphenolic fraction isolated from grape seeds in human prostate carcinoma DU145 cells: Modulation of mitogenic signaling and cell-cycle regulators and induction of G1 arrest and apoptosis. *Mol. Carcinog.* **28:** 129–138.
20. Pezzuto, J.M. 2006. Resveratrol as a cancer chemopreventive agent. In *Resveratrol in Health and Disease*. B.B. Aggarwal & S. Shishodia, Eds.: 233–383. Marcel Dekker, Inc. New York, New York.
21. Narayanan, B.A., N.K. Narayanan, G.D. Stoner & B.P. Bullock. 2002. Interactive gene expression pattern in prostate cancer cells exposed to phenolic antioxidants. *Life Sci.* **70:** 1821–1839.
22. Yu, R., V. Hebbar, D.W. Kim, *et al.* 2001. Resveratrol inhibits phorbol ester and UV-induced activator protein 1 activation by interfering with mitogen-activated protein kinase pathways. *Mol. Pharmacol.* **60:** 217–224.
23. Francy-Guilford, J. & J.M. Pezzuto. 2008. Mechanisms of cancer chemopreventive agents: A perspective. *Planta Med.* **74:** 1644–1650.
24. Hoshino, J., E.-J. Park, T.P. Kondratyuk, *et al.* 2010. Selective synthesis and biological evaluation of sulfate-conjugated resveratrol metabolites. *J. Med. Chem.* **53:** 5033–5043.
25. Calamini, B., K. Ratia, M.G. Malkowski, *et al.* 2010. Pleiotropic mechanisms facilitated by resveratrol and its metabolites. *Biochem. J.* **429:** 273–282.
26. Buryanovskyy, L., Y. Fu, M. Boyd, *et al.* 2004. Crystal structure of quinone reductase 2 in complex with resveratrol. *Biochemistry* **43:** 11417–11426.
27. Howitz, K.T., K.J. Bitterman, H.Y. Cohen, *et al.* 2003. Small molecule activators of sirtuins extend *Saccharomyces cerevisiae* lifespan. *Nature* **425:** 191–196.
28. Pacholec, M., J.E. Bleasdale, B. Chrunyk, *et al.* 2010. SRT1720, SRT2183, SRT1460, and resveratrol are not direct activators of SIRT1. *J. Biol. Chem.* **285:** 8340–8351.
29. Miller, R.A., D. Harrison, C.M. Astle, *et al.* 2010. Rapamycin, but not resveratrol or simvastatin, extends lifespan of genetically heterogeneous mice. *J. Gerontol. A Biol. Sci. Med. Sci.* Advance online publication. doi: 10.1093/gerona/glq178.
30. Pezzuto, J.M., V. Venkatasubramanian, M. Hamad, & K.R. Morris. 2009. Unraveling the relationship between grapes and health. *J. Nutr.* **139:** 1783S–1787S.

Ann. N.Y. Acad. Sci. ISSN 0077-8923

ANNALS OF THE NEW YORK ACADEMY OF SCIENCES
Issue: *Resveratrol and Health*

Safety of resveratrol with examples for high purity, *trans*-resveratrol, resVida®

J. A. Edwards, M. Beck, C. Riegger, and J. Bausch
DSM Nutritional Products Ltd., Kaiseraugst, Switzerland

Address for correspondence: James A. Edwards, DSM Nutritional Products Ltd, NRD/CH Product Safety, Wurmisweg 576, Kaiseraugst CH-4303, Switzerland. james-a.edwards@dsm.com

Studies with resVida® (a high purity *trans*-resveratrol) show that *trans*-resveratrol is a substance of low oral toxicity. An acceptable daily intake (ADI) in food of 450 mg/day has been defined, a level well beyond natural dietary intake of *trans*-resveratrol. The ADI was based on no-observed-adverse-effect-levels (NOAELs) of 750 mg/kg bw/day in 13-week developmental toxicity studies by the dietary route and a standard safety margin of 100. In studies by gavage, the kidney and bladder are target organs at very high dosages (2,000–3,000 mg/kg bw/day). Six-month studies in rat and rabbit models show no significant increase in toxicity in comparison to 4-week studies. Lower quoted NOAELs in gavage studies (ca. 300 mg/kg bw/day) potentially reflect more rapid bioavailability, but different dosage regimes complicate comparisons. Short-term studies show no genotoxicity *in vivo*. A 6-month mouse carcinogenicity model showed no increase in tumors. Clinical data support an ADI of at least 450 mg/day, and kinetic data from the DSM 13-week toxicity study also support the expectation of no increase in toxicity with longer term intake.

Keywords: *trans*-resveratrol; stilbene; safety; toxicology; development; kidney

Introduction

A stilbene with three hydroxyl groups, and as such a polyphenol, *trans*-resveratrol (CAS 501–36-0, Fig. 1) belongs to a group of substances known as phytoalexins, which are low molecular weight secondary metabolites produced by plants as a defensive or stress response. In natural food sources, *trans*-resveratrol is primarily found in red wine and grapes.

DSM Nutritional Products Ltd. produces *trans*-resveratrol in a multiple-step process from chemical precursors. Each step of the chemical process is carefully controlled to ensure a consistent purity and yield of the end-product. The end-product, resVida®, has a purity of not less than 99% and is good manufacturing process grade. resVida® has been thoroughly tested for safe oral use;[1] these studies were designed to specifically support the safety of resVida® at dosages that might be used in food or as a supplement. The studies were not intended to demonstrate safety at dosages of >1 g, as has recently been of interest for potential pharmaceutical or chemo-preventative usage of *trans*-resveratrol. In this paper we compare published safety studies undertaken by DSM Nutritional Products Ltd. with other relevant studies in the scientific literature.

History of use considerations

An initial starting point in the safety evaluation of a naturally occurring food substance is its natural intake. In natural food sources, *trans*-resveratrol is primarily found in red wine and grapes and at lower levels in white wine, peanuts, chocolate (cacao), and mulberries. Humans are exposed to *trans*-resveratrol by consuming those foods or dietary supplements containing resveratrol. The daily intake of dietary resveratrol (from naturally occurring resveratrol) is mainly from the consumption of wine and grapes and foods derived from them. Among plants high levels of resveratrol are found in the roots of the Japanese knotweed, *P. cuspidatum*, which is not a normal dietary source but can be used for extraction of *trans*-resveratrol. Besides the

doi: 10.1111/j.1749-6632.2010.05855.x

trans-isomer, *cis*-resveratrol is also present in grapes and other resveratrol sources, but in much lower concentrations.

An unpublished DSM evaluation gives an average daily intake for *trans*-resveratrol of about 0.01–0.45 mg/day for Europe, with upper intake levels of 1–2 mg/day. These values, however, are potentially conservative. The glycoside *trans*-piceid has been suggested to be a potential additional natural form of *trans*-resveratrol exposure;[2] *trans*-piceid (also named *trans*-polydatin) is a natural dietary precursor of resveratrol (Fig. 1) found at up to 10 times higher levels than resveratrol in wine and up to five times higher levels in dark chocolate.[3] It has been shown in a number of studies that *trans*-resveratrol can be measured in plasma when *trans*-piceid—without *trans*-resveratrol present—is taken orally. Most likely, *trans*-resveratrol is formed in the intestinal tract by the activity of beta-glycosidase and/or lactase phloridzin hydrolase in the lumen or enterocytes of the intestinal tract.[4] Therefore, the estimate of a human intake of up to approximately 2 mg/day is probably conservative; it would be more accurate to derive an intake taking into account all direct and indirect (metabolic precursor) sources. In general, however, it is expected that such a derived intake value would still be relatively low in comparison to reported human supplement dosages already in use. An evaluation of combined resveratrol and piceid intake in a Spanish population gave a mean of 0.93 mg/day.[5]

In comparison to a number of other naturally consumed polyphenols, an intake level for *trans*-resveratrol of up to 2 mg/day is rather low. For example, for the EGCG polyphenol from green tea, based on green tea consumption for men and women, average consumption in Japan is in the range of 120–380 mg.[6] This is a level of intake in a similar range to that of dietary supplementation in some other countries. In the case of *trans*-resveratrol, the proposed safe dosages based on safety testing are two orders of magnitude above the European upper dietary intake, so the history of human use is, relatively speaking, of more limited value to the overall safety evaluation.

DSM safety package for resVida®

resVida® has been thoroughly tested for safe oral use.[1] The studies undertaken were not only intended to evaluate safety for human food use but also included safety for workers in the manufacturing process (e.g., lymph node assay for sensitization). Environmental studies were also performed but are not presented here or in the publication referred to.[1]

trans-Piceid *trans*-Resveratrol

Figure 1. Glycoside (*trans*-piceid) and aglycone (*trans*-resveratrol).

The safety studies were performed using representative material from the manufacturing process and thereby cover the trace by-products that are present. Such studies were considered necessary to enable registration and regulatory approval for use in food and also to support the safety evaluation necessary prior to the conduct of planned clinical studies for resVida®. Preclinical testing was performed according to internationally recognized guidelines of Good Laboratory Practice, following the relevant OECD study design guidelines and performed at specialist research organizations with experience in undertaking such studies. The preclinical safety studies for resVida® as published[1] included:

- worker safety studies;
- genotoxicity studies;
- developmental toxicity;
- subchronic toxicity;
- kinetic data; and
- ADME (absorption, distribution, metabolism, and excretion) studies

The worker safety studies for irritation and sensitization were sufficient for the DSM manufacturing process and no further information was added.

A bacterial reverse mutation assay (Ames test) with *Salmonella typhimurium* and *Eschericha coli* performed by DSM showed no mutagenic potential *in vitro*. There was existing published data indicating positive clastogenic activity *in vitro*, and this was confirmed in a DSM *in vitro* study (chromosome aberration test, with human lymphocytes) in which threshold concentrations (with and without metabolizing fraction) for the clastogenic response were also defined. More importantly, an *in vivo* study with clastogenicity as an end point (micronucleus test in the rat, with a dose up to 2,000 mg/kg bw) was negative.[1]

The mechanism by which *trans*-resveratrol induces a clastogenic resonse is not completely clear. Potential activity *in vitro* through copper ion interaction has been shown.[7] In addition, anti-oxidant activity, which *trans*-resveratrol has, can be prooxidative and lead to clastogenic activity *in vitro*. This has been shown, for example, for the catechin EGCG, which induces hydrogen peroxide production in culture media.[8] However, a DSM mechanistic study showed only a 20% potency for *trans*-resveratrol in comparison to EGCG.[9] Such a hydrogen peroxide response is most unlikely *in vivo* due to counter activity by catalase and other *in vivo* components. The *in vitro* micronucleus test is also known for a high rate of false positive results; the more important result is the negative *in vivo* response. The absence of genotoxicity concern has also been referenced for the *trans*-resveratrol product SRT501, for which various short-term genotoxicity studies were performed.[10]

A 6-month mouse carcinogenicity model (p53 knockout mouse study) has also been performed.[11] The mice, which are deficient in the tumor suppressor gene p53, showed no increase in incidence of either benign or malignant tumors at the dose of 1,000 mg/kg bw/day by gavage. Evaluation of higher dosages of 2,000, or 4,000 mg/kg/day was precluded due to toxicity and mortality. At these high dosages histopathology identified the kidney (hydronephrosis) and urinary bladder (epithelial hyperplasia) as target tissues. The kidney was also identified as the target organ at high gavage dosages in the rat.[12] With respect to the main purpose of this p53 knockout mouse study, *trans*-resveratrol demonstrated no evidence of oncogenic potential through a genetic based mechanism. In contrast, the positive control p-cresidine induced urinary bladder cancer.

Reproductive toxicology

There is potential concern that *trans*-resveratrol has an estrogenic activity due to its stilbene structure and structural similarity to the potent estrogen diethylstilbestrol. However, it has been shown that the estrogenic activity of stilbenes vary significantly according to structure.[13] Importantly for *trans*-resveratrol, a definitive study for estrogenic activity in mammals (i.e., an *in vivo* uterotrophic assay) with immature female Wistar rats dosed subcutaneously with *trans*-resveratrol (18, 58, and 575 mg/kg) was negative.[14] Published information on phyto-estrogenic activity *in vitro* contrasts with the negative *in vivo* uterotrophic assay. Nevertheless, in view of the concern of potential activity, sensitive end points for estrogenic influence were incorporated into the DSM developmental and 13-week toxicology studies. No effects on these end points of ano-genital distance in fetuses or on seminology or estrous cycling in adults were observed.[1]

No significant pre-existing adverse data on the potential developmental toxicity of *trans*-resveratrol were available and DSM undertook a regulatory study, following guideline OECD 414, to evaluate effects on embryo-fetal survival and development.[1] Dietary concentrations of resVida® were used to achieve an intake up to 750 mg/kg bw/day, this dose having induced slight food and weight gain reduction in the 13-week toxicity study. No treatment-related effect on embryo-fetal survival, fetal weight, structural abnormalities, or ano-genital distance was observed; the study concluded that 750 mg/kg bw/day was the no-observed-adverse-effect-level (NOAEL). Since this study, a further developmental toxicity study (Table 1) with *trans*-resveratrol and administration to pregnant female rats by gavage has been referred to in the literature; again, no evidence of embryo-fetal toxicity was observed.[10]

General toxicology

Taking into account a published 4-week study with *trans*-resveratrol in the rat by gavage,[12] in which the kidney was identified as a target organ at the high dosage of 3,000 mg/kg bw/day and a NOAEL of 300 mg/kg bw/day was defined, DSM undertook a dietary 4-week preliminary rat study with resVida® at dietary concentration rasulting in intake of up to 500 mg/kg bw/day, determined to be the NOAEL.[1] In a subsequent dietary 13-week (subchronic) rat study, a high dose of 750 mg/kg bw/day was selected. This dose was defined as NOAEL, as the minor effects seen were considered not to be adverse. Other general toxicology studies (Table 2) with *trans*-resveratrol up to 13 weeks in duration have been cited in the literature. [10,15] All of these other studies have been by gavage administration.

Table 1. No Observed Adverse Effect Levels (NOAELs) in regulatory developmental toxicity studies

Species	Publication	Route	Doses (mg/kg bw/day)	Duration	NOAEL (mg/kg bw/day)
Rat	Williams 2009[1] DSM study	Dietary	Up to 750	Days 5–20 gestation	750: Maternal and Developmental
Rat	Elliott 2009[10]	Gavage	0, 300, 1,000 and 3,000	Days 7–17 gestation	300: Maternal 1,000: Developmental

In addition to the standard toxicological end points, the 13-week rat toxicity study with resVida® included evaluation of endpoints for potential estrogenic activity, which were unaffected. No preneoplastic changes were observed histopathologically.[1] Kinetic evaluation involving plasma sampling at three time points during the study (after 1, 4, and 13 weeks of treatment) were undertaken to assess potential accumulation of *trans*-resveratrol with time. A dose-dependent increase in plasma concentration of *trans*-resveratrol (both unconjugated and total *trans*-resveratrol were measured) was similar at all 3 time points, with no indication of accumulation with treatment duration.[1]

In other rat toxicology studies up to subchronic (13-weeks) duration,[10,15] NOAELs of the same order of magnitude as in the DSM studies with resVida® were observed, but were approximately half the DSM value of 750 mg/kg bw/day. Such a difference could be due to the different kinetic profile following dietary administration in comparison to gavage. Bolus gavage would normally be expected to produce more rapid bioavailability and higher plasma C_{max} values, which can often correlate with higher toxicity, than the more extended plasma profile following dietary administration. However, different dosage regimes were used in the different studies, and it is possible the differences between dietary and gavage administration are less than might be apparent from face value comparison of NOAELs. The purity and impurity content of the specific source of *trans*-resveratrol used could also have a bearing.

The NOAELs for the *trans*-resveratrol gavage studies in different species (Table 2) show good inter-species concordance and there is no indication that the rat is less sensitive than other species, in terms of toxicological response. In the rabbit the target organ of toxicity identified at high dosages was also the kidney. The species tested included the dog with dosages up to 300 mg/kg bw/day and in this study functional measurements including Q-T interval were included.[10] Functional cardiac measurements were also measured in a single dose study with telemetered dogs at a dosage up to 1,000 mg/kg body

Table 2. No Observed Adverse Effect Levels (NOAELs) in different general toxicity studies up to subchronic in duration

Species Strain	Study type	Publication	Route	Doses mg/kg bw/day	Duration weeks	NOAEL mg/kg bw/day
Rat Sprague-Dawley	Subacute	Crowell 2004[12]	Gavage	0, 300, 1,000 and 3,000	4	300
Rat Wistar	Subacute	Williams 2009[1] DSM study	Dietary	Up to 500	4	≥ 500
Rat Wistar	Subchronic	Williams 2009[1] DSM study	Dietary	Up to 750	13	750
Rat Sprague-Dawley	Subchronic	McCormick 2008[15]	Gavage	0, 200, 400 and 1,000	13	200
Dog	Subacute	Elliott 2009[10]	Gavage	Up to 300	4	300
Rabbit	Subacute	Elliott 2009[10]	Gavage	Up to 750	4	500: Males 250: Females

Table 3. Comparison of NOAELs between subacute and chronic studies referred to by Elliott *et al.*[10]

Study type	Route	Duration	Species doses (mg/kg bw/day)	Result NOAEL (mg/kg bw/day)
Subacute	Gavage	28 days	Rat: Up to 1,000	300
	Gavage	28 days	Rabbit: Up to 750	500; Males: 250: Females
Chronic	Gavage	6 months	Rat: 0, 300, 1,000 and 2,000	300
	Gavage	6 months	Rabbit: 0, 100, 300 and 500	500

weight (bw). No adverse effect on cardiac function was observed.[10] This additional cardiac information is reassuring when it is considered that the history of safe human use for *trans*-resveratrol relates to low intake, and that in the sector of pharmaceutical development functional cardiac effects are a common reason for compound attrition.

In addition to the p53 mouse study, 6-month studies in the rat and rabbit have been performed, and for both of these species there are corresponding shorter term 4-week studies with the same source of *trans*-resveratrol.[10] For both species (Table 3), the NOAEL was the same or greater than in the 1-month study, indicating that no clear increase in the toxic potential of *trans*-resveratrol had occurred with the longer treatment.

trans-Resveratrol has since been shown to extend the lifespan of evolutionarily distant lower order species, e.g., *Drosophila melanogaster*. Also, there are studies in mice with treatment from one year of age (50 weeks) up to full life span (up to death) such that the mice on average were treated for over a year.[16] For mice on high-calorie (HC) diet, *trans*-resveratrol increased survival of those middle-aged and shifted the physiology towards that of mice on standard diet.[16] Although there is information in the mouse supporting the longer term safety of *trans*-resveratrol, the studies were not performed according to design criteria of regulatory toxicology studies, for example with lower dosage regimen and testing in only one gender. However, the data are supporting information to the more conventional studies indicating no expected increase in the toxicity of *trans*-resveratrol with chronic intake.

Acceptable daily intake (ADI) and safety factor

In both of the main toxicity studies in the rat undertaken for resVida®, 750 mg/kg bw/day was defined as the NOAEL.[1] Using this NOAEL and a standard default safety factor of 100, an ADI of 450 mg/day was defined for a 60 kg adult with respect to use in food.[1]

resVida® (with specified manufacturing process and known reproducibility) obtained self-GRAS (generally regarded as safe) in the U.S. in 2008 following independent toxicological expert evaluation with a defined ADI of 450 mg/day. The marketing of resVida® in the U.S. since 2008 has not been associated with safety issues (adverse reports). The use of a dose as high as the ADI is, however, not necessary when efficacy data indicates benefits at lower dosages and there is evidence to indicate that beneficial effects can be seen at a dosage as low as 30 mg/day.[17] It is expected that further clinical information of this sort will be available in the future so that appropriate human dosages as a nutritional supplement can be better defined.

The standard default safety factor of 100 (applied to data from toxicity studies) incorporates the uncertainty factors of 10 for intra-species differences and 10 for inter-species differences. This has been refined with further subdivision of each factor into toxicokinetic (TK) and toxicodynamic (TD) elements. It is recognized within the underlying concept of 100 fold safety factor that there may be justifications to support lower factors when appropriate information is known for a substance. Based on the interspecies similarity that is shown in the kinetics, metabolism, and excretion (mainly in urine) of *trans*-resveratrol, it can be argued that the default interspecies safety factor for *trans*-resveratrol could be set lower than 10. Such arguments can be used to justify a lower safety factor.[18] There is also no evidence of accumulation in plasma with time and no pathological changes that might be expected to progress with time, such as preneoplastic changes, were observed.[1] An additional safety factor of 2 (giving an overall 200 fold factor) is sometimes advocated by some authorities (EFSA, ECHA) for extrapolation from subchronic data to life-time exposure when no chronic data exists. However, for *trans*-resveratrol, this can be argued to be unnecessary based on the available toxicological and kinetic data,[1] and taking into account the general underlying concepts behind the standard default safety factor.

The ADI is defined for adults and assumes reasonable health status. There are published human studies with higher dosages, up to 5 g/day,[10,19] and in disease conditions that indicate *trans*-resveratrol as a possible therapeutic or co-therapeutic agent. These studies are of limited relevance to *trans*-resveratrol found in food at a dosage an order of magnitude lower. In one of these clinical studies in patients with multiple myeloma, kidney disease (cast nephropathy) has been reported.[20] The kidney has been identified as the target organ or *trans*-resveratrol in preclinical studies, but cast nephropathy is a condition closely associated with multiple myeloma, so the finding in this study is of doubtful significance outside of this disease condition. In other clinical studies it seems that 5 g/day has been tolerated for up to a month with only minor side effects, including clinical chemistry of hematological, cardiovascular, liver and renal parameters.[10] There is, however, recent information indicating that *trans*-resveratrol at a dose of 1 g/day may inhibit the metabolism of certain drugs metabolized by certain CYP iso-enzymes, in particular CYP2C9.[21] This is an aspect of safety that is relevant for people on medication and needs further clarification whether it is a relevant consideration for people consuming lower amounts of *trans*-resveratrol as a dietary supplement.

Discussion and conclusion

The history of safe human use of *trans*-resveratrol seems to be restricted to relatively low intake of up to 2 mg/day, although this may be conservative if the precursor *trans*-piceid is taken into account. This intake of up to 2 mg/day is relatively low in comparison to the level of safe oral intake that is derived from oral preclinical studies, which for the high purity *trans*-resveratrol, resVida®, was determined to be 450 mg/day.

Potential concerns regarding reproductive toxicology of *trans*-resveratrol due to the stilbene structure and potential estrogenic action are not supported by the *in vivo* data from preclinical testing. Also, with the recent use of *trans*-resveratrol in anti-cancer studies and, thus, the possibility that *trans*-resveratrol has an anti-proliferative effect on cells, embryo-fetal development could be a potential unwanted a target of action. However, there is no indication of activity on embryo-fetal development in two guideline developmental toxicity studies up to high dosages by the oral route.

Studies with resVida® show that *trans*-resveratrol is a substance of low oral toxicity. An ADI in food of 450 mg/day has been defined, a level well beyond natural dietary intake of *trans*-resveratrol. The ADI was based on NOAELs of 750 mg/kg bw/day in 13-week and developmental toxicity studies[1] by the dietary route and a standard safety margin of 100. In studies by gavage, the kidney and bladder are target organs at very high dosages (2,000–3,000 mg/kg bw/day). Six-month studies in rat and rabbit show no significant increase in toxicity in comparison to four-week studies.[10] Lower quoted NOAELs in gavage studies (ca. 300 mg/kg bw/day) potentially reflect more rapid bioavailability, but different dosage regimes complicate comparisons. Short term studies show no genotoxicity *in vivo*. A six-month mouse carcinogenicity model showed no increase in tumors.[11] Clinical data support an ADI of at least 450 mg/day and kinetic data from the DSM 13-week toxicity study also support the expectation of no increase in toxicity with longer term intake.[1]

Human clinical studies have been performed with dosages up to 5 g/day for 28 days and generally support the expectation of safety at the ADI defined for resVida®.[1] A recent study with a high dosage of *trans*-resveratrol of 1 g/day showed a potential influence on drugs metabolized by certain CYP iso-enzymes.[21] This is an aspect of safety that is relevant for people on medication and needs further clarification as to whether it is relevant for people consuming lower dosages of *trans*-resveratrol as a dietary supplement. The expectation based upon unpublished *in vitro* CYP interaction studies undertaken by DSM was that at the lower dosage of dietary supplementation, significant interaction was not expected.

Conflicts of Interest

All authors are employees of DSM Nutritional Products Ltd.

References

1. Williams, L.D., G.A. Burdock, J.A. Edwards, *et al.* 2009. Safety studies conducted on high-purity *trans*-resveratrol in experimental animals. *Food Chem. Toxicol.* **47:** 2170–2182.
2. Burkon, A. & V. Somoza. 2008. Quantification of free and protein-bound trans-resveratrol metabolites and identification of trans-resveratrol-C/O-conjugated diglucuronides—two novel resveratrol metabolites in human plasma. *Mol. Nutr. Food Res.* **52:** 549–557.
3. Hurst, W.J., J.A. Glinski, K.B. Miller, *et al.* 2008. Survey of the trans-resveratrol and trans-piceid content of cocoa-containing and chocolate products. *J. Agric. Food Chem.* **56:** 8374–8378.
4. Henry-Vitrac, C., A. Desmouliere, D. Girard, *et al.* 2006. Transport, deglycosylation, and metabolism of trans-piceid by small intestinal epithelial cells. *Eur. J. Nutr.* **45:** 376–382.
5. Zamora-Ros, R., C. Andres-Lacueva1, R. Lamuela-Raventos, *et al.* 2008. Concentrations of resveratrol and derivatives in foods and estimation of dietary intake in a Spanish population: European Prospective Investigation into Cancer and Nutrition (EPIC)-Spain cohort. *Br. J. Nutr.* **100:** 188–196.
6. Tsubono, Y., T. Takahashi, Y. Iwase, *et al.* 1997. Dietary differences with green tea intake among middle-aged Japanese men and women. *Prev. Med.* **26:** 704–710.
7. Fukuhara, K., M. Nagakawa, I. Nakanishi, *et al.* 2006. Structural basis for DNA-cleaving activity of resveratrol in the presence of Cu(II). *Bioorg. Med. Chem.* **14:** 1437–1443.
8. Long, L., D. Kirkland, J. Whitwell & B. Halliwell. 2007. Different cytotoxic and clastogenic effects of epigallocatechin gallate in various cell-culture media due to variable rates of its oxidation in the culture medium. *Mutat. Res.* **634:** 177–183.
9. Pappa, G., T. Wöhrle, & M. Török. 2008. Automated in vitro Micronucleus testing of natural compounds in correlation with hydrogen peroxide detection. *SOT Annu. Congr.* Poster 2250.
10. Elliott, P.J., S. Walpole, L. Morelli, *et al.* 2009. Resveratrol/SRT-501. *Drugs Future* **34:** 291–295.
11. Horn, T.L., M.J. Cwik, R.L. Morrissey, *et al.* 2007. Oncogenicity evaluation of resveratrol in p53(+/-) (p53 knockout) mice. *Food Chem. Toxicol.* **45:** 55–63.
12. Crowell, J.A., P.J. Korytko, R.L. Morrissey, *et al.* 2004. Resveratrol associated renal toxicity. *Toxicol. Sci.* **82:** 614–619.
13. Sanoh, S., S. Kitamura, K. Sugihara, *et al.* 2003. Estrogenic activity of stilbene derivatives. *J. Health Sci.* **49:** 359–367.
14. Freyberger, A., E. Hartmann,H. Hildebrand & F. Krötlinger. 2001. Differential response of immature rat uterine tissue to ethinylestradiol and the red wine constituent resveratrol. *Arch. Toxicol.* **74:** 709–715.
15. McCormick, D.L., W.D. Johnson, R.L. Morrissey, et al. 2008. Cardiopreotective activity of resveratrol in rats. *SOT Annu. Congr.* Poster 711.
16. Baur, J.A., K.J. Pearson, N.L. Price, *et al.* 2006. Resveratrol improves health and survival of mice on a high-calorie diet. *Nature* **444:** 337–342.
17. Wong, R.H., P.R. Howe, J.D. Buckley, *et al.* 2010. Acute resveratrol supplementation improves flow-mediated dilatation in overweight/obese individuals with mildly elevated blood pressure. *Nutr. Metab. Cardiovasc. Dis.* 29 Jul 2010. Epub ahead of print.
18. Renwick, A.G. & N.R. Lazarus. 1998. Human variability and noncancer risk assessment—an analysis of the default uncertainty factor. *Regul. Toxicol. Pharmacol.* **27:** 3–20.
19. Boocock, D., G. Faust, K. Patel, *et al.* 2007. Phase I dose escalation pharmacokinetic study in healthy volunteers of resveratrol, a potential cancer chemopreventive agent. *Cancer Epidemiol. Biomarkers Prev.* **16:** 1246–1252.
20. http://www.clinicaltrials.gov/ct2/show/NCT00920556?term=SRT501&rank=2. 22 April 2010.
21. Chow, H., L. Garland, C. Hsu, *et al.* 2010. Resveratrol modulates drug- and carcinogen-metabolizing enzymes in a healthy volunteer study. *Cancer Prev. Res.* **9:** 1168–1175.

Ann. N.Y. Acad. Sci. ISSN 0077-8923

ANNALS OF THE NEW YORK ACADEMY OF SCIENCES
Issue: *Resveratrol and Health*

Resveratrol and life extension

Beamon Agarwal and Joseph A. Baur
Institute for Diabetes, Obesity, and Metabolism, and Department of Physiology, University of Pennsylvania School of Medicine, Philadelphia, Pennsylvania

Address for correspondence: Joseph A. Baur, Institute for Diabetes, Obesity, and Metabolism, University of Pennsylvania School of Medicine, Philadelphia, PA 19104. baur@mail.med.upenn.edu

Age is the most important risk factor for diseases affecting the Western world, and slowing age-related degeneration would greatly improve the quality of human life. In rodents, caloric restriction (CR) extends lifespan by up to 50%. However, attempts to mimic the effects of CR pharmacologically have been limited by our poor understanding of the mechanisms involved. SIRT1 is proposed to mediate key aspects of CR, and small molecule activators may therefore act as CR mimetics. The polyphenol resveratrol activates SIRT1 in an *in vitro* assay, and produces changes that resemble CR *in vivo*, including improvements in insulin sensitivity, endurance, and overall survival in obese mice. However, resveratrol has numerous other targets that could contribute to its health benefits. Moreover, unlike *bona fide* CR, resveratrol has not been shown to extend lifespan in lean mice. Overexpression of SIRT1 or treatment with a novel activator is sufficient to improve metabolism, supporting the idea that resveratrol could act through this pathway. However, the poor phenotype of SIRT1 null mice has thus far precluded a more definitive test.

Keywords: longevity; sirtuins; caloric restriction; SIRT1; resveratrol

Caloric restriction

Caloric restriction (CR), defined as a reduction of energy intake in the absence of malnutrition, is the most robust and reproducible way known to improve health and extend lifespan across species.[1] The seminal experiment linking energy intake to heath and longevity in rats was reported by Clive McKay, Mary Crowell, and Leonard Maynard in 1935,[2] and since that time, benefits of CR in mammals have been shown to include the prevention of major age-related diseases, including neoplasms, diabetes, cardiovascular diseases, and neurodegenerative disorders. This is accompanied by the suppression of chronic inflammatory responses and a general maintenance of a more youthful physiology. CR from a young age slows growth, delays the onset of puberty, and extends lifespan even in strains of rodents that die for different reasons. Because of the wide variety of age-related changes and diseases that are prevented, and since maximum lifespan potential is increased in parallel to mean lifespan, CR has been hypothesized to slow the basic processes that lead to aging, rather than counteract specific diseases.

Despite the passage of 73 years, and intense interest in the subject, no specific molecular mechanism is widely accepted to account for the beneficial effects of CR, and the question of whether similar benefits could be obtained in humans remains open. Although CR has been found to increase lifespan by ~20–50% in most species that have been tested, exceptions have been noted, including the DBA2 strain of mice,[3] and a number of simpler species.[4] More recently, a survey of lab mice highlighted the fact that there are many strains that fail to respond or have decreased longevity in response to the standard CR protocol (40% reduction in caloric intake).[5] Nevertheless, a survival benefit has been demonstrated in rhesus monkeys, along with major improvements in age-related diseases.[6] Short-term studies in humans have shown improvements in insulin sensitivity[7] and cardiovascular disease risk factors,[8] suggesting that regardless of its ultimate effects on longevity, CR has the potential to improve human health. Since it is unlikely that a large

doi: 10.1111/j.1749-6632.2010.05850.x
Ann. N.Y. Acad. Sci. 1215 (2011) 138–143 © 2011 New York Academy of Sciences.

proportion of the population would be willing or able to adopt a CR lifestyle, and because of lingering concerns about potential tradeoffs such as decreased immune or reproductive function, there has been great interest in developing drugs to recapitulate the beneficial effects of CR.[9] A number of observations suggest that this might be feasible, including the existence of mutations that block effects of CR,[10–12] and the observation that a similar calorie deficit induced by exercise is not sufficient to extend life to the same degree as CR.[13] These observations support the idea that at least some of the benefits provided by CR are the result of an active response, and could theoretically be obtained in the absence of reduced energy intake.

Sirtuins

Sirtuins are an evolutionarily conserved family of deacetylases and ADP-ribosyltransferases[14] that derive their name from *Saccharomyces cerevisiae SIR2* (Silent Information Regulator 2) the first member of the family to be described. *SIR2* homologs extend life when overexpressed in yeast, worms, and flies.[15–17] Unlike other deacetylases, the activity of sirtuin enzymes is linked to metabolic processes within the cell through a requirement for nicotinamide adenine dinucleotide (NAD^+).[18,19] This observation led to the proposal that sirtuins might respond to CR, and mediate some of its benefits. Indeed, the initial experiments in multiple organisms provided strong support for this hypothesis.[16,20,21] However, it has since become clear that in yeast and worms, and likely other organisms, restricting energy intake in different ways can extend longevity through multiple genetically distinct pathways. The involvement of specific factors, including sirtuins, depends on the protocol employed.[22,23] The potential of the seven mammalian sirtuins (SIRT1–7) to positively influence longevity is not yet clear, although CR enhances expression of SIRT1–3 and SIRT6 in various tissues. SIRT1-null mice are short-lived and show no increase in activity level or longevity when subjected to CR,[24,25] and overexpression of SIRT1 induces phenotypes that resemble CR.[26] On the other hand, there are at least two cases in which CR is though to decrease sirtuin activity—SIRT4 in the pancreas[27] and SIRT1 in the liver.[28] In addition, hepatic SIRT1 is dispensable for the metabolic effects of CR, and despite their shorter lifespans, SIRT1-null mice have reduced damage to lipids and proteins in the brain,[29] indicating that this enzyme could have both positive and negative effects. These results make sirtuins intriguing components of the response to nutritional stress, and promising candidates to mediate beneficial effects of CR, but much remains to be elucidated about their regulation and functions.

Resveratrol and longevity in lower organisms

Based on the idea that sirtuins might mediate the beneficial effects of CR, Howitz *et al.* performed a screen for small molecule activators of SIRT1 and identified resveratrol,[30] a small polyphenol that was already known to have chemopreventive activity[31] and suspected to contribute to the health benefits of red wine. As discussed in more detail below, the assay used in this screen has become a major source of controversy in the field, since it relies on a fluorescent substrate, and non-fluorescent assays have produced differing results. Nevertheless, resveratrol was subsequently shown to extend life in yeast, worms, and flies in a *SIR2* (SIRT1 homolog)-dependent manner[30,32] (Fig. 1). Notably, lifespan extensions in yeast and flies, as well as the SIR2-dependence of the effect in worms, have been challenged,[34,36] while at the same time, other groups have provided supporting evidence for the original observations.[33,37,39,45] The explanation for these conflicting reports is not clear, but they may indicate that lab-to-lab variations in diets or that genetic backgrounds can play a larger role in the response to resveratrol than has been appreciated.[46] Further support for the ability of resveratrol to improve health and extend life was provided by a series of experiments in *N. furzeri*, a short-lived species of fish. Resveratrol not only extended average lifespan by up to 56%, but also improved motor function and delayed neurodegenerative processes in these animals.[41] These experiments in lower organisms set the stage for testing the effects of resveratrol on longevity in mammals.

Resveratrol and longevity in mice

The ability of resveratrol to mimic lower energy intake was originally tested in two parallel experiments in mice.[42,43] In the first, animals were fed a resveratrol-containing diet to test whether it would recapitulate effects of CR. In the second, animals were fed a high-fat diet with or without resveratrol to test whether it could prevent some of the

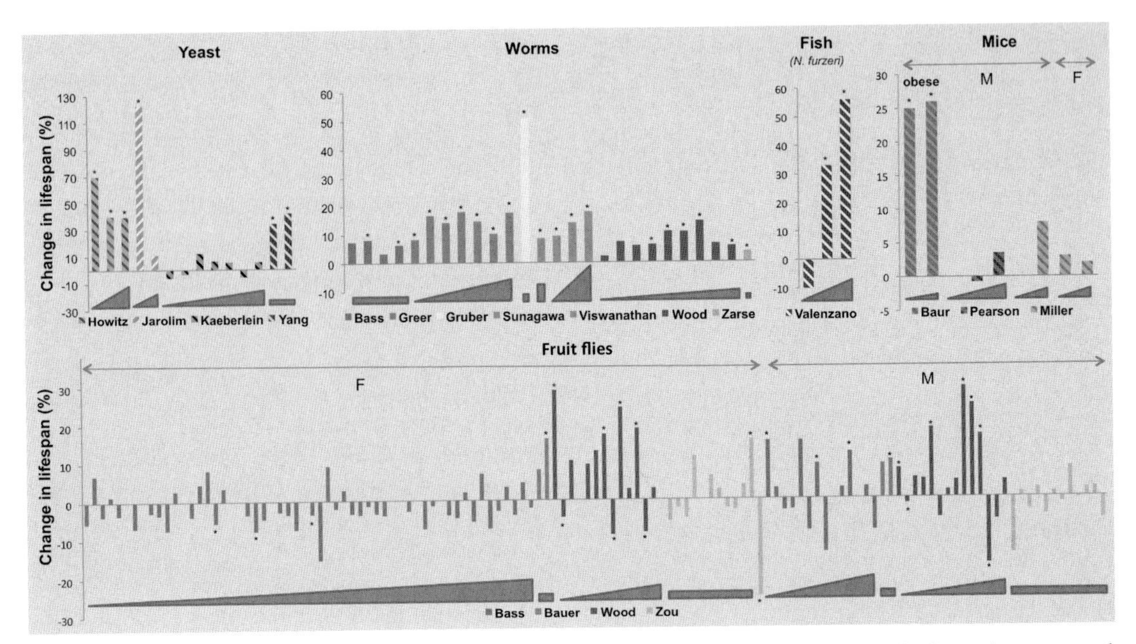

Figure 1. The effect of resveratrol on mean or median lifespan in model organisms. Data are organized by first author name, and correspond to the following references: Howitz,[30] Jarolim,[33] Kaeberlein,[34] Yang,[35] Bass,[36] Greer,[22] Gruber,[37] Sunagawa,[38] Viswanathan,[39] Wood,[32] Zarse,[40] Valenzano,[41] Baur,[42] Pearson,[43] Miller,[44] Bauer,[45] and Zou.[46] Data from Jarolim, Bass, Wood, Valenzano, and Miller are based on median lifespan, while the remainder are based on means. Relative resveratrol dose is approximated by the blue wedges under each graph. Other variables, such as the specific strain and the composition of the diets, varied considerably between and within papers, and in cases where explicit calculations were not provided, results were approximated from graphs. In the Baur reference, the mice being treated were obese due to a high-fat diet. In flies and mice, males are indicated by "M" and females by "F". Asterisks indicate that the result was considered statistically significant by the authors.

detrimental consequences of obesity. In each case, improvements in physiology were noted, ranging from increased insulin sensitivity in the obese mice, to better motor coordination in both groups, to fewer cataracts and higher bone strength in the lean mice. Strikingly, resveratrol was found to mimic the transcriptional response to CR, both in this study and in a parallel independent study.[47,48] Resveratrol increased survival in the obese mice such that their lifespans were equivalent to those of lean, untreated animals. However, adding resveratrol to the diet of mice on a normal diet did not produce any further increase, suggesting that resveratrol might primarily be counteracting the negative consequences of obesity, rather than slowing aging in a CR-like manner.

Several follow-up studies have confirmed that resveratrol does not increase lifespan in healthy mice. Increasing the dose of resveratrol to approximately 200 mg/kg had no effect on survival, and delivering about 1.5 g/kg appeared to be toxic.[43] A larger study conducted as part of the National Institute on Aging's Intervention Testing Program also found no significant increase in survival in either gender (prior tests were performed in males only).[44] An important caveat to note is that all of these studies began with middle-age mice (12 months old), and CR has a larger effect on longevity when the treatment begins earlier.[49] For this reason, several groups are currently studying cohorts that were started at a younger age, although no positive results have been reported to date. Post-mortem analyses of the animals fed high-fat diets support the hypothesis that the life extension that has been observed results primarily from alleviating vascular complications and fatty liver, which are not major contributors to mortality in lean mice.

Importance of SIRT1 in the effects of resveratrol

Resveratrol was hypothesized to be a CR mimetic based on the observation that it can activate SIRT1 in an *in vitro* assay.[30] Many of the effects that have been observed in resveratrol-treated animals are consistent with the modulation of SIRT1's targets, particularly the transcriptional coactivator peroxisome proliferator activator gamma coactivator 1 alpha (PGC-1α).[42,50] However, resveratrol is known to produce a wide array of effects in mammalian cells, including activation of AMP-activated protein kinase (AMPK), an energy sensor that is involved in some of the same pathways as SIRT1 and phosphorylates PGC-1α directly.[42,51] Although SIRT1 can activate the kinase upstream of AMPK, this pathway does not appear to be necessary for AMPK stimulation by resveratrol,[52] and in fact, there is some evidence to suggest that SIRT1 might be downstream of AMPK.[53,54] Further, the validity of the *in vitro* SIRT1 assay has been challenged by several groups, undermining the presumption that resveratrol activates SIRT1 directly. Therefore, it will be important to carefully test the role of SIRT1 in the physiological changes induced by resveratrol.

Several lines of evidence support the assertion that SIRT1 is activated by resveratrol treatment, and that this could plausibly explain some of its beneficial effects *in vivo*. First, there are numerous reports of SIRT1-dependent effects in resveratrol treated cells. These include transcriptional changes (e.g. Refs. 55, 56) as well as macroscopic effects such as the inhibition of adipogenesis.[57] Second, SIRT1 substrates are consistently deacetylated in resveratrol-treated animals, suggesting increased SIRT1 activity.[42,50] Finally, treating mice with a novel SIRT1 activator, SRT1720, recapitulates many of the effects of resveratrol *in vivo*,[58] and overexpressing SIRT1 is sufficient to improve motor coordination and preserve insulin sensitivity in mice fed a high-fat diet.[26,59,60] These observations suggest that SIRT1 may well mediate some of the salient effects of resveratrol *in vivo*.

What has become more controversial is whether resveratrol activates SIRT1 through the direct binding mechanism that was originally proposed. The *in vitro* screen that was used to identify resveratrol as a SIRT1 activator employed a fluorescent substrate. It has been known since 2005 that removing the fluorophore eliminates the activation,[34,61] and this has alternately been interpreted to mean that the original result was an artifact, or that having the fluorophore present somehow mimics the intracellular situation better (e.g., that endogenous substrates are bulkier than the peptides used *in vitro*, or that residues from a SIRT1 binding partner interact with substrates). A recent report showed that resveratrol fails to activate SIRT1 *in vitro* even when full-length endogenous substrates are used, and importantly, showed that the novel SIRT1 activator SRT1720 also exhibits substrate-dependent effects on SIRT1 activity.[62] This was interpreted as strong evidence that both of the molecules that have been used to probe the function of SIRT1 could be working through alternative mechanisms, and influencing SIRT1 only indirectly. The report was followed closely by another study in which SRT1720 was shown to activate SIRT1 against certain substrates that contain only natural amino acids, reasserting the view that activation of SIRT1 *in vivo* occurs through a direct binding mechanism.[63] It was also noted that the similar effects of resveratrol and SRT1720, which are not structurally related, are difficult to reconcile in the absence of a common target, presumed to be SIRT1.

While the biochemical mechanism accounting for SIRT1 activation is likely to remain contentious for some time, testing the importance of SIRT1 in resveratrol's effects is feasible and should be a high priority for the field. Constitutive deletion of the SIRT1 gene causes developmental defects that compromise viability on some backgrounds, and results in smaller mice with altered metabolism.[64,65] Although these mice lack some of the normal responses to CR,[29] and are not as well protected from tumorigenesis by resveratrol,[66] their poor phenotype makes the interpretation of such experiments difficult. Using a floxed SIRT1 allele developed by the Alt lab at Harvard Medical School,[64] it should be possible to avoid many of the complications of constitutive SIRT1 null mice, and to provide a more definitive test of the importance of SIRT1 in the benefits of resveratrol.

Conclusion

Resveratrol extends life in multiple species, although some discrepancies between labs remain unexplained. In mice, resveratrol prevents the early mortality associated with obesity, but there is currently no experimental evidence to suggest that it can prolong life in lean, healthy animals. The deaths that have been prevented in obese mice seem to be primarily those that result from vascular complications, which are not a significant cause of mortality in lean mice but are associated with morbidity and mortality in humans. Therefore, the mouse studies provide a good rationale for studying the effects of resveratrol on human health. However, the influence of factors such as interspecies differences in metabolism, genetic variation, diet, physical activity, disease, and mental health should not be underestimated when extrapolating from rodent models. Further experimental evidence is needed to clarify the importance of SIRT1 and other potential mechanisms in the effects of resveratrol.

Conflicts of interest

The author declares no conflicts of interest.

References

1. Weindruch, R. & R.L. Walford. 1988. *The Retardation of Aging and Disease by Dietary Restriction*. Charles C. Thomas. Springfield, IL.
2. McCay, C.M., M.F. Crowell & L.A. Maynard. 1935. The effect of retarded growth upon the length of life span and upon the ultimate body size. *J. Nutr.* **10:** 63–79.
3. Forster, M.J., P. Morris & R.S. Sohal. 2003. Genotype and age influence the effect of caloric intake on mortality in mice. *Faseb J.* **17:** 690–692.
4. Shanley, D.P. & T.B. Kirkwood. 2006. Caloric restriction does not enhance longevity in all species and is unlikely to do so in humans. *Biogerontology* **7:** 165–168.
5. Liao, C.Y. *et al.* 2010. Genetic variation in the murine lifespan response to dietary restriction: from life extension to life shortening. *Aging Cell* **9:** 92–95.
6. Colman, R.J. *et al.* 2009. Caloric restriction delays disease onset and mortality in rhesus monkeys. *Science* **325:** 201–204.
7. Weiss, E.P. *et al.* 2006. Improvements in glucose tolerance and insulin action induced by increasing energy expenditure or decreasing energy intake: a randomized controlled trial. *Am. J. Clin. Nutr.* **84:** 1033–1042.
8. Lefevre, M. *et al.* 2009. Caloric restriction alone and with exercise improves CVD risk in healthy non-obese individuals. *Atherosclerosis* **203:** 206–213.
9. Ingram, D.K. *et al.* 2006. Calorie restriction mimetics: an emerging research field. *Aging Cell* **5:** 97–108.
10. Anderson, R.M. *et al.* 2003. Nicotinamide and PNC1 govern lifespan extension by calorie restriction in Saccharomyces cerevisiae. *Nature* **423:** 181–185.
11. Bonkowski, M.S. *et al.* 2006. Targeted disruption of growth hormone receptor interferes with the beneficial actions of calorie restriction. *Proc. Natl. Acad. Sci. USA* **103:** 7901–7905.
12. Panowski, S.H. *et al.* 2007. PHA-4/Foxa mediates diet-restriction-induced longevity of C. elegans. *Nature* **447:** 550–555.
13. Holloszy, J.O. *et al.* 1985. Effect of voluntary exercise on longevity of rats. *J. Appl. Physiol.* **59:** 826–831.
14. Frye, R.A. 2000. Phylogenetic classification of prokaryotic and eukaryotic Sir2-like proteins. *Biochem. Biophys. Res. Commun.* **273:** 793–798.
15. Kaeberlein, M., M. McVey & L. Guarente. 1999. The SIR2/3/4 complex and SIR2 alone promote longevity in Saccharomyces cerevisiae by two different mechanisms. *Genes Dev.* **13:** 2570–2580.
16. Rogina, B. & S.L. Helfand. 2004. Sir2 mediates longevity in the fly through a pathway related to calorie restriction. *Proc. Natl. Acad. Sci. USA* **101:** 15998–16003.
17. Tissenbaum, H.A. & L. Guarente. 2001. Increased dosage of a sir-2 gene extends lifespan in Caenorhabditis elegans. *Nature* **410:** 227–230.
18. Imai, S. *et al.* 2000. Transcriptional silencing and longevity protein Sir2 is an NAD-dependent histone deacetylase. *Nature* **403:** 795–800.
19. Vaziri, H. *et al.* 2001. hSIR2(SIRT1) functions as an NAD-dependent p53 deacetylase. *Cell* **107:** 149–159.
20. Lin, S.J., P.A. Defossez & L. Guarente. 2000. Requirement of NAD and SIR2 for life-span extension by calorie restriction in Saccharomyces cerevisiae. *Science* **289:** 2126–2128.
21. Wang, Y. & H.A. Tissenbaum. 2006. Overlapping and distinct functions for a Caenorhabditis elegans SIR2 and DAF-16/FOXO. *Mech. Ageing Dev.* **127:** 48–56.
22. Greer, E.L. & A. Brunet. 2009. Different dietary restriction regimens extend lifespan by both independent and overlapping genetic pathways in C. elegans. *Aging Cell.* **8:** 113–127.
23. Lamming, D.W. *et al.* 2006. Response to Comment on "HST2 mediates SIR2-independent life-span extension by calorie restriction". *Science* **312:** 1312c.
24. Chen, D. *et al.* 2005. Increase in activity during calorie restriction requires Sirt1. *Science* **310:** 1641.
25. Boily, G. *et al.* 2008. SirT1 regulates energy metabolism and response to caloric restriction in mice. *PLoS ONE* **3:** e1759.
26. Bordone, L. *et al.* 2007. SIRT1 transgenic mice show phenotypes resembling calorie restriction. *Aging Cell.* **6:** 759–767.
27. Haigis, M.C. *et al.* 2006. SIRT4 inhibits glutamate dehydrogenase and opposes the effects of calorie restriction in pancreatic beta cells. *Cell* **126:** 941–954.
28. Chen, D. *et al.* 2008. Tissue-specific regulation of SIRT1 by calorie restriction. *Genes Dev.* **22:** 1753–1757.
29. Li, Y. *et al.* 2008. SirT1 inhibition reduces IGF-I/IRS-2/Ras/ERK1/2 signaling and protects neurons. *Cell Metab.* **8:** 38–48.
30. Howitz, K.T. *et al.* 2003. Small molecule activators of sirtuins extend Saccharomyces cerevisiae lifespan. *Nature* **425:** 191–196.
31. Jang, M. *et al.* 1997. Cancer chemopreventive activity of resveratrol, a natural product derived from grapes. *Science* **275:** 218–220.
32. Wood, J.G. *et al.* 2004. Sirtuin activators mimic caloric restriction and delay ageing in metazoans. *Nature* **430:** 686–689.
33. Jarolim, S. *et al.* 2004. A novel assay for replicative lifespan in Saccharomyces cerevisiae. *FEMS Yeast Res.* **5:** 169–177.
34. Kaeberlein, M. *et al.* 2005. Substrate specific activation of sirtuins by resveratrol. *J. Biol. Chem.* **280:** 17038–17045.
35. Yang, H. *et al.* 2007. Design and synthesis of compounds that extend yeast replicative lifespan. *Aging Cell* **6:** 35–43.
36. Bass, T.M. *et al.* 2007. Effects of resveratrol on lifespan in Drosophila melanogaster and Caenorhabditis elegans. *Mech. Ageing Dev.* **128:** 546–552.

37. Gruber, J., S.Y. Tang & B. Halliwell. 2007. Evidence for a trade-off between survival and fitness caused by resveratrol treatment of Caenorhabditis elegans. *Ann. N.Y. Acad. Sci.* **1100:** 530–542.
38. Sunagawa, T. *et al.* Procyanidins from Apples (Malus pumila Mill.) Extend the Lifespan of Caenorhabditis elegans. *Planta Med.* 2010 Aug 17 [Epub ahead of print] doi: 10.1055/s-0030-1250204.
39. Viswanathan, M. *et al.* 2005. A role for SIR-2.1 regulation of ER stress response genes in determining C. elegans life span. *Dev Cell.* **9:** 605–615.
40. Zarse, K. *et al.* 2010. Differential effects of resveratrol and SRT1720 on lifespan of adult Caenorhabditis elegans. *Horm. Metab. Res.* **42:** 837–839.
41. Valenzano, D.R. *et al.* 2006. Resveratrol prolongs lifespan and retards the onset of age-related markers in a short-lived vertebrate. *Curr. Biol.* **16:** 296–300.
42. Baur, J.A. *et al.* 2006. Resveratrol improves health and survival of mice on a high-calorie diet. *Nature* **444:** 337–342.
43. Pearson, K.J. *et al.* 2008. Resveratrol delays age-related deterioration and mimics transcriptional aspects of dietary restriction without extending life span. *Cell Metab.* **8:** 157–168.
44. Miller, R.A., D.E. Harrison, C. Astle, *et al.* 2010. Rapamycin, but not resveratrol or simvastatin, extends lifespan of genetically heterogeneous mice. *J. Gerontology: Biol. Sci.* In press.
45. Bauer, J.H. *et al.* 2004. An accelerated assay for the identification of lifespan-extending interventions in Drosophila melanogaster. *Proc. Natl. Acad. Sci. USA* **101:** 12980–12985.
46. Zou, S. *et al.* 2009. The prolongevity effect of resveratrol depends on dietary composition and calorie intake in a tephritid fruit fly. *Exp. Gerontol.* **44:** 472–476.
47. Pearson, K.J. *et al.* 2008. Resveratrol delays age-related deterioration and mimics transcriptional aspects of dietary restriction without extending life Span. *Cell Metab.* **8:** 157–168.
48. Barger, J.L. *et al.* 2008. A low dose of dietary resveratrol partially mimics caloric restriction and retards aging parameters in mice. *PLoS ONE* **3:** e2264.
49. Weindruch, R. & R.L. Walford. 1982. Dietary restriction in mice beginning at 1 year of age: effect on life-span and spontaneous cancer incidence. *Science* **215:** 1415–1418.
50. Lagouge, M. *et al.* 2006. Resveratrol improves mitochondrial function and protects against metabolic disease by activating SIRT1 and PGC-1alpha. *Cell* **127:** 1109–1122.
51. Zang, M. *et al.* 2006. Polyphenols stimulate AMP-activated protein kinase, lower lipids, and inhibit accelerated atherosclerosis in diabetic LDL receptor-deficient mice. *Diabetes* **55:** 2180–2191.
52. Dasgupta, B. & J. Milbrandt. 2007. Resveratrol stimulates AMP kinase activity in neurons. *Proc Natl Acad Sci USA* **104:** 7217–7222.
53. Canto, C. *et al.* 2010. Interdependence of AMPK and SIRT1 for metabolic adaptation to fasting and exercise in skeletal muscle. *Cell Metab.* **11:** 213–219.
54. Fulco, M. *et al.* 2008. Glucose restriction inhibits skeletal myoblast differentiation by activating SIRT1 through AMPK-mediated regulation of Nampt. *Dev Cell* **14:** 661–673.
55. Csiszar, A. *et al.* 2009. Resveratrol induces mitochondrial biogenesis in endothelial cells. *Am. J. Physiol. Heart Circ. Physiol.* **297:** H13–H20.
56. Gracia-Sancho, J. *et al.* 2010. Activation of SIRT1 by resveratrol induces KLF2 expression conferring an endothelial vasoprotective phenotype. *Cardiovasc. Res.* **85:** 514–519.
57. Picard, F. *et al.* 2004. Sirt1 promotes fat mobilization in white adipocytes by repressing PPAR-gamma. *Nature* **429:** 771–776.
58. Feige, J.N. *et al.* 2008. Specific SIRT1 activation mimics low energy levels and protects against diet-induced metabolic disorders by enhancing fat oxidation. *Cell Metab.* **8:** 347–358.
59. Banks, A.S. *et al.* 2008. SirT1 gain of function increases energy efficiency and prevents diabetes in mice. *Cell Metab.* **8:** 333–341.
60. Pfluger, P.T. *et al.* 2008. Sirt1 protects against high-fat diet-induced metabolic damage. *Proc. Natl. Acad. Sci. USA* **105:** 9793–9798.
61. Borra, M.T., B.C. Smith & J.M. Denu. 2005. Mechanism of human SIRT1 activation by resveratrol. *J. Biol. Chem.* **280:** 17187–17195.
62. Pacholec, M. *et al.* 2010. SRT1720, SRT2183, SRT1460, and resveratrol are not direct activators of SIRT1. *J. Biol. Chem.* **285:** 8340–8351.
63. Dai, H. *et al.* 2010. SIRT1 activation by small molecules – kinetic and biophysical evidence for direct interaction of enzyme and activator. *J. Biol. Chem.* **285:** 32695–32703.
64. Cheng, H.L. *et al.* 2003. Developmental defects and p53 hyperacetylation in Sir2 homolog (SIRT1)-deficient mice. *Proc. Natl. Acad. Sci. USA* **100:** 10794–10799.
65. McBurney, M.W. *et al.* 2003. The mammalian SIR2alpha protein has a role in embryogenesis and gametogenesis. *Mol. Cell Biol.* **23:** 38–54.
66. Boily, G. *et al.* 2009. SirT1-null mice develop tumors at normal rates but are poorly protected by resveratrol. *Oncogene* **28:** 2882–2893.

Ann. N.Y. Acad. Sci. ISSN 0077-8923

ANNALS OF THE NEW YORK ACADEMY OF SCIENCES
Issue: *Resveratrol and Health*

Resveratrol in cancer management: where are we and where we go from here?

Mary Ndiaye,[1] Raj Kumar,[2] and Nihal Ahmad[1,3]

[1]Department of Dermatology, University of Wisconsin, Madison, Wisconsin. [2]Department of Basic Sciences, The Commonwealth Medical College, Scranton, Pennsylvania. [3]University of Wisconsin Carbone Cancer Center, Madison, Wisconsin

Address for correspondence: Nihal Ahmad, Department of Dermatology, University of Wisconsin, 425 Medical Sciences Center, 1300 University Avenue, Madison, WI 53706. nahmad@wisc.edu

Resveratrol has been shown to afford protection against several diseases. A plethora of studies have suggested that resveratrol imparts cancer chemopreventive and therapeutic responses. However, an important issue with the future development of resveratrol for disease management is its low bioavailability due to its rapid metabolism in mammals. Therefore, efforts are needed to enhance its bioavailability in humans. In this direction, some possible scenarios include enhancing the bioavailability of resveratrol by novel mechanism-based combinations with agents that can inhibit the *in vivo* metabolism of resveratrol, nanoparticle-mediated delivery, use of naturally occurring or synthetic analogues of resveratrol, and use of conjugated metabolites of resveratrol, though these need to be carefully evaluated as they may need to be deconjugated from resveratrol at the target organ to elicit a biological response. Thus, concerted and multidisciplinary efforts are needed to take reseratrol to the next level, that is, from the "bench-to-bedside."

Keywords: resveratrol; bioavailability; nanoparticles; piceatannol

Introduction

Resveratrol (3,5,4′-trihydroxystilbene), a phytoalexin antioxidant found in grapes, red wines, berries, and peanuts, has been shown to afford protection against several diseases, including cancer.[1–8] In 1992, Renaud and Lorgeril presented the concept of the "French Paradox."[9] In most countries, intake of saturated fat has been positively correlated with a high mortality from coronary heart disease (CHD); however, the scenario in France is somewhat paradoxical as despite a high intake of saturated fat, the observed mortality from CHD is low there.[9] It is thought that the high consumption of red wine in France is responsible for this paradox, and that resveratrol is the active therapeutic agent in red wine. Resveratrol is produced by some plants upon exposure to certain environmental stresses, usually a fungus or microbe, as a defense mechanism, which classifies it as a phytoalexin. This property naturally leads itself to exploitation as a disease-fighting agent. The use of resveratrol-containing herbs has been known for long time. The root of *Polygonum cuspidatum* contains an active constituent of resveratrol and has been used in traditional Japanese and Chinese medicine to fight dermatitis, favus, hyperlipemia, and gonorrhea.[10,11]

Since the publication of the first paper on cancer chemoprevention by resveratrol from Dr. John Pezzuto's laboratory,[12] a plethora of preclinical studies, including *in vitro* and *in vivo* studies from our laboratory, have suggested that resveratrol imparts chemopreventive as well as therapeutic response against several cancers.[2,4,7,13–15] The use of resveratrol is continuously increasing as a supplement and complementary and alternative medicine (CAM) approach toward the management of several diseases and conditions, including cancer, diabetes, obesity, and Alzheimer's disease.[1,8,13–21] Several dose-finding clinical trials have been undertaken to look at the bioavailability of resveratrol following its ingestion.[7,22–24] In addition, at present, there are several ongoing clinical trials aimed at

doi: 10.1111/j.1749-6632.2010.05851.x

evaluating the health-beneficial effects of resveratrol, including cancers of colon, multiple myeloma, and follicular lymphoma (http://clinicaltrials.gov). However, an important issue with the future development of resveratrol for disease management appears to be its low bioavailability due to its rapid metabolism in mammals. Because of the proven efficacy of resveratrol in preclinical settings, it is necessary that concerted multidisciplinary efforts are undertaken to enhance its bioavailability and activity in order to take this promising agent from the bench to the bedside.

Metabolism and bioavailability

Resveratrol is found in either *cis-* or *trans-*configurations, with *trans*-resveratrol being the dominant chemical form occurring in nature, as well as being the most thoroughly studied by scientists. The *cis* form can be synthesized in the laboratory by exposing *trans*-resveratrol to ultraviolet light or heat. The most abundant metabolites of *trans*-resveratrol are glucuronides and sulfates, and at least seven forms have been identified.[25] Based on studies, resveratrol has a low toxicity and can be given in relatively high doses without adverse effects in humans, a quality that makes it an excellent candidate for disease management.[22,26,27] It has been shown that resveratrol is rapidly metabolized, usually within 30–60 min following oral administration, although the peak plasma concentration can be delayed slightly by changing certain variables, including eating certain kinds of food with the drug.[26,28] The avenue of administration, including administering resveratrol in wine, grape juice, or capsules, appears to impact the availability of resveratrol and its metabolites.[29] Several studies have been done to assess the metabolism of resveratrol in a variety of systems. For example, Walle *et al.* used C^{14}-labeled *trans*-resveratrol (given either orally or via *intravenous* injection), and found that resveratrol was very quickly absorbed but also extremely quickly metabolized, with most of the oral dose being recovered in the urine.[23] The plasma concentration of resveratrol when given *intravenously* rapidly declined over a 1 h period of time and thereafter followed a similar pace as the oral dose. This study also found that when given orally, some of the resveratrol was metabolized by hydrogenation, which the authors speculated was done by intestinal microflora, not by the body itself.[23] Other studies have also found similar results.[29–31]

Enhancing resveratrol bioavailability

Because the rapid metabolism seems to be a limiting factor in translating resveratrol's promising chemopreventive and chemotherapeutic effect in humans, researchers have begun to focus on different means of enhancing the bioavailability of resveratrol, as well as discovering and synthesizing novel resveratrol analogues and metabolites with superior efficacy and bioavailability. Some efforts in this direction are summarized later.

Inhibiting the metabolism: mechanism-based combinations

One approach toward enhancing bioavailability of resveratrol could be to combine it with other agent(s) that can inhibit its metabolism *in vivo*. Studies have shown that during metabolism, the majority of resveratrol undergoes (i) glucuronide conjugation by uridine diphosphate glucoronosyltransferases (UGTs) to result in glucuronides (3-*O*-glucuronide and 4′-*O*-glucuronide), and (ii) sulfate conjugation by sulfotransferases (SULTs) to form resveratrol 3-*O*-sulfate in humans and several other sulfate conjugates (4′-sulfate, 3,5-disulfate, 3,4′-disulfate, 3,4′,5-trisulfate) in rodents.[4] This gives rise to the opportunity to use resveratrol in combination with other agents that are known to inhibit these reactions in order to prolong resveratrol's presence *in vivo*. Studies have shown that piperine, an alkaloid derived from black pepper (*Piper* spp.), inhibits glucuronidation of several agents, including another polyphenol, curcumin, *in vivo*.[32–34] Another study has shown that piperine enhances the bioavailability of the tea polyphenol (–)-epigallocatechin-3-gallate (which is also subjected to glucuronidation) in mice.[35] Further studies have shown that some other naturally occurring flavonoids, including quercetin and myricetin, may also inhibit glucuronidation and sulfation of resveratrol.[36–38] However, these studies have not gone further to determine the effects on bioavailability of resveratrol. There has been initial work done using a combination of resveratrol with quercetin that has shown that there is a synergistic effect on oxidative injury in human erythrocytes, but no bioavailability data was collected for that study.[39] Another group recently published data from a small human subject trial ($n = 8$) that looked at resveratrol and quercetin. The authors determined *trans*-resveratrol pharmacokinetics when given alone along with either a standard or high-fat breakfast, and with or without quercetin and/or alcohol.[40] It was found that the maximum serum concentration of resveratrol was delayed with the high-fat breakfast, but no effect was seen with the quercetin.[40] However, the authors pointed out that their studies did not take into account metabolites of resveratrol and only measured *trans*-resveratrol levels, so quercetin could have had effects on the metabolites and not resveratrol, which would not have been detected using the study parameters.[40]

Using nanotechnology

Nanoparticle-mediated delivery could serve as a useful tool in enhancing the bioavailability of resveratrol. Nanotechnology is an emerging area of research that operates at the crossroads of biology, chemistry, physics, engineering, and medicine. Design of nanoparticles for an effective and regulated delivery of anti-cancer agents is being considered as one of the most exciting applications of nanotechnology.[41] Nanoparticles used in drug technology are small objects sized between 1 and 100 nm that are proposed to allow superior administration of the test object or chemical into the cell of interest. These particles can be made of many different materials, including lipids, metals, and proteins. It is conceivable that nanoparticle-encapsulated resveratrol, because of its significantly enhanced bioavailability, could provide a superior response *in vivo*.

A few studies have attempted to enhance the efficacy of resveratrol via its encapsulation in liposomal or nanoparticles.[42–46] In a recent study, Teskac and Kristl have shown that solid lipid nanoparticles loaded with resveratrol can cross the cell membrane in keratinocytes, thereby enhancing the cells' time of exposure to resveratrol.[44] The authors found that the resveratrol binds tightly within the lipid nanoparticles and forms a layer on the outside. These outer layers become diffused upon initial exposure, and then the inner layers are released slowly, with a delay up to 5 h following initial exposure.[44] Another study used bovine serum albumin-bound resveratrol nanoparticles in primary ovarian cancer xenograft mice.[45] The authors found that not only was the nanoparticle-bound resveratrol easier to work with due to better solubility, but it had a higher impact on inhibition of tumor growth, compared to resveratrol alone.[45] It is expected that nanoparticle-mediated delivery will provide a dose-advantage in a therapeutic setting. This is supported by a recent *in vitro* study by Shao *et al.* demonstrating that incorporating resveratrol into mPEG-PCL (methoxy poly(ethylene glycol)-poly(caprolactone))-based nanoparticles not only caused cell death, but was found to lower the dose of resveratrol needed to have the same cytotoxicity, with 4 μM nanoparticle-bound resveratrol having similar response as 8 μM unbound resveratrol.[43] They also showed that this encapsulation resulted in a sustained release of resveratrol, even 10 h following the initial exposure. Thus, nanoparticle mediated delivery of resveratrol appears to be a promising strategy for enhancing the efficacy of resveratrol.

Synthetic manipulation: analogues of resveratrol

A complementary strategy that needs to be aggressively pursued is to discover and define novel analogues of resveratrol, both naturally occurring and synthetic. These analogues should have the same structural backbone of resveratrol, with chemical modifications resulting into superior efficacy.[47] An example of a naturally occurring resveratrol analogue is piceatannol (monohydroxylated resveratrol; 3,3′,4,5′-tetrahydroxy-*trans*-stilbene), which is found in wine and grapes in low concentrations. The anti-proliferative effects of piceatannol were first described in 1984 by Ferrigni *et al.*[48] Piceatannol can be synthesized by the addition of a hydroxyl group to one end of the resveratrol molecule, as is done *in vivo* by the enzyme CYP1B1.[49] A few reports have studied the effects of piceatannol in certain cancer cell types, including leukemia, prostate, breast, and bladder.[50–53] One group has looked at piceatannol in human breast epithelial cells and found that this resveratrol analogue inhibits phorbol ester-induced NF-κB activation and cyclooxygenase-2 expression, as well as blocking migration and transformation of the cells.[54] In a different study designed to test cytotoxic, cytoprotective, and antioxidant properties of resveratrol analogues, it was shown that piceatannol was the only analogue tested that had a cytoprotective effect in C6 cells.[55] Based on these studies, we believe that piceatannol needs to be carefully studied for its chemopreventive effects, alone as well as in combination with resveratrol.

At present, researchers are attempting to synthesize a wide variety of resveratrol analogues with superior efficacy.[56–58] In a recent study, nine resveratrol analogues were synthesized by replacing a section of the resveratrol molecule with pseudoheterocyclic or heterocyclic scaffolds.[59] The authors then tested the biological activities of these compounds and found that, compared to resveratrol, several of these new compounds had a reduced IC_{50}.[59] Another group of investigators designed a set of resveratrol analogues with the purpose of inhibiting key enzymes involved in the conversion of androgens to estrogen, in an attempt to create a therapeutic agent for some breast cancers. These researchers were able to find two products that had superior efficacy against the specific enzymes as compared to resveratrol.[58] In a different study, glucosyl groups were added to resveratrol to increase its solubility.[60] This study showed that, when given to rats, these glucosyl-analogues formed metabolites similar to resveratrol and showed a better *in vivo* retention, with the blood concentration *versus* time curve being shifted to longer times.[60] By actively exploring modifications to the resveratrol backbone, it is hoped that further progress can be made on increasing the bioavailability and activity of this drug *in vivo*.

Chemoprevention by resveratrol metabolites

As noted earlier, resveratrol is metabolized very quickly in mammals; it is glucuronated and sulfonated in the liver and duodenum. In spite of this, some promise can be seen in the *in vivo* trials that have been undertaken.[16] It is thought that the effects on cancer in these *in vivo* trials could be a result of these metabolites working directly on the tumors themselves. Traditionally, it is thought that when substances are glucuronated and sulfonated, the resulting compounds are less able to cross the cell membrane. However, the undeniable efficacy of resveratrol *in vitro* and in animal models, despite its quick metabolism, leads some researchers to believe that these particular metabolites may be exhibiting biological activity that is separate from the activity of resveratrol. It is increasingly believed that resveratrol is metabolized by the body and its metabolite conjugates are transported to different tissues, where the compounds are deconjugated at the target site of action, thereby releasing the substrate to elicit biological activity on its own.[24,61] This area of research is very much in its infancy, with much of the research being focused on determining the exact metabolites. One recent study done in humans by Patel and colleagues measured the amount of resveratrol and six of its metabolites in a set of colorectal cancer patients.[30] They found that the levels of resveratrol, as well as those of its metabolites, were low in serum but much higher in colon tissue, especially in the right side of the colon.[30] This study supports the conjugation-deconjugation hypothesis, as described earlier. In another study aimed at elucidating the utility of particular metabolites, Hoshino *et al.* synthesized five different sulfate metabolites of resveratrol and tested their activities on several biological processes *in vitro*. They found that the metabolites were not as effective at killing the cells as resveratrol was, but some exhibited better activity on specific biological targets, including radical scavenging.[62] This information can be used to exclude some of the metabolites from further research so a more directed focus can be made on the candidates that are more likely to have a biological effect when used *in vivo*.

Conclusion

Resveratrol has shown much promise in preclinical trials, and because of its good safety profile resveratrol could be an ideal chemopreventive and/or chemotherapy agent. However, the rapid metabolism of resveratrol has been a continuing challenge. There are several approaches currently underway to try to overcome this problem, which appears to be a major hindrance in the translational relevance of resveratrol. Indeed, multidisciplinary efforts of scientists from diverse disciplines (chemistry, biology, engineering, and clinics) will hopefully be able to take this promising agent from the bench to the bedside.

Conflicts of interest

The authors declare no conflicts of interest.

References

1. Aziz, M.H. *et al.* 2005. Chemoprevention of skin cancer by grape constituent resveratrol: relevance to human disease? *FASEB J.* **19:** 1193–1195.
2. Reagan-Shaw, S., H. Mukhtar & N. Ahmad. 2008. Resveratrol imparts photoprotection of normal cells and enhances the efficacy of radiation therapy in cancer cells. *Photochem. Photobiol.* **84:** 415–421.

3. Afaq, F. *et al.* 2002. Botanical antioxidants for chemoprevention of photocarcinogenesis. *Front. Biosci.* **7:** d784–d792.
4. Shankar, S., G. Singh & R.K. Srivastava. 2007. Chemoprevention by resveratrol: molecular mechanisms and therapeutic potential. *Front. Biosci.* **12:** 4839–4854.
5. Bhat, K.P. & J.M. Pezzuto. 2002. Cancer chemopreventive activity of resveratrol. *Ann. N. Y. Acad. Sci.* **957:** 210–229.
6. Gullett, N.P. *et al.* 2010. Cancer prevention with natural compounds. *Semin. Oncol.* **37:** 258–281.
7. Aziz, M.H., R. Kumar & N. Ahmad. 2003. Cancer chemoprevention by resveratrol: *in vitro* and *in vivo* studies and the underlying mechanisms (review). *Int. J. Oncol.* **23:** 17–28.
8. Baur, J.A. & D.A. Sinclair. 2006. Therapeutic potential of resveratrol: the *in vivo* evidence. *Nat. Rev. Drug Discov.* **5:** 493–506.
9. Renaud, S. & M. de Lorgeril. 1992. Wine, alcohol, platelets, and the French paradox for coronary heart disease. *Lancet* **339:** 1523–1526.
10. Nonomura, S., H. Kanagawa & A. Makimoto. 1963. [Chemical Constituents of Polygonaceous Plants. I. Studies on the Components of Ko-J O-Kon. (Polygonum Cuspidatum Sieb. Et Zucc.)]. *Yakugaku Zasshi.* **83:** 988–990.
11. Clement, M.V. *et al.* 1998. Chemopreventive agent resveratrol, a natural product derived from grapes, triggers CD95 signaling-dependent apoptosis in human tumor cells. *Blood* **92:** 996–1002.
12. Jang, M. *et al.* 1997. Cancer chemopreventive activity of resveratrol, a natural product derived from grapes. *Science* **275:** 218–220.
13. Aziz, M.H., F. Afaq & N. Ahmad. 2005. Prevention of ultraviolet-B radiation damage by resveratrol in mouse skin is mediated via modulation in surviving. *Photochem. Photobiol.* **81:** 25–31.
14. Aziz, M.H. *et al.* 2006. Resveratrol-caused apoptosis of human prostate carcinoma LNCaP cells is mediated via modulation of phosphatidylinositol 3′-kinase/Akt pathway and Bcl-2 family proteins. *Mol. Cancer Ther.* **5:** 1335–1341.
15. Reagan-Shaw, S. *et al.* 2004. Modulations of critical cell cycle regulatory events during chemoprevention of ultraviolet B-mediated responses by resveratrol in SKH-1 hairless mouse skin. *Oncogene* **23:** 5151–5160.
16. Bishayee, A. 2009. Cancer prevention and treatment with resveratrol: from rodent studies to clinical trials. *Cancer Prev. Res. (Phila Pa).* **2:** 409–418.
17. Afaq, F., V.M. Adhami & N. Ahmad. 2003. Prevention of short-term ultraviolet B radiation-mediated damages by resveratrol in SKH-1 hairless mice. *Toxicol. Appl. Pharmacol.* **186:** 28–37.
18. Ahmad, K.A. *et al.* 2004. Resveratrol inhibits drug-induced apoptosis in human leukemia cells by creating an intracellular milieu nonpermissive for death execution. *Cancer Res.* **64:** 1452–1459.
19. Ahmad, K.A. *et al.* 2007. Protein kinase CK2 modulates apoptosis induced by resveratrol and epigallocatechin-3-gallate in prostate cancer cells. *Mol. Cancer Ther.* **6:** 1006–1012.
20. Dong, Z. 2003. Molecular mechanism of the chemopreventive effect of resveratrol. *Mutat. Res.* **523–524:** 145–150.
21. Saiko, P. *et al.* 2008. Resveratrol and its analogs: defense against cancer, coronary disease and neurodegenerative maladies or just a fad? *Mutat. Res.* **658:** 68–94.
22. Boocock, D.J. *et al.* 2007. Phase I dose escalation pharmacokinetic study in healthy volunteers of resveratrol, a potential cancer chemopreventive agent. *Cancer Epidemiol. Biomarkers Prev.* **16:** 1246–1252.
23. Walle, T. *et al.* 2004. High absorption but very low bioavailability of oral resveratrol in humans. *Drug Metab. Dispos.* **32:** 1377–1382.
24. Wenzel, E. & V. Somoza. 2005. Metabolism and bioavailability of trans-resveratrol. *Mol. Nutr. Food Res.* **49:** 472–481.
25. Burkon, A. & V. Somoza. 2008. Quantification of free and protein-bound trans-resveratrol metabolites and identification of trans-resveratrol-C/O-conjugated diglucuronides – two novel resveratrol metabolites in human plasma. *Mol. Nutr. Food Res.* **52:** 549–557.
26. Almeida, L. *et al.* 2009. Pharmacokinetic and safety profile of trans-resveratrol in a rising multiple-dose study in healthy volunteers. *Mol. Nutr. Food Res.* **53**(Suppl 1): S7–S15.
27. Nunes, T. *et al.* 2009. Pharmacokinetics of trans-resveratrol following repeated administration in healthy elderly and young subjects. *J. Clin. Pharmacol.* **49:** 1477–1482.
28. Vaz-da-Silva, M. *et al.* 2008. Effect of food on the pharmacokinetic profile of trans-resveratrol. *Int. J. Clin. Pharmacol. Ther.* **46:** 564–570.
29. Cottart, C.H. *et al.* 2010. Resveratrol bioavailability and toxicity in humans. *Mol. Nutr. Food Res.* **54:** 7–16.
30. Patel, K.R. *et al.* 2010. Clinical pharmacology of resveratrol and its metabolites in colorectal cancer patients. *Cancer Res.* **70**: 7392–7399.
31. Brown, V.A. *et al.* 2010. Repeat dose study of the cancer chemopreventive agent resveratrol in healthy volunteers: safety, pharmacokinetics and effect on the insulin-like growth factor axis. *Cancer Res.* Advance online publication. doi: 10.1158/0008-5472.CAN-10-2364.
32. Reen, R.K. *et al.* 1993. Impairment of UDP-glucose dehydrogenase and glucuronidation activities in liver and small intestine of rat and guinea pig *in vitro* by piperine. *Biochem. Pharmacol.* **46:** 229–238.
33. Atal, C.K., R.K. Dubey & J. Singh. 1985. Biochemical basis of enhanced drug bioavailability by piperine: evidence that piperine is a potent inhibitor of drug metabolism. *J. Pharmacol. Exp. Ther.* **232:** 258–262.
34. Srinivasan, K. 2007. Black pepper and its pungent principle-piperine: a review of diverse physiological effects. *Crit. Rev. Food. Sci. Nutr.* **47:** 735–748.
35. Lambert, J.D. *et al.* 2004. Piperine enhances the bioavailability of the tea polyphenol (-)-epigallocatechin-3-gallate in mice. *J. Nutr.* **134:** 1948–1952.
36. de Santi, C. *et al.* 2000. Glucuronidation of resveratrol, a natural product present in grape and wine, in the human liver. *Xenobiotica* **30:** 1047–1054.
37. De Santi, C. *et al.* 2000. Sulphation of resveratrol, a natural compound present in wine, and its inhibition by natural flavonoids. *Xenobiotica* **30:** 857–866.
38. De Santi, C. *et al.* 2000. Sulphation of resveratrol, a natural product present in grapes and wine, in the human liver and duodenum. *Xenobiotica* **30:** 609–617.

39. Mikstacka, R., A.M. Rimando & E. Ignatowicz. 2010. Antioxidant effect of trans-resveratrol, pterostilbene, quercetin and their combinations in human erythrocytes *in vitro*. *Plant. Foods Hum. Nutr.* **65:** 57–63.
40. la Porte, C. *et al.* 2010. Steady-State pharmacokinetics and tolerability of trans-resveratrol 2000 mg twice daily with food, quercetin and alcohol (ethanol) in healthy human subjects. *Clin. Pharmacokinet.* **49:** 449–454.
41. Siddiqui, I.A. *et al.* 2010. Nanochemoprevention: sustained release of bioactive food components for cancer prevention. *Nutr. Cancer.* **62:** 883–890.
42. Lu, X. *et al.* 2009. Resveratrol-loaded polymeric micelles protect cells from Abeta-induced oxidative stress. *Int. J. Pharm.* **375:** 89–96.
43. Shao, J. *et al.* 2009. Enhanced growth inhibition effect of resveratrol incorporated into biodegradable nanoparticles against glioma cells is mediated by the induction of intracellular reactive oxygen species levels. *Colloids Surf. B. Biointerfaces* **72:** 40–47.
44. Teskac, K. & J. Kristl. 2010. The evidence for solid lipid nanoparticles mediated cell uptake of resveratrol. *Int. J. Pharm.* **390:** 61–69.
45. Guo, L. *et al.* 2010. Anticancer activity and molecular mechanism of resveratrol-bovine serum albumin nanoparticles on subcutaneously implanted human primary ovarian carcinoma cells in nude mice. *Cancer Biother. Radiopharm.* **25:** 471–477.
46. Narayanan, N.K. *et al.* 2009. Liposome encapsulation of curcumin and resveratrol in combination reduces prostate cancer incidence in PTEN knockout mice. *Int. J. Cancer.* **125:** 1–8.
47. Szekeres, T. *et al.* 2010. Resveratrol and resveratrol analogues–structure-activity relationship. *Pharm. Res.* **27:** 1042–1048.
48. Ferrigni, N.R. *et al.* 1984. Use of potato disc and brine shrimp bioassays to detect activity and isolate piceatannol as the antileukemic principle from the seeds of Euphorbia lagascae. *J. Nat. Prod.* **47:** 347–352.
49. Potter, G.A. *et al.* 2002. The cancer preventative agent resveratrol is converted to the anticancer agent piceatannol by the cytochrome P450 enzyme CYP1B1. *Br. J. Cancer* **86:** 774–778.
50. Kim, E.J. *et al.* 2009. The grape component piceatannol induces apoptosis in DU145 human prostate cancer cells via the activation of extrinsic and intrinsic pathways. *J. Med. Food* **12:** 943–951.
51. Lee, Y.M. *et al.* 2009. Piceatannol, a natural stilbene from grapes, induces G1 cell cycle arrest in androgen-insensitive DU145 human prostate cancer cells via the inhibition of CDK activity. *Cancer Lett.* **285:** 166–173.
52. Vo, T.P. *et al.* 2009. Pro- and anti-carcinogenic mechanisms of piceatannol are activated dose-dependently in MCF-7 breast cancer cells. *Carcinogenesis.* Advance online publication. doi: 10.1093/carcin/bgp199.
53. Kuo, P.L. & Y.L. Hsu. 2008. The grape and wine constituent piceatannol inhibits proliferation of human bladder cancer cells via blocking cell cycle progression and inducing Fas/membrane bound Fas ligand-mediated apoptotic pathway. *Mol. Nutr. Food Res.* **52:** 408–418.
54. Son, P.S. *et al.* 2010. Piceatannol, a catechol-type polyphenol, inhibits phorbol ester-induced NF-{kappa}B activation and cyclooxygenase-2 expression in human breast epithelial cells: cysteine 179 of IKK{beta} as a potential target. *Carcinogenesis* **31:** 1442–1449.
55. Ruweler, M. *et al.* 2009. Cytotoxic, cytoprotective and antioxidant activities of resveratrol and analogues in C6 astroglioma cells *in vitro*. *Chem. Biol. Interact.* **182:** 128–135.
56. Horvath, Z. *et al.* 2007. Novel resveratrol derivatives induce apoptosis and cause cell cycle arrest in prostate cancer cell lines. *Anticancer Res.* **27:** 3459–3464.
57. Saiko, P. *et al.* 2007. N-hydroxy-N′-(3,4,5-trimethoxyphenyl)-3,4,5-trimethoxy-benzamidine, a novel resveratrol analog, inhibits ribonucleotide reductase in HL-60 human promyelocytic leukemia cells: synergistic antitumor activity with arabinofuranosylcytosine. *Int. J. Oncol.* **31:** 1261–1266.
58. Sun, B. *et al.* 2010. Design, synthesis, and biological evaluation of resveratrol analogues as aromatase and quinone reductase 2 inhibitors for chemoprevention of cancer. *Bioorg. Med. Chem.* **18:** 5352–5366.
59. Bertini, S. *et al.* 2010. Synthesis of heterocycle-based analogs of resveratrol and their antitumor and vasorelaxing properties. *Bioorg. Med. Chem.* **18:** 6715–6724.
60. Biasutto, L. *et al.* 2009. Soluble polyphenols: synthesis and bioavailability of 3,4′,5-tri(alpha-D-glucose-3-O-succinyl) resveratrol. *Bioorg. Med. Chem. Lett.* **19:** 6721–6724.
61. Reagan-Shaw, S., M. Nihal & N. Ahmad. 2008. Dose translation from animal to human studies revisited. *FASEB J.* **22:** 659–661.
62. Hoshino, J. *et al.* 2010. Selective synthesis and biological evaluation of sulfate-conjugated resveratrol metabolites. *J. Med. Chem.* **53:** 5033–5043.

Ann. N.Y. Acad. Sci. ISSN 0077-8923

ANNALS OF THE NEW YORK ACADEMY OF SCIENCES
Issue: *Resveratrol and Health*

Chemosensitization of tumors by resveratrol

Subash C. Gupta, Ramaswamy Kannappan, Simone Reuter, Ji Hye Kim, and Bharat B. Aggarwal

Cytokine Research Laboratory, Department of Experimental Therapeutics, The University of Texas MD Anderson Cancer Center, Houston, Texas

Address for correspondence: Bharat B. Aggarwal, Cytokine Research Laboratory, Department of Experimental Therapeutics, The University of Texas MD Anderson Cancer Center, Houston, TX 77030. aggarwal@mdanderson.org

Because tumors develop resistance to chemotherapeutic agents, the cancer research community continues to search for effective chemosensitizers. One promising possibility is to use dietary agents that sensitize tumors to the chemotherapeutics. In this review, we discuss that the use of resveratrol can sensitize tumor cells to chemotherapeutic agents. The tumors shown to be sensitized by resveratrol include lung carcinoma, acute myeloid leukemia, promyelocytic leukemia, multiple myeloma, prostate cancer, oral epidermoid carcinoma, and pancreatic cancer. The chemotherapeutic agents include vincristine, adriamycin, paclitaxel, doxorubicin, cisplatin, gefitinib, 5-fluorouracil, velcade, and gemcitabine. The chemosensitization of tumor cells by resveratrol appears to be mediated through its ability to modulate multiple cell-signaling molecules, including drug transporters, cell survival proteins, cell proliferative proteins, and members of the NF-κB and STAT3 signaling pathways. Interestingly, this nutraceutical has also been reported to suppress apoptosis induced by paclitaxel, vincristine, and daunorubicin in some tumor cells. The potential mechanisms underlying this dual effect are discussed. Overall, studies suggest that resveratrol can be used to sensitize tumors to standard cancer chemotherapeutics.

Keywords: apoptosis; cancer therapy; chemoresistance; chemosensitization; resveratrol; tumor

Introduction

Since the inception of the national war on cancer in 1971, significant progress has been made in understanding risk factors, treatments, and prognosis of the disease. Despite this fact, the overall cancer death rate has not decreased appreciably, and the disease remains one of the leading causes of mortality worldwide. The therapies available to date for cancer treatment are surgery, radiotherapy, and chemotherapy. Chemotherapy is often used as a main regimen in the overall treatment of most cancers. However, the development of tumor resistance to chemotherapy (chemoresistance) presents a major hurdle in cancer therapy.[1] Further, complications emerge when cancer cells develop chemoresistance by multiple mechanisms. Therefore, there is an urgent need to identify a strategy that can overcome chemoresistance and sensitize tumor cells to chemotherapeutic agents.

Chemosensitization is one strategy to overcome chemoresistance. It is based on the use of one drug to enhance the activity of another by modulating one or more mechanisms of resistance. Among the potential chemosensitizers are natural agents such as resveratrol. Over the years, more and more natural products have been discovered to be effective against cancer drugs because of their mutitargeting property, low cost, low toxicity, and immediate availability. More than 60% of the anticancer drugs available to date in the market are of natural origin.[2,3]

Resveratrol (3,4′,5-trihydroxy-trans-stilbene) was first isolated in 1940 as an ingredient of the roots of white hellebore (*Veratrum grandiflorum* O. Loes) and since then it has been identified in extracts from more than 70 other plant species (Fig. 1).[4–7] It is one of the main constituents of red wine, the consumption of which has been associated with 40% lower incidence of heart infarction in France than in other comparable countries, the so-called French paradox.[8] Resveratrol is a phytoalexin (antimicrobial) that has

doi: 10.1111/j.1749-6632.2010.05852.x

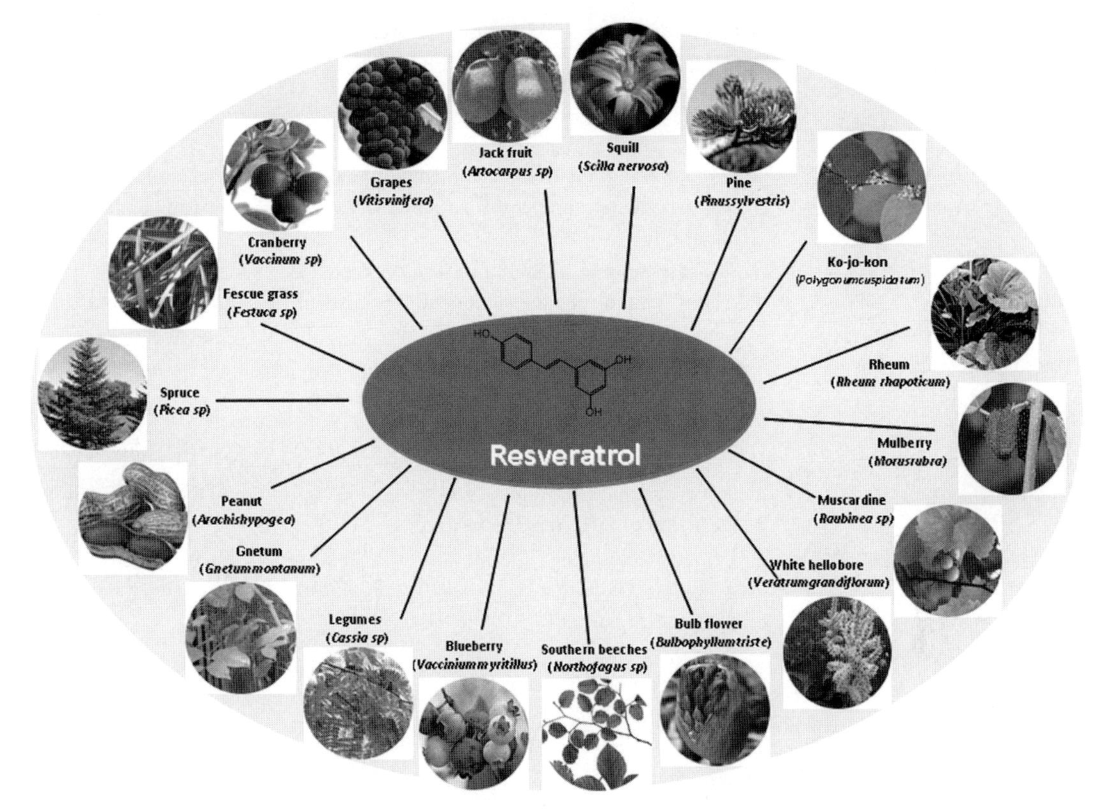

Figure 1. Molecular structure and sources of resveratrol. The sources of resveratrol include fruits, vegetables, legumes, and weeds.

broad-spectrum beneficial health effects including anti-infective, antioxidant, and cardioprotective functions.[5] Since Jang *et al.* published the first article on the anticancer potential of resveratrol in 1997, great interest from cancer researchers has continued in this molecule.[9]

The anticancer activities of resveratrol are mediated through modulation of several cell-signaling molecules that regulate cell cycle progression, inflammation, proliferation, apoptosis, invasion, metastasis, and angiogenesis of tumor cells.[6,10,11] It has been shown that resveratrol can sensitize resistant cells to chemotherapeutic agents by overcoming one or more mechanisms of chemoresistance. In some tumor cells, however, resveratrol has been shown to act as chemoprotector. The potential mechanisms underlying this dual effect are described in this review.

Mechanisms of chemoresistance in tumor cells

The mechanisms of chemoresistance in tumor cells can be intrinsic (cells are resistant before the treatment) or acquired (resistance is developed during the treatment). The most important mechanisms of chemoresistance, as listed in Table 1, are described below.

Drug influx and efflux

Because the antitumor agent must reach into the cancer cell in a concentration sufficient to exert its effect, alterations in the uptake or efflux of the drug could be responsible for the acquisition of chemoresistance.[12] Drug efflux from cells is mediated by the activity of transporter proteins called ATP-dependent multidrug transporters. The most important drug transporters associated with chemoresistance are multidrug resistance protein (MDR; P-glycoprotein, P-gp),[13] multi-drug resistance associated protein (MRP1),[14] lung resistance-related protein (LRP), [15] and breast cancer resistance protein (BRCP). [16]

Inactivation of chemotherapeutic agents

Another chemoresistance mechanism is mediated by the glutathione/glutathione S-transferase (GSH/GST) system.[17] A number of reports have

Table 1. Mechanism of chemoresistance of tumor cells

Mechanism	Effect
Increased drug efflux	
P-glycoprotein ↑	Transports lipophilic and natural compounds out of the resistant tumor cells[13]
MRP ↑	Transports chemotherapeutic drugs out of the cells in association with cellular glutathione[14]
LRP ↑	Transports compounds out of the cells by exocytosis[15]
BRCP ↑	Resistant to mitoxantrone, daunorubicin, doxorubicin, and topoisomerase I inhibitors[16]
Increased drug inactivation	
GSH/GST ↑	Detoxification of drugs to an inactive compound[17]
DPD ↑	Catabolism of 5-FU in the liver[19]
Altered target molecules	An increase, an inactivation, or a decrease in the target molecules[20,21]
DNA repair ↑	Removal of drug-DNA adduct and repair of drug-induced DNA lesions through the action of DNA repair enzymes[66]
Altered growth factor signaling	
EGFR ↑	Activation of EGFR that leads to activation of intracellular cell survival pathways such as PI3K/Akt, ERK1/2 and STAT3[25]
Altered cell death signaling	
p53 ↓	Enhanced tumor cell survival [67]
Bcl-2, Bcl-xL ↑	Enhancement of cell survival[30]
Bax ↓	An increase in cell survival[68]
c-FLIP↑	Inactivation of caspase-8[31]
XIAP ↑	Inhibition of activities of caspases[32,69]
Survivin ↑	Inhibition of caspase-9 activity[70]
NF-κB ↑	Upregulation of genes involved in growth regulation, inflammation, carcinogenesis, and apoptosis[34,35]
STAT3 ↑	Upregulation of Bcl-xL, Mcl-1, survivin, and cyclins[71]

Bax, Bcl-2-associated X protein; Bcl-2, B cell lymphoma-2; Bcl-xL, B cell lymphoma extra large; BRCP, breast cancer resistance protein; c-FLIP, FLICE/caspase 8 inhibitory protein; GSH/GST, glutathione/glutathione S-transferase; DPD, dihydropyrimidine dehydrogenase; 5-FU, 5-fluorouracil; EGFR, epidermal growth factor receptor; ERK, extracellular signal-regulated kinase; LRP, lung-resistance-related protein; Mcl-1, myeloid cell leukemia 1; MRP, multi-drug resistance associated protein; NF-κB, nuclear factor kappa B, PI3K, phosphoionositide 3 kinase; Akt, AKT8 virus oncogene cellular homolog; STAT3, signal transducers and activators of transcription protein-3; XIAP, X-linked inhibitor of apoptosis; ↑ up-regulation, ↓ down-regulation.

shown that the resistance of tumor cells to chemotherapeutic drugs is associated with an increase in the expression of GSH or GST.[18] Dihydropyrimidine dehydrogenase (DPD), which is responsible for catabolizing 5-fluorouracil (5-FU) primarily in the liver, has been shown to confer resistance to 5-FU in colorectal cancer cells through this mechanism.[19]

Alterations in target molecules

During the course of chemotherapy, the target may be modified or decreased to a level where it ceases to have any significant cellular influence. The best example of a molecular target susceptible to modifications is topoisomerase II. Modifications in the topoisomerase II contribute to the development of resistance to various classes of drugs.[20,21]

Enhanced DNA repair

Direct or indirect DNA alteration is the basis of the mechanism of action of many cytostatic drugs. Tumor cells have gained enhanced ability to repair damaged DNA by multiple mechanisms, viz., direct repair, mismatch repair (MMR), base excision repair

Table 2. Chemosensitization of tumors by resveratrol

Chemotherapeutic agent	Tumor type	Mechanism
Cytotoxic drugs	Tumor cells	Cell cycle arrest, survivin↓ [44]
Cisplatin, Gefitinib, Paclitaxel	NSCLC	Survivin ↓ [45]
Paclitaxel	NHL and MM	Bcl-xL ↓ [46]
5-FU	Colon cancer cells	Centrosome amplification[56] Caspase-6 ↑[a47]
Velcade	MM and T cell leukemia	Fas/CD95 recruitment [48]
Doxorubcin	Melanoma	Cyclin D1 ↓ and p53 ↑ [49]
Vincristine, adriamycin, Paclitaxel	Oral carcinoma KBv200 cells	P-glycoprotein ↓, Bcl-2 ↓ [51]
Doxorubicin	AML cells	MRP1 ↓ [52]
Paclitaxel	Lung cancer cells	$p21^{waf1}$ ↑ [53]
Velcade, thalidomide	MM	NF-κB ↓ and STAT3 ↓ [42]
Gemcitabine	Pancreatic cancer	NF-κB ↓ [57]
Anticancer agents	HL60/VCR[58]	
Doxorubicin, cisplatin	Ovarian and uterine cancer cells[59]	
Gemcitabine	Pancreatic cancer cells[60]	

AML, acute myeloid leukemia; Bcl-2, B cell lymphoma-2; Bcl-xL, B cell lymphoma extra large; 5-FU, 5-fluorouracil; MM, multiple myeloma; MRP, multi-drug resistance associated protein; NF-κB, nuclear factor kappa B; NHL, Non-Hodgkin Lymphoma; NSCLC, non small cell lung cancer; STAT3, signal transducers and activators of transcription protein-3; ↑, upregulation; ↓, down-regulation; ↑[a], activation.

(BER), and nucleotide excision repair (NER). For example, BER appears to be important in resistance to some lesions induced by chemotherapy.[22] One of the key elements in the NER process is the excision repair cross-complementing 1 protein (ERCC1).[23] Upregulation of ERCC1 has been associated with chemoresistance of various tumors including colon cancer cells.[23]

Growth factor signaling

The epidermal growth factor receptor (EGFR) activates molecular pathways involved in a variety of cellular processes, such as differentiation, proliferation, survival, and transformation.[24] Overexpression of EGFR in tumor cells has been associated with an increase in resistance to chemotherapeutic agents.[25] It has been proposed that constitutive activation of pathways downstream of EGFR such as the PI3K/Akt and MAPK pathways can also confer resistance to EGFR-targeted therapy.[26]

Altered cell death regulation

Acquisition of resistance to cell death mechanisms, in particular to apoptotic cell death, is one of the hallmarks of tumor cells and plays an important role in the development of resistance to anticancer agents.[27] Tumor cells acquire resistance to apoptosis by such mechanisms as down-regulating pro-apoptotic proteins (p53 and Bax), up-regulating antiapoptotic proteins (Bcl-2; Bcl-xL), and activating survival pathways.

A key factor in the induction of apoptosis in response to chemotherapy is the tumor suppressor protein p53. In approximately 50% of cancer tumors, p53 is found to be mutated.[28] A large study carried out in 1997 by the National Cancer Institute (NCI) revealed a positive correlation between p53 status and sensitivity to cytotoxic drugs.[29]

Bcl-2 family proteins regulate both pro-apoptotic and anti-apoptotic proteins. Bcl-2 itself is an antiapoptotic protein that is inappropriately overexpressed in a number of tumors and has been reported to contribute to chemoresistance.[30] Some other cell survival proteins that have been associated with development of chemoresistance include c-FLIP,[31] X-linked inhibitor of apoptosis (XIAP),[32] and survivin.[33]

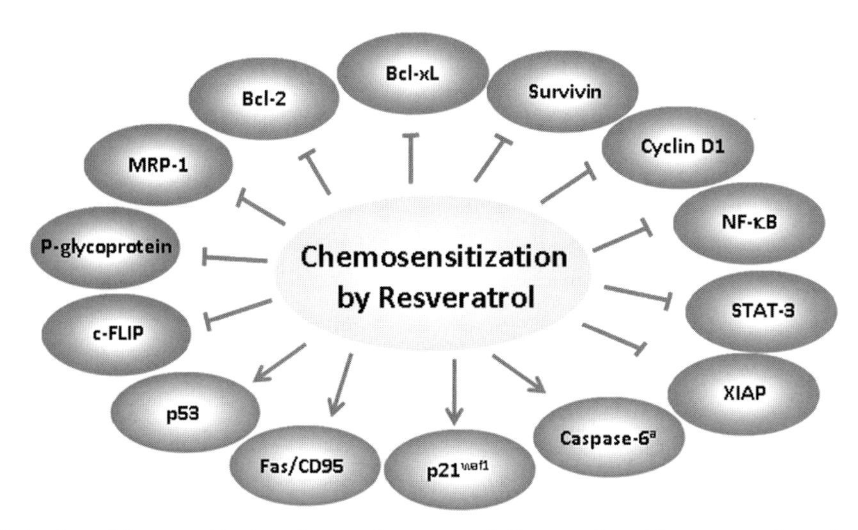

Figure 2. Mechanism of chemosensitization of tumors by resveratrol. Resveratrol sensitizes tumor cells to chemotherapeutic agents by targeting proteins involved in cell survival, cell proliferation, and drug transport. ↓, down-regulation, ↑, up-regulation, [a]activation.

NF-κB activation pathway

The transcription factor nuclear factor (NF)-κB, originally identified in 1986, is a ubiquitously expressed transcription factor that regulates over 500 genes involved in immunoregulation, growth regulation, inflammation, carcinogenesis, and apoptosis.[34,35] Various *in vitro* and *in vivo* studies have shown that constitutive activation of NF-κB inhibits chemotherapy-induced apoptosis in a number of tumor types.[36–38]

STAT3 activation pathway

Signal transducer and activator of transcription 3 (STAT3) is a ubiquitously expressed member of the STAT family of transcription factors that is activated by tyrosine phosphorylation via upstream receptors such as epidermal growth factor (EGF) and platelet-derived growth factor and cytokines such as interleukin-6 (IL-6).[39] Work from our laboratory as well as others have shown that STAT3 can confer tumor resistance to chemotherapeutic agents.[40–43]

Chemosensitization of tumor cells by resveratrol

Resveratrol exerts its sensitization effect by modulating one or more mechanisms of chemoresistance (Table 2, Fig. 2). Research from *in vitro* and *in vivo* studies indicate that resveratrol can overcome chemoresistance in tumor cells by modulating apoptotic pathways, down-regulating drug transporters, and down-modulating proteins involved in tumor cell proliferation. In addition, resveratrol has also been shown to overcome chemoresistance by inhibiting NF-κB and STAT3 pathway.

Most of the reports indicate that resveratrol sensitizes tumor cells to chemotherapeutic agents by modulating cell survival proteins. For example, it has been shown to sensitize human cancer cell lines (neuroblastoma, glioblastoma, breast carcinoma, prostate carcinoma, leukemia, and pancreatic carcinoma) to such chemotherapeutic agents as doxorubicin, cytarabine (AraC), actinomycin D, taxol, and methotrexate by down-regulating survivin expression and increasing apoptosis.[44] Similarly, in another study, resveratrol potentiated apoptosis induced by chemotherapeutic agents such as cisplatin, gefitinib, and paclitaxel in multidrug-resistant non-small-cell lung cancer cells that was associated with a decrease in survivin expression.[45] In non-Hodgkin's lymphoma (NHL) and multiple myeloma cells, inhibition of Bcl-xL expression by resveratrol was critical for its sensitization effect on paclitaxel-induced apoptosis.[46] Resveratrol has also been shown to sensitize tumor cells by inducing caspase activation. For example, in a study using HCT116 human colon cancer cells, resveratrol exerted synergistic effects with 5-FU in a caspase-6 dependent manner. Furthermore, the nature of resveratrol's effect was dependent on concentration. At higher concentrations, resveratrol synergistically promoted

5-FU-mediated apoptosis, while it inhibited 5-FU-triggered apoptosis at lower concentrations.[47] In one study, resveratrol was shown to enhance the apoptotic potential of perifosine and velcade in multiple myeloma and T cell leukemia cells by enhancing recruitment of Fas/CD95 death receptor in the extrinsic pathway of apoptosis.[48]

In a few cases, resveratrol has been shown to sensitize tumor cells to chemotherapeutic agents through modulation of the tumor suppressor gene p53. For example, it was shown to enhance the chemotherapeutic potential of doxorubicin in chemoresistant B16 melanoma cells through upregulation of p53.[49] Interestingly, in another study an enhancement in the sensitization effect of resveratrol on apoptosis induced by various drugs in cancer cell lines was found to be p53 independent.[44]

In some tumor cells, resveratrol has been shown to eliminate chemoresistance by decreasing the activity of membrane transporters such as P-gp and multidrug resistance-associated protein. One study showed that resveratrol induced accumulation of daunorubicin in human multidrug-resistant carcinoma KB-C2 cells in a concentration-dependent manner. Although the authors did not examine the effect of resveratrol on chemosensitivity, they observed that the increase in daunorubicin accumulation was associated with a decrease in P-gp level.[50] In a recent study, Quan *et al.* examined the potential of resveratrol to overcome chemoresistance of oral epidermoid carcinoma KBv200 cells to vincristine, adriamycin, and paclitaxel.[51] Resveratrol produced a synergistic effect when combined with the chemotherapeutic agents and reversed the multidrug-resistant phenotype of KBv200 cells. The reversal of chemoresistance phenotype was associated with a decrease in the expression of P-gp. Using three doxorubicin-resistant acute myeloid leukemia cells (AML-2/DX30, AML-2/DX100, AML-2/DX300), Kweon *et al.*, recently investigated the potential of resveratrol to reverse the resistance phenotype of tumor cells.[52] They found an enhanced expression of the MRP1 gene in chemoresistant cells. When resveratrol was administered, the expression of MRP1 was downregulated, while cellular uptake of doxorubicin into the resistant cells was increased. Based on these observations, the group concluded that resveratrol may facilitate the cellular uptake of doxorubicin via an induced downregulation of MRP1 expression, and that resveratrol may prove useful in overcoming doxorubicin resistance, or in sensitizing doxorubicin-resistant AML cells to antileukemic agents.

Resveratrol has also been shown to enhance chemosensitivity of tumor cells by arresting cells at different stages of the cell cycle and by downregulating the genes involved in cell proliferation. For example, in chemoresistant B16 melanoma, resveratrol enhanced doxorubicin induced cytotoxicity that was accompanied by a downregulation of cyclin D1.[49] Combined treatment with resveratrol was also associated with an increase in the cell cycle arrest at the G(1)-S phase. In another study, when resveratrol was administered prior to paclitaxel treatment in lung cancer cells (A549, EBC-1, Lu65), a significant enhancement in the antiproliferative potential of paclitaxel was observed. Resveratrol also induced the expression of cell cycle regulator p21/waf1 expression and lowered the paclitaxel threshold required for killing tumor cells.[53]

In a few cases, resveratrol has been shown to overcome chemoresistance through downregulation of NF-κB and STAT3 pathways. In our laboratory, resveratrol was shown to enhance the apoptotic and antiproliferative potential of velcade and thalidomide in multiple myeloma cells.[42] Such an enhancement was associated with inhibition of NF-κB and STAT3 activation pathways. Resveratrol administration was also associated with accumulation of sub-G(1) population, increase in Bax release, and activation of caspase-3. This was further correlated with downregulation of various proliferative and antiapoptotic gene products, including cyclin D1, cIAP-2, XIAP, survivin, Bcl-2, Bcl-xL, Bfl-1/A1, and TRAF2. Investigation of the mechanism revealed that resveratrol inhibited NF-κB activation through inhibition of IκBα phosphorylation and IKK activation. These observations were further supported by an inhibition of NF-κB and STAT3 in multiple myeloma patients.

Centrosome amplification, defined as the presence of three or more centrosomes in a cell, is a common feature in cancer cells. Centrosome amplification is often associated with genetic instability, resulting in the accumulation of deleterious genomic stresses in the cells. Therefore, cells with centrosome amplification are likely to be more efficiently eliminated by cell death mechanisms.[54] This has been appropriately described as "apoptosis limits centrosome amplification and genomic instability."[55] Recently, Lee *et al.* investigated the relationship between centrosome amplification and apoptosis sensitivity in colon cancer cells. They found that centrosome amplification is a spontaneous process in cancer cells and could also be induced by subtoxic concentrations of 5-FU. The group further found that cancer cells with centrosome amplification, whether spontaneous or 5-FU-induced, were more sensitive to apoptosis induction by resveratrol.[56] Based on their observations, the group concluded that centrosome amplification might be a novel chemosensitization approach to cancer therapy.

In some cases, resveratrol has been shown to overcome chemoresistance by more than one mechanism. For example, the reversal of chemoresistace of KBv200 cells to vincristine, adriamycin, and paclitaxel by reseveratrol was shown not to be only due to down-regulation of P-gp but also due to a decrease in Bcl-2 expression.[51] Similarly, in a number of cancer cell lines, resveratrol was shown to overcome chemoresistance by inducing cell cycle arrest at S-phase and down-regulating survivin to enhance the effect of a variety of chemotherapeutic agents.[44] Another study using lung cancer cells showed that resveratrol has potential to enhance chemotherapeutic potential of paclitaxel by enhancing both pro-apoptotic and antiproliferative potential.[53]

Although most of the reports are based on the *in vitro* data, some *in vivo* studies also support the chemosensitizing potential of resveratrol. Zhao *et al.* examined the efficacy of resveratrol in a nude mice bearing multidrug-resistant human non-small-cell lung cancer cells (NSCLC). The cells were treated with resveratrol at a concentration of 25, 50, or 100 μM in *in vitro* studies and nude mice were implanted with NSCLC and fed a special diet that included resveratrol at a dose of either 1 g/kg/day or 3 g/kg/day. The rate of cell proliferation, apoptosis ratio, cell cycle phase distribution, IC_{50} values of cisplatin, gefitinib, and paclitaxel, implanted tumor volume, and expression of survivin in resveratrol-treated and control mice were determined. Resveratrol significantly inhibited the proliferation of cancer cells, induced apoptosis, arrested the cell cycle between G0-G1 and S-phase or at the G2/M phase, decreased the IC_{50} values of chemotherapeutic drugs, and showed antitumor effects in nude mice that were implanted with cancer cells. In addition, resveratrol affected the proliferation of cancer cells in a dose- and time-dependent manner. Expression of survivin in cancer cells decreased even at low doses of resveratrol, but the effects were dose-dependent.[45]

Whether resveratrol can sensitize pancreatic cancer (PaCa) cells to gemcitabine *in vitro* and *in vivo* was investigated recently in our laboratory.[57] In *in vitro*, resveratrol inhibited the proliferation of human PaCa cell lines, synergized the apoptotic effects of gemcitabine, inhibited the constitutive activation of NF-κB and expression of Bcl-2, Bcl-xL, COX-2, cyclin D1, MMP-9, and VEGF. We then established PaCa xenografts in nude mice, randomized them into four groups, and treated them with vehicle, gemcitabine, resveratrol, or a combination. Resveratrol significantly suppressed the growth of the tumor, and this effect was further enhanced by gemcitabine. The markers of proliferation index, Ki-67, and micro vessel density, CD31, were significantly down-regulated in tumor tissue by the combination of gemcitabine and resveratrol. As compared to vehicle control, resveratrol also suppressed NF-κB activation and expression of cyclin D1, COX-2, ICAM-1, MMP-9, and survivin. Overall the results demonstrated that resveratrol can potentiate the effects of gemcitabine through suppression of markers of proliferation, invasion, angiogenesis, and metastasis.

In a few cases, resveratrol has been shown to sensitize tumor cells to chemotherapy; however, the molecular mechanism for this action is not known. For example, resveratrol exhibited diverse anticancer activities with doxorubicin, cycloheximide, busulfan, gemcitabine, and paclitaxel in multidrug-resistant variant HL60/VCR (P-gp positive) cells.[58] Similarly, in human ovarian (OVCAR-3) and uterine (Ishikawa) cancer cells, resveratrol was shown to enhance the growth inhibitory/anticancer activity of cisplatin and doxorubicin.[59] A polymethoxylated resveratrol analogue, *N*-hydroxy-*N*′-(3,4,5-trimethoxphenyl)-3,4,5-trimethoxy-benzamidine (KITC), also showed synergistic effect with gemcitabine as an anticancer agent in the human pancreatic cancer cell lines AsPC-1 and BxPC-3.[60] In these studies, the molecular mechanism of resveratrol action was not elucidated.

Table 3. Chemoprotection of tumors by resveratrol

Chemotherapeutic agent	Tumor type	Mechanism
Paclitaxel	SH-SY5Y neuroblastoma cells	Caspase-7 ↓, PARP cleavage ↓ [61]
Paclitaxel	Neuroblastoma cells	S-phase block[62]
Paclitaxel	Bladder cancer cells	Caspase-3 activation ↓, PARP cleavage ↓, G_2/M cell cycle arrest ↓[63]
Vincristine, Daunorubicin H_2O_2	Leukemia cells	NADPH oxidase ↑[a64]
Paclitaxel	Breast cancer cells	Bcl-2 active, ROS ↓ [65]

Bcl-2, B cell lymphoma-2; NADPH, nicotinamide adenine dinucleotide phosphate reduced; PARP, poly (ADP-ribose) polymerase; ROS, reactive oxygen species; ↓, down-regulation; ↑[a], activation.

Resveratrol as a chemoprotector

Chemosensitization strategy aims at using one drug to enhance the activity of another selectively in the tumor cells, while limiting any undesired toxicity or side effects in normal cells. Although resveratrol has been shown to increase the efficacy of chemotherapeutic agents, under certain circumstances, in some cell types, it has also been reported to reduce the efficacy of chemotherapeutic agents (Table 3). Similarly, evidence is also emerging showing that such normal cells as endothelial cells, lymphocytes, and chondrocytes are vulnerable to resveratrol.[10] These contradictory reports present a major caveat for use of resveratrol as a chemosensitizer.

The potential of resveratrol as an antiapoptotic agent was investigated in human neuroblastoma cell line SH-SY5Y.[61] When neuroblastoma cells were treated with paclitaxel, a significant increase in apoptosis was observed. The induction in paclitaxel-induced apoptosis was abrogated by resveratrol treatment. Investigation of the mechanism revealed that resveratrol was able to inhibit the activation of caspase 7 and PARP cleavage that occurred in SH-SY5Y cells exposed to paclitaxel. In a subsequent study, the group evaluated the antiapoptotic effect of resveratrol by studying its effect on cell cycle progression in neuroblastoma cells.[62] Resveratrol was found to induce S-phase cell cycle arrest that was associated with an increase of cyclin E and cyclin A, a downregulation of cyclin D1 with no alteration in cyclin B1, and cdk 1 activation. The resveratrol-induced S-phase block prevented neuroblastoma cells from entering mitosis, the phase of the cell cycle in which paclitaxel exerts its activity.

In another study, resveratrol negatively affected paclitaxel-induced apoptosis in 5637 bladder cancer cells.[63] The antagonism was associated with a decrease in paclitaxel-induced caspase-3 activation, PARP cleavage, and G2/M cell cycle arrest. The antagonism of paclitaxel-induced apoptosis by resveratrol seemed to involve several intracellular signal pathways including Akt, MAPK, and NF-κB.[63] When human leukemia cells were exposed to low concentrations of resveratrol (4–8 μM), a decrease in apoptosis induced by hydrogen peroxide or the anticancer drugs vincristine, and daunorubicin was observed.[64] Resveratrol exerted its effect by acting as antioxidant and reduced caspase activation, DNA fragmentation, and translocation of cytochrome *c* induced by the agents.[64] The authors showed that NADPH oxidase-dependent elevation of intracellular superoxide that blocks mitochondrial hydrogen peroxide production was essential for apoptosis inhibition by resveratrol.

The sensitization effect of resveratrol on paclitaxel-induced cell death in breast cancer cells was investigated.[65] Resveratrol strongly diminished the susceptibility of MDA-MB-435s, MDA-MB-231, and SKBR-3 cells to paclitaxel-induced cell death in culture, although the effect was not observed in MCF-7 cells. Using MDA-MB-435s cells as a representative model, the authors made a similar observation in athymic nude mice. Mechanistically, the modulating effect of resveratrol was attributable to its inhibition of paclitaxel-induced G(2)/M cell cycle arrest, together with an accumulation of cells in the S-phase. In addition, resveratrol was also found to suppress paclitaxel-induced accumulation of reactive oxygen species (ROS) and subsequently the inactivation of antiapoptotic Bcl-2 family proteins. The group concluded that concomitant use of resveratrol with paclitaxel is detrimental in certain types of human cancers.

Summary, conclusions, and future perspective

Acquisition of chemoresistance remains one of the promising problems of chemotherapy failure in cancer patients. Cancer cells are very intelligent: they will do whatever they need to survive. The fact that tumor cells develop multiple chemoresistance mechanisms and that more than one mechanism may work simultaneously complicates the success of chemotherapy. In this regard, resveratrol seems to be an ideal candidate as an adjunct to reverse chemoresistance. However, additional research is required to clarify the following issues.

First, despite more than 1100 publications on cancer chemotherapeutic potential of resveratrol, with around 20 reports on its chemosensitization potential, only a few studies using clinically relevant animal models have been done. Second, combinations may be synergistically cytotoxic to normal cells as well. Most of the studies done to date have focused on tumor cells only. Whether and how resveratrol discriminates between normal and cancer cells is largely unknown. Third, resveratrol has been shown to act as chemoprotector in some cancer cells. Fourth, it remains to be seen whether the expected chemosensitization will be reproduced in humans. Would such a combination increase the patient's life?

Future studies should focus on careful and accurate characterization of molecular mechanisms of chemosensitization, determination of efficacy of resveratrol combinations by clinically relevant *in vivo* studies, and, finally, demonstration of safety and effectiveness of combinations in patients.

Acknowledgments

We thank Walter Pagel from the Department of Scientific Publications for carefully proofreading the manuscript and providing valuable comments. Dr. Aggarwal is the Ransom Horne, Jr., Professor of Cancer Research. This work was supported by a grant from the Clayton Foundation for Research (B.B.A.), a core grant from the National Institutes of Health (CA-16 672), a program project grant from National Institutes of Health (NIH CA-124787-01A2), and a grant from the Center for Targeted Therapy of M.D. Anderson Cancer Center.

Conflicts of interest

The authors declare no conflicts of interest.

References

1. Higgins, C.F. 2007. Multiple molecular mechanisms for multidrug resistance transporters. *Nature* **446:** 749–757.
2. Gupta, S.C. *et al.* 2010. Regulation of survival, proliferation, invasion, angiogenesis, and metastasis of tumor cells through modulation of inflammatory pathways by nutraceuticals. *Cancer Metastasis Rev.* **29:** 405–434.
3. Newman, D.J., G.M. Cragg & K.M. Snader. 2003. Natural products as sources of new drugs over the period 1981–2002. *J. Nat. Prod.* **66:** 1022–1037.
4. Takaoka, M. 1940. Of the phenolic substances of white hellebore (Veratrum grandiflorum Loes. Fil.) *J. Faculty Sci. Hokkaido Imperial University* **3:** 1–16.
5. Baur, J.A. & D.A. Sinclair. 2006. Therapeutic potential of resveratrol: the in vivo evidence. *Nat. Rev. Drug Discov.* **5:** 493–506.
6. Aggarwal, B.B. *et al.* 2004. Role of resveratrol in prevention and therapy of cancer: preclinical and clinical studies. *Anticancer Res.* **24:** 2783–2840.
7. Aggarwal, B.B., Shishodia, S. 2006. *Resveratrol in Health and Disease.* CRC Press Taylor & Francis Group. Los Angeles.
8. Richard, J.L. 1987. [Coronary risk factors. The French paradox]. Arch Mal Coeur Vaiss. 80 Spec No: 17–21.
9. Jang, M. *et al.* 1997. Cancer chemopreventive activity of resveratrol, a natural product derived from grapes. *Science* **275:** 218–220.
10. Shakibaei, M., K.B. Harikumar & B.B. Aggarwal. 2009. Resveratrol addiction: to die or not to die. *Mol. Nutr. Food Res.* **53:** 115–128.
11. Harikumar, K.B. & B.B. Aggarwal. 2008. Resveratrol: a multitargeted agent for age-associated chronic diseases. *Cell Cycle* **7:** 1020–1035.
12. Huang, Y. *et al.* 2004. Membrane transporters and channels: role of the transportome in cancer chemosensitivity and chemoresistance. *Cancer Res.* **64:** 4294–4301.
13. Endicott, J.A. & V. Ling. 1989. The biochemistry of P-glycoprotein-mediated multidrug resistance. *Annu. Rev. Biochem.* **58:** 137–171.
14. Deeley, R.G. & S.P. Cole. 1997. Function, evolution and structure of multidrug resistance protein (MRP). *Semin. Cancer Biol.* **8:** 193–204.
15. Izquierdo, M.A. *et al.* 1995. Drug resistance-associated marker Lrp for prediction of response to chemotherapy and prognoses in advanced ovarian carcinoma. *J. Natl. Cancer Inst.* **87:** 1230–1237.
16. Lee, J.S. *et al.* 1997. Reduced drug accumulation and multidrug resistance in human breast cancer cells without associated P-glycoprotein or MRP overexpression. *J. Cell. Biochem.* **65:** 513–526.
17. Batist, G. *et al.* 1986. Overexpression of a novel anionic glutathione transferase in multidrug-resistant human breast cancer cells. *J. Biol. Chem.* **261:** 15544–15549.

18. Chao, C.C. *et al.* 1992. Overexpression of glutathione S-transferase and elevation of thiol pools in a multidrug-resistant human colon cancer cell line. *Mol. Pharmacol.* **41:** 69–75.
19. Diasio, R.B. & B.E. Harris. 1989. Clinical pharmacology of 5-fluorouracil. *Clin Pharmacokinet.* **16:** 215–237.
20. Cole, S.P. *et al.* 1991. Non-P-glycoprotein-mediated multidrug resistance in a small cell lung cancer cell line: evidence for decreased susceptibility to drug-induced DNA damage and reduced levels of topoisomerase II. *Cancer Res.* **51:** 3345–3352.
21. Webb, C.D. *et al.* 1991. Attenuated topoisomerase II content directly correlates with a low level of drug resistance in a Chinese hamster ovary cell line. *Cancer Res.* **51:** 6543–6549.
22. Liu, L. *et al.* 1999. Pharmacologic disruption of base excision repair sensitizes mismatch repair-deficient and -proficient colon cancer cells to methylating agents. *Clin. Cancer Res.* **5:** 2908–2917.
23. Youn, C.K. *et al.* 2004. Oncogenic H-Ras up-regulates expression of ERCC1 to protect cells from platinum-based anticancer agents. *Cancer Res.* **64:** 4849–4857.
24. Yarden, Y. & M.X. Sliwkowski. 2001. Untangling the ErbB signalling network. *Nat. Rev. Mol. Cell Biol.* **2:** 127–137.
25. Chen, X., T.K. Yeung & Z. Wang. 2000. Enhanced drug resistance in cells coexpressing ErbB2 with EGF receptor or ErbB3. *Biochem. Biophys. Res. Commun.* **277:** 757–763.
26. Magne, N. *et al.* 2002. Influence of epidermal growth factor receptor (EGFR), p53 and intrinsic MAP kinase pathway status of tumour cells on the antiproliferative effect of ZD1839 ("Iressa"). *Br. J. Cancer* **86:** 1518–1523.
27. Hanahan, D. & R.A. Weinberg. 2000. The hallmarks of cancer. *Cell* **100:** 57–70.
28. Levine, A.J. 1997. p53, the cellular gatekeeper for growth and division. *Cell* **88:** 323–331.
29. O'Connor, P.M. *et al.* 1997. Characterization of the p53 tumor suppressor pathway in cell lines of the National Cancer Institute anticancer drug screen and correlations with the growth-inhibitory potency of 123 anticancer agents. *Cancer Res.* **57:** 4285–4300.
30. Chun, E. & K.Y. Lee. 2004. Bcl-2 and Bcl-xL are important for the induction of paclitaxel resistance in human hepatocellular carcinoma cells. *Biochem. Biophys. Res. Commun.* **315:** 771–779.
31. Longley, D.B. *et al.* 2006. c-FLIP inhibits chemotherapy-induced colorectal cancer cell death. *Oncogene* **25:** 838–848.
32. Sasaki, H. *et al.* 2000. Down-regulation of X-linked inhibitor of apoptosis protein induces apoptosis in chemoresistant human ovarian cancer cells. *Cancer Res.* **60:** 5659–5666.
33. Altieri, D.C. 2003. Validating survivin as a cancer therapeutic target. *Nat. Rev. Cancer* **3:** 46–54.
34. Gupta, S.C. *et al.* Inhibiting NF-kappaB activation by small molecules as a therapeutic strategy. *Biochim. Biophys. Acta.* Advance online publication. doi: 10.1016/j.bbagrm.2010.05.004.
35. Sen, R. & D. Baltimore. 1986. Multiple nuclear factors interact with the immunoglobulin enhancer sequences. *Cell* **46:** 705–716.
36. Bharti, A.C. & B.B. Aggarwal. 2002. Chemopreventive agents induce suppression of nuclear factor-kappaB leading to chemosensitization. *Ann. N.Y. Acad. Sci.* 973: 392–395.
37. Bharti, A.C. & B.B. Aggarwal. 2002. Nuclear factor-kappa B and cancer: its role in prevention and therapy. *Biochem. Pharmacol.* **64:** 883–888.
38. Wang, C.Y., M.W. Mayo & A.S. Baldwin, Jr. 1996. TNF- and cancer therapy-induced apoptosis: potentiation by inhibition of NF-kappaB. *Science* **274:** 784–787.
39. Levy, D.E. & J.E. Darnell, Jr. 2002. Stats: transcriptional control and biological impact. *Nat. Rev. Mol. Cell Biol.* **3:** 651–662.
40. Bharti, A.C. *et al.* 2004. Nuclear factor-kappaB and STAT3 are constitutively active in CD138+ cells derived from multiple myeloma patients, and suppression of these transcription factors leads to apoptosis. *Blood* **103:** 3175–3184.
41. Real, P.J. *et al.* 2002. Resistance to chemotherapy via Stat3-dependent overexpression of Bcl-2 in metastatic breast cancer cells. *Oncogene* **21:** 7611–7618.
42. Bhardwaj, A. *et al.* 2007. Resveratrol inhibits proliferation, induces apoptosis, and overcomes chemoresistance through down-regulation of STAT3 and nuclear factor-kappaB-regulated antiapoptotic and cell survival gene products in human multiple myeloma cells. *Blood* **109:** 2293–2302.
43. Ahn, K.S. *et al.* 2008. Guggulsterone, a farnesoid X receptor antagonist, inhibits constitutive and inducible STAT3 activation through induction of a protein tyrosine phosphatase SHP-1. *Cancer Res.* **68:** 4406–4415.
44. Fulda, S. & K.M. Debatin. 2004. Sensitization for anticancer drug-induced apoptosis by the chemopreventive agent resveratrol. *Oncogene* **23:** 6702–6711.
45. Zhao, W. *et al.* Resveratrol down-regulates survivin and induces apoptosis in human multidrug-resistant SPC-A-1/CDDP cells. *Oncol Rep.* **23:** 279–286.
46. Jazirehi, A.R. & B. Bonavida. 2004. Resveratrol modifies the expression of apoptotic regulatory proteins and sensitizes non-Hodgkin's lymphoma and multiple myeloma cell lines to paclitaxel-induced apoptosis. *Mol. Cancer Ther.* **3:** 71–84.
47. Chan, J.Y. *et al.* 2008. Resveratrol displays converse dose-related effects on 5-fluorouracil-evoked colon cancer cell apoptosis: the roles of caspase-6 and p53. *Cancer Biol Ther.* **7:** 1305–1312.
48. Reis-Sobreiro, M., C. Gajate & F. Mollinedo. 2009. Involvement of mitochondria and recruitment of Fas/CD95 signaling in lipid rafts in resveratrol-mediated antimyeloma and antileukemia actions. *Oncogene* **28:** 3221–3234.
49. Gatouillat, G. *et al.* Resveratrol induces cell-cycle disruption and apoptosis in chemoresistant B16 melanoma. *J. Cell. Biochem.* **110:** 893–902.
50. Nabekura, T., S. Kamiyama & S. Kitagawa. 2005. Effects of dietary chemopreventive phytochemicals on P-glycoprotein function. *Biochem. Biophys. Res. Commun.* **327:** 866–870.
51. Quan, F. *et al.* 2008. Reversal effect of resveratrol on multidrug resistance in KBv200 cell line. *Biomed. Pharmacother.* **62:** 622–629.
52. Kweon, S.H., J.H. Song & T.S. Kim. Resveratrol-mediated reversal of doxorubicin resistance in acute myeloid leukemia

cells via downregulation of MRP1 expression. *Biochem. Biophys. Res. Commun.* **395:** 104–110.
53. Kubota, T. *et al.* 2003. Combined effects of resveratrol and paclitaxel on lung cancer cells. *Anticancer Res.* **23:** 4039–4046.
54. Nigg, E.A. 2002. Centrosome aberrations: cause or consequence of cancer progression? *Nat. Rev. Cancer* **2:** 815–825.
55. Cuomo, M.E. *et al.* 2008. p53-Driven apoptosis limits centrosome amplification and genomic instability downstream of NPM1 phosphorylation. *Nat. Cell Biol.* **10:** 723–730.
56. Lee, S.C., J.Y. Chan & S. Pervaiz. Spontaneous and 5-fluorouracil-induced centrosome amplification lowers the threshold to resveratrol-evoked apoptosis in colon cancer cells. *Cancer Lett.* **288:** 36–41.
57. Harikumar, K.B. *et al.* Resveratrol, a multitargeted agent, can enhance antitumor activity of gemcitabine *in vitro* and in orthotopic mouse model of human pancreatic cancer. *Int. J. Cancer* **127:** 257–268.
58. Duraj, J. *et al.* 2006. Diverse resveratrol sensitization to apoptosis induced by anticancer drugs in sensitive and resistant leukemia cells. *Neoplasma* **53:** 384–392.
59. Rezk, Y.A. *et al.* 2006. Use of resveratrol to improve the effectiveness of cisplatin and doxorubicin: study in human gynecologic cancer cell lines and in rodent heart. *Am. J. Obstet. Gynecol.* **194:** e23–e26.
60. Bernhaus, A. *et al.* 2009. Antitumor effects of KITC, a new resveratrol derivative, in AsPC-1 and BxPC-3 human pancreatic carcinoma cells. *Invest. New Drugs* **27:** 393–401.
61. Nicolini, G. *et al.* 2001. Anti-apoptotic effect of trans-resveratrol on paclitaxel-induced apoptosis in the human neuroblastoma SH-SY5Y cell line. *Neurosci. Lett.* **302:** 41–44.
62. Rigolio, R. *et al.* 2005. Resveratrol interference with the cell cycle protects human neuroblastoma SH-SY5Y cell from paclitaxel-induced apoptosis. *Neurochem. Int.* **46:** 205–211.
63. Mao, Q.Q. *et al.* 2010. Resveratrol confers resistance against taxol via induction of cell cycle arrest in human cancer cell lines. *Mol. Nutr. Food Res.* **54:** 1574–1584.
64. Ahmad, K.A. *et al.* 2004. Resveratrol inhibits drug-induced apoptosis in human leukemia cells by creating an intracellular milieu nonpermissive for death execution. *Cancer Res.* **64:** 1452–1459.
65. Fukui, M., N. Yamabe & B.T. Zhu. Resveratrol attenuates the anticancer efficacy of paclitaxel in human breast cancer cells *in vitro* and *in vivo*. *Eur. J. Cancer* **46:** 1882–1891.
66. Heim, M.M. *et al.* 2000. Differential modulation of chemosensitivity to alkylating agents and platinum compounds by DNA repair modulators in human lung cancer cell lines. *J. Cancer Res. Clin. Oncol.* **126:** 198–204.
67. Miyashita, T. *et al.* 1994. Tumor suppressor p53 is a regulator of bcl-2 and bax gene expression *in vitro* and *in vivo*. Oncogene **9:** 1799–1805.
68. Hayward, R.L. *et al.* 2004. Enhanced oxaliplatin-induced apoptosis following antisense Bcl-xl down-regulation is p53 and Bax dependent: genetic evidence for specificity of the antisense effect. *Mol. Cancer Ther.* **3:** 169–178.
69. Li, J. *et al.* 2001. Human ovarian cancer and cisplatin resistance: possible role of inhibitor of apoptosis proteins. *Endocrinology* **142:** 370–380.
70. Salvesen, G.S. & C.S. Duckett. 2002. IAP proteins: blocking the road to death's door. *Nat. Rev. Mol. Cell Biol.* **3:** 401–410.
71. Aoki, Y., G.M. Feldman & G. Tosato. 2003. Inhibition of STAT3 signaling induces apoptosis and decreases survivin expression in primary effusion lymphoma. *Blood* **101:** 1535–1542.

Ann. N.Y. Acad. Sci. ISSN 0077-8923

ANNALS OF THE NEW YORK ACADEMY OF SCIENCES
Issue: *Resveratrol and Health*

Clinical trials of resveratrol

Ketan R. Patel, Edwina Scott, Victoria A. Brown, Andreas J. Gescher, William P. Steward, and Karen Brown
Department of Cancer Studies and Molecular Medicine, University of Leicester, Leicester, United Kingdom

Address for correspondence: Karen Brown, 502 RKCSB, Department of Cancer Studies and Molecular Medicine, University of Leicester, Leicester, United Kingdom. kb20@le.ac.uk

An expanding body of preclinical evidence suggests resveratrol has the potential to impact a variety of human diseases. To translate encouraging experimental findings into human benefits, information is first needed on the safety, pharmacokinetics, pharmacodynamics, and, ultimately, clinical efficacy of resveratrol. Published clinical trials have largely focused on characterizing the pharmacokinetics and metabolism of resveratrol. Recent studies have also evaluated safety and potential mechanisms of activity following multiple dosing, and have found resveratrol to be safe and reasonably well-tolerated at doses of up to 5 g/day. However, the occurrence of mild to moderate side effects is likely to limit the doses employed in future trials to significantly less than this amount. This review describes the available clinical data, outlines how it supports the continuing development of resveratrol, and suggests what additional information is needed to increase the chances of success in future clinical trials.

Keywords: resveratrol; clinical; pharmacokinetics; colon; pharmacodynamic; safety

Introduction

An increasing body of evidence from preclinical studies suggests resveratrol has the potential to impact a variety of human diseases. To translate encouraging experimental findings into human benefit, information is first needed on the safety, pharmacokinetics, pharmacodynamics, and ultimately, clinical efficacy of resveratrol. Valuable data are accumulating on the safety and tolerability of resveratrol in humans, while the clinical pharmacokinetic and metabolism profiles are becoming reasonably well-defined. However, there are currently no published demonstrations of therapeutic or protective effects of resveratrol in appropriately designed clinical trials. Furthermore, there is presently very little evidence of pharmacological activity in terms of molecular or biochemical changes in humans that might be useful as surrogate efficacy biomarkers; this is an area that urgently needs addressing. A review of the database (http://clinicaltrials.gov/) reveals a total of 16 studies involving resveratrol that are either active or recruiting, plus six more that have recently been completed.[1] These include trials aimed at investigating the potential role of resveratrol in the management of type 2 diabetes, obesity, Alzheimer's disease, and cancer. The following paper reviews the available clinical data, outlines how it supports the continuing development of resveratrol, and suggests what additional information is needed.

Resveratrol pharmacokinetics and metabolism in humans

A number of clinical investigations have assessed the pharmacokinetics and metabolism of resveratrol in humans following oral ingestion of either the single synthetic agent or as a constituent of a particular food or drink.[2–7] These studies, which are summarized in Table 1, have employed a range of doses and administration schedules, as well as a variety of formulations. An understanding of the clinical pharmacokinetics and metabolism of resveratrol is essential to enable appropriate doses to be chosen for larger efficacy trials, for defining relevant concentrations that should be used in preclinical mechanistic studies, and for identifying metabolites that may contribute to or diminish the activity of the parent compound.

doi: 10.1111/j.1749-6632.2010.05853.x

Table 1. Summary of published clinical trials involving resveratrol. The figures in parentheses (column 1) refer to the number of participants in each study

Cohort	Form of resveratrol	Resveratrol dose	Dosing schedule	Study outcome	Reference
Healthy males (12)	Delivered in white wine, white grape juice or vegetable juice	25 mg/70 kg	Single	Resveratrol absorption was similar in all three matrices	8
Healthy males (3)	Dissolved in 5 mL whisky mixed with 50 mL water	0.03, 0.5, or 1 mg/kg	Single	Pharmacokinetic and metabolite profile	9
Healthy males (3)	Delivered in grape juice (200, 400, 600 or 1200 mL)	0.32, 0.64, 0.96, or 1.92 mg	Single	Pharmacokinetic and metabolite profile	9
Healthy males (3) and females (3)	^{14}C-resveratrol taken orally and intravenously	25 mg	Single	Pharmacokinetic and metabolite profile	10
Healthy males (11)	250 mL red wine	5.38 mg	Single	Resveratrol and metabolites identified in low-density lipoprotein after moderate wine intake	2
Healthy males (14) and females (11)	300 mL or 600 mL red wine, consumed after fasting, or with meals of varying lipid content	246 μg, 480 μg, or 1.92 mg	Single	Resveratrol bioavailability was not influenced by food, or lipid content	3
Healthy males (18) and females (22)	500 mg capsules	0.5, 1, 2.5, or 5.0 g	Single	Pharmacokinetic and metabolite profile. Resveratrol did not cause serious adverse events	11
Healthy males (9)	Dissolved in 100 mL of 15% ethanol, made up in low-fat milk to a total volume of 500 mL	85.5 mg/70 kg	Single	Pharmacokinetic and metabolite profile. Resveratrol and its metabolites shown to have a high affinity for protein binding	4
Healthy males (11)	250 mL red wine, 1 L grape juice, or 10 tablets	14 μg/kg	Single	Bioavailability of resveratrol from wine and grape juice sixfold higher than that from tablets	5
Healthy males (4) and females (20)	Capsules	250 or 500 mg	Single; once daily on three separate days	Single doses of resveratrol can modulate cerebral blood flow variables	29
Healthy males (12 young) and females (12 elderly)	Capsules	200 mg	Single followed by multiple doses thrice daily (2 days) and a final single dose	Pharmacokinetic and metabolite profile. Resveratrol was well-tolerated by young and elderly subjects	22

Continued

Table 1. *Continued*

Cohort	Form of resveratrol	Resveratrol dose	Dosing schedule	Study outcome	Reference
Healthy males (20) and females (20)	Capsules	25, 50, 100, or 150 mg	Multiple; six times/day, for 13 doses	Pharmacokinetic and metabolite profile. Resveratrol was well-tolerated, but with some mild adverse events reported	21
Healthy males (3) and females (5)	Capsules; taken with food, quercetin or 100 mL 5% alcohol	2 g	Multiple; twice daily	Pharmacokinetic and metabolite profile. A high-fat meal reduced AUC and C_{max}. Resveratrol was well-tolerated, although diarrhea was frequently observed	6
Healthy males (22) and females (18)	500 mg caplets	0.5, 1, 2.5, or 5.0 g	Multiple; once daily for 29 days	Pharmacokinetic and metabolite profile. Resveratrol caused a reduction in IGF-1 and IGFBP-3 plasma levels. 2.5 and 5.0 g caused mild to moderate gastrointestinal symptoms	12
Male (9) and female (11) colorectal cancer patients	500 mg caplets	0.5 or 1 g	Multiple; once daily for 8 days	Pharmacokinetic and metabolite profile in colon/tumor tissue	15
Healthy males (11) and females (31)	500 mg caplets	1 g	Multiple; once daily for 28 days	Resveratrol shown to modulate enzyme systems involved in carcinogen activation and detoxification. Resveratrol was well-tolerated	20
Healthy males (10) and females (10)	300 mL sparkling wine or 200 mL either white wine or red wine	0.357, 0.398 or 2.56 mg/day	Multiple; once daily for 28 days	Resveratrol metabolites in urine may be useful biomarkers of wine intake in epidemiological and intervention studies	7

Among the first of these pharmacokinetic investigations was that by Goldberg *et al.*, which involved oral administration of resveratrol (25 mg/70 kg) to healthy male subjects in three different matrices, white wine, white grape juice, and vegetable juice, to examine the influence on absorption.[8] Blood samples were taken up to 4 h after ingestion, and the highest recorded serum resveratrol/metabolite levels were consistently achieved after 30 min. Concentrations then declined rapidly, reaching baseline levels within 4 h. Free resveratrol accounted for only a small fraction of the total dose in plasma (1.7–1.9%), with glucuronide and sulfate conjugates dominating the profile in both plasma and urine. The total absorption and peak concentrations of resveratrol related species were broadly equivalent in aqueous and alcoholic matrices. However, the low levels of free resveratrol attained (<40 nmol/L) led the authors to conclude that circulating concentrations generated through consumption of dietary sources would be inadequate to elicit biological effects, based on that needed for activity in cultured cells *in vitro* (5–100 μM).[8]

A study comparing resveratrol pharmacokinetics after administration of grape juice, which predominantly contains glucosides, or as the pure free aglycone dissolved in a small amount of whisky and diluted with water,[9] essentially reinforced the findings of the Goldberg investigation. In addition, the results suggested that glycoside forms of resveratrol are absorbed to a lesser extent than the aglycone.[8]

In a more comprehensive study, the absorption, bioavailability, and metabolic fate of resveratrol was traced through the use of radioisotope labeling.[10] Following oral ingestion of a dietary relevant amount (25 mg) of [^{14}C]-resveratrol by six healthy volunteers, at least 70% of the dose was absorbed, which is unusually high for a dietary polyphenol.[10] However, despite relatively efficient absorption, the bioavailability of unchanged resveratrol was very low, due to rapid and extensive metabolism. Peak plasma levels of radioactivity equated to ~2 μM total [^{14}C]-labeled species, but the vast majority of this was due to the presence of phase II metabolites; only trace amounts of unchanged resveratrol (<5 ng/mL) could be detected. A secondary peak, observed at 6 h after oral dosing was indicative of enterohepatic recirculation of conjugated metabolites by reabsorption after intestinal hydrolysis.

In the same study, urinary analysis revealed that most of the radioactivity following oral dosing was recovered in the urine (53–85%) with the proportion excreted in the feces highly variable (0.3–38%).[10] Administration of a higher unlabeled oral dose (100 mg) to one subject allowed characterization of metabolites eliminated via the urine by high performance liquid chromatography-mass spectrometry (HPLC-MS) and demonstrated the involvement of three transformation pathways. Five major metabolites were identified: two isomeric resveratrol monoglucuronides, a dihydroresveratrol monoglucuronide, a resveratrol monosulfate, and a dihydroresveratrol sulfate. Intestinal microflora were suggested as the potential source of hydrogenated resveratrol metabolites formed by saturation of the aliphatic double bond. Sulfate metabolites excreted in the urine accounted for approximately 24% of the dose administered, and this conjugation reaction was proposed as the rate limiting step governing the poor bioavailability of resveratrol, whereas glucuronides contributed about half this much. The remainder seemed to consist largely of early eluting, polar derivatives of unknown structure.[10]

A recent phase I dose escalation study evaluated the safety and pharmacokinetics of resveratrol administered as a single dose (0.5, 1.0, 2.5, or 5.0 g) in healthy volunteers.[11] Plasma and urine concentrations of resveratrol and metabolites were determined by HPLC with UV detection and structurally identified using HPLC-tandem mass spectrometry (LC-MS/MS). In addition, fecal samples were collected for analysis. At the highest dose, the average peak plasma concentration (C_{max}) of resveratrol was 539 ± 384 ng/mL (mean ± SD, $n = 10$), which equates to ~2.4 μM. In comparison, the C_{max} levels of two monoglucuronides (resveratrol-3-*O*-glucuronide and resveratrol-4′-*O*-glucuronide) plus resveratrol-3-*O*-sulfate were three- to eight-fold higher (Fig. 1). Furthermore, the area under the plasma concentration versus time curve (AUC) for these metabolites was up to 23-fold greater than for resveratrol, indicating considerably higher systemic exposure to the conjugates. Additional metabolites detected were resveratrol-4′-*O*-sulfate and resveratrol disulfate, although these were relatively low in abundance. Renal excretion of resveratrol and its metabolites was rapid, with 77% of all urinary agent-derived species excreted within 4 h of consuming the 0.5 g dose. The appearance of a secondary resveratrol peak in the plasma, together with the detection of a large amount of parent compound in the feces compared to metabolites, is again consistent with enterohepatic circulation suggested in the earlier [^{14}C]-resveratrol study by Walle *et al.*[10]

In a follow up study, resveratrol was administered at the same doses for 29 days to assess pharmacokinetics following repeated dosing.[12] Resveratrol and metabolite levels were measured by HPLC-UV in plasma pre-dose and at 0.25, 1.0, 1.5, 5, 12, and 24 h post-dose on a day between the 21st and the 28th day of dosing. The mean average plasma concentration (C_{av}) and C_{max} values of parent resveratrol across the four groups increased with dose and ranged from 0.04–0.55 μM and 0.19–4.24 μM,

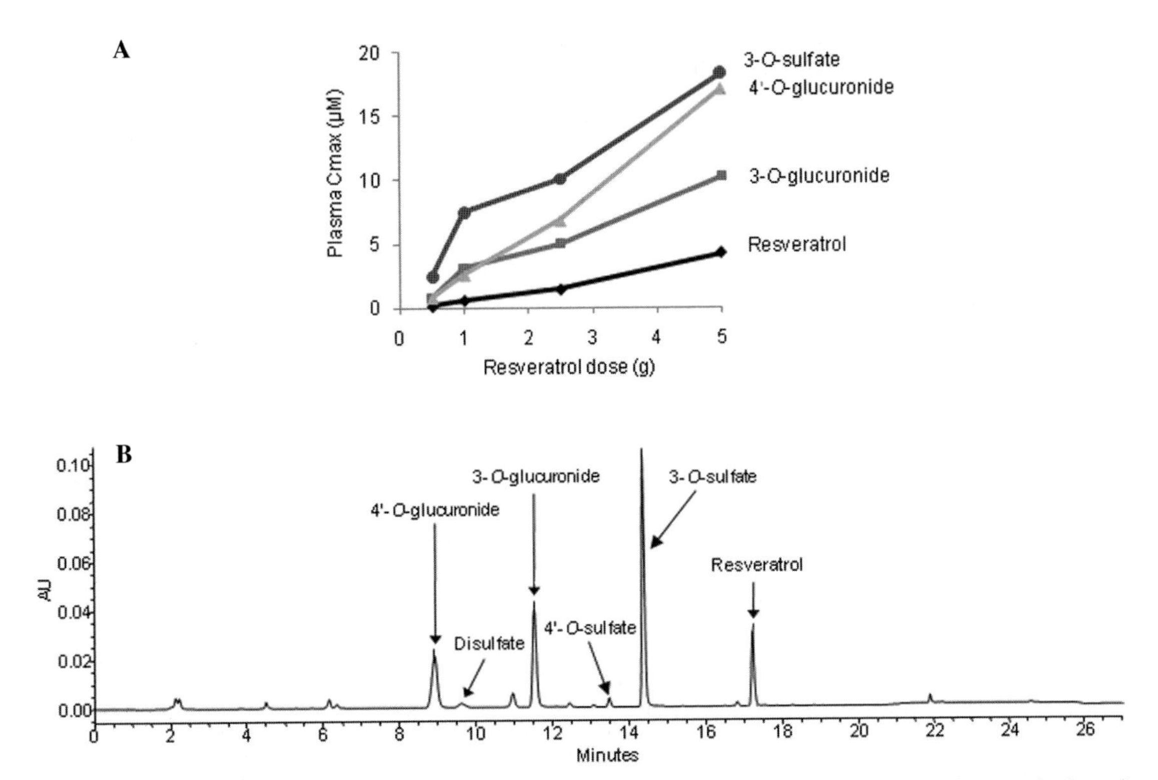

Figure 1. Resveratrol pharmacokinetics in healthy volunteers following multiple dosing. (A) Relationship between the dose of resveratrol ingested and maximum plasma concentration of the parent compound and its main metabolites, the 3-*O*-glucuronide, 4′-*O*-glucuronide, and 3-*O*-sulfate. Volunteers took resveratrol caplets at a dose of 0.5, 1.0, 2.5, or 5.0 g/day for 29 days, and pharmacokinetic profiling was performed between days 21 and 28. Values are the mean of 10 subjects for each dose group. (B) Typical profile of plasma metabolites in a healthy volunteer receiving 2.5 g resveratrol per day. Figure produced from data presented in Ref. 12.

respectively. Using this particular regimen, similar plasma concentrations of resveratrol and its main metabolites were attained following multiple or single dosing;[11,12] significant accumulation with repeated ingestion was only evident at the highest dose. Resveratrol-3-*O*-sulfate, resveratrol-4′-*O*-glucuronide, and resveratrol-3-*O*-glucuronide were the major plasma metabolites, with C_{max} values between 2.4- and 13-fold greater than resveratrol itself. AUC values for these metabolites also exceeded those of the parent, in the case of resveratrol-3-*O*-sulfate by up to 20-fold. The apparent total body clearances and volumes of distribution were calculated following urinary analysis and were consistent with the rapid metabolism and low bioavailability of resveratrol.

An important aspect of resveratrol pharmacokinetics is assessing whether it reaches the proposed sites of action after oral ingestion in humans and determining the concentrations attained in target tissues. Although this has been studied in detail in rodents,[13,14] it clearly presents challenges in humans. A recent clinical trial conducted in colorectal cancer patients has begun to address these issues.[15] Recruited individuals (10 per group) consumed either one or two 500 mg resveratrol caplets daily for eight consecutive days prior to resection. During surgery, samples of tumor and adjacent sections of apparently normal colon tissue were obtained for analysis. In addition to resveratrol, six metabolites were identified in the tissue by HPLC-UV and/or LC-MS/MS: resveratrol-3-*O*-glucuronide, resveratrol-4′-*O*-glucuronide, resveratrol-3-*O*-sulfate, resveratrol-4′-*O*-sulfate, resveratrol disulfate, and resveratrol sulfate glucuronide. In contrast to the pattern in plasma of healthy volunteers[12] where the sulfate glucuronide is a minor component or more commonly absent, it was a predominant metabolite in tissue from 14 out of 20 patients.[15] There was substantial variation in the tissue concentrations measured, both in different samples from the same patient and among individuals. When all ten patients on each dose were considered together, the highest mean concentration of resveratrol was found in normal tissue localized proximal to the tumor on the right side, where it reached ~19 and 674 nmol/g, after the 0.5 and 1.0 g interventions, respectively. The corresponding values in tumor tissue were lower at ~8 and 94 nmol/g (or μmol/L assuming 1 mL has a mass of 1 g). Importantly, the levels of free resveratrol detected in many cases exceeded the concentrations that have been widely reported to have activity in numerous preclinical systems.[16] Maximal mean tissue concentrations determined for resveratrol metabolites (in nmol resveratrol equivalents/g) were 86 for resveratrol-3-*O*-glucuronide at the 0.5 g dose level and 67 for resveratrol-3-*O*-sulfate in patients on 1.0 g resveratrol, both observed in normal right-sided colorectal tissue proximal to the tumor.[15] Interestingly, levels of resveratrol and its metabolites were consistently higher in tissues originating in the right side of the colon compared to the left. This difference was best exemplified by one patient who had two tumors removed; the levels of resveratrol-derived species were considerably higher in the cecal (right-sided) tumor than the one excised from the sigmoid colon (left-sided). This phenomenon may be due to a higher resveratrol content of fecal matter passing through the right side of the colon relative to the left. In addition, during this passage the luminal contents become more solid, which may hinder the absorption of resveratrol.

Although high levels of resveratrol are achievable in the colon, the profile in other internal tissues might be expected to more closely reflect the situation in plasma. Therefore, whether resveratrol is active in these organs may be more dependent on the intrinsic activity of the metabolites present and/or their ability to regenerate the parent compound. Emerging data on the properties of resveratrol metabolites suggests that the 3- and 4′-*O*-sulfates engage several mechanisms consistent with anticancer activity.[17,18] Although reports to date indicate that resveratrol glucuronides are inactive *in vitro*, they may provide a pool for local or systemic regeneration of resveratrol *in vivo* through β-glucuronidase mediated hydrolysis.[19] In evaluating the potential health benefits of resveratrol it is therefore of key importance to determine the distribution and metabolite profile in other human target tissues and advance understanding of the inherent activity and pharmacokinetics of resveratrol metabolites.

Resveratrol safety and tolerability

Only a small number of clinical trials using resveratrol as a single-agent formulated as a medicinal product have formally addressed and reported on safety and tolerability.[6,11,12,20,21] Only one of these studies included a placebo control group,[21] so it is difficult to completely ascribe the adverse effects experienced to ingestion of resveratrol; however, a consistent pattern is evident that allows some conclusions to be drawn. The dose escalation study reported by Brown *et al.* describes toxicity data from 44 healthy volunteers (10–12 per group) who consumed resveratrol for 29 days at a daily dose of 0.5, 1.0, 2.5, or 5.0 g.[12] Resveratrol was deemed safe, as borne out by the lack of serious adverse events detected by clinical, biochemical, or hematological indices, during both the intervention and 2 week follow-up phase. Overall, 28 participants reported at least one adverse effect that was considered possibly due to resveratrol, with the majority occurring in people taking the two highest doses. The most common toxicity was gastrointestinal, particularly diarrhea, nausea, and abdominal pain, all of which occurred only in individuals taking in excess of 1 g resveratrol per day. Typically, the onset of gastrointestinal symptoms was ~1 h after caplet ingestion with improvement during the course of the day. Approximately, 90% of all events reported were graded as mild, according to the National Cancer Institute Common Terminology Criteria for Adverse Events (CTCAE). In addition, there were four cases of moderate severity diarrhea. Based on the overall findings from this study, the authors recommended that daily doses of resveratrol for subsequent clinical evaluation should not exceed 1 g.[12]

Results from a similarly designed 4-week study by Chow *et al.* reinforce these data in that 1 g resveratrol taken once daily was generally well-tolerated in healthy participants.[20] All reported adverse events were CTC grade 1 or 2 with many being mild and transient; the frequency of side effects experienced were consistent with those observed in the trial described by Brown *et al.*[12] and shorter-term studies involving fractionated daily doses of resveratrol.[6,21,22]

Pharmacodynamic effects of resveratrol in humans

Considering the wealth of preclinical data on the potential mechanisms of action engaged by resveratrol there is relatively little published evidence of pharmacodynamic effects in humans. In the phase I study described by Brown *et al.*, ingestion of resveratrol for four weeks by healthy volunteers caused a small but significant decrease in circulating insulin-like growth factor-1 (IGF-1) and IGF binding protein 3 (IGFBP-3), compared to pre-dosing values.[12] The IGF signaling system influences malignant development; IGFs, which are mitogenic and antiapoptotic,[23,24] can affect cell differentiation, neoplastic transformation, and metastasis.[24–26] The IGF system is regulated by IGF binding proteins, particularly IGFBP-3, which binds IGFs in the extracellular milieu, reducing circulating levels and the potential for interaction with IGF receptors. The potential importance of IGF-1 in cancer is illustrated by several studies suggesting a direct relationship between levels of IGF-1 and risk of colorectal, prostate, breast, or lung cancer.[27] The role of IGFBP-3 is less clear. Conventionally, this peptide is thought to have a protective effect due to sequestration of IGF-1; however, there are also data indicating that higher concentrations are associated with an increased rather than reduced risk of premenopausal breast cancer.[28] Furthermore, although high levels have been inversely correlated with lung cancer risk in a meta-regression analysis, there was no such association with other cancers such as colon and prostate.[28] In the volunteer study, when results from all trial participants were combined, consumption of resveratrol reduced IGF-1 and IGFBP-3 levels weakly compared to baseline, albeit significantly.[12] However, when each dose group was considered separately, those taking 2.5 g resveratrol experienced the most prominent and consistent reductions in plasma IGF-1, whereas no significant changes were apparent with lower or higher doses (0.5, 1.0, 5.0 g). Mean IGFBP-3 concentrations in individuals on 1.0 or 2.5 g resveratrol were also significantly reduced but the other doses had no effect. The ability of resveratrol to decrease circulating IGF-1 and IGFBP-3 in humans may constitute an anti-carcinogenic mechanism. However, further studies are required to confirm or refute this association and, if appropriate, elucidate why the relationship appears to be nonlinear; an understanding of the underlying mechanisms would strengthen the potential value of IGF-1 and/or IGFBP-3 as biomarkers to monitor resveratrol efficacy.

Several other potential markers of activity were investigated in blood samples from the volunteers. Ingestion of resveratrol for 29 days failed to significantly affect circulating levels of prostaglandin E-2, reflecting perturbation of the arachidonic acid cascade, or influence leukocyte levels of the malondialdehyde-DNA adduct M_1dG, which is a measure of DNA oxidation.[12] With respect to activity in target tissues, resveratrol ingestion for 8 days reduced the proliferation of colorectal epithelial cells in cancer tissue compared to pre-dose levels, as assessed by Ki-67 positivity detected by immunohistochemistry.[15] The effect was small and was only significant when data were combined from patients in both dose groups (0.5 and 1.0 g/day); however, it suggests resveratrol has the potential to favorably alter cell proliferation in humans.

The ability of resveratrol to protect against DNA damage and resulting initiating mutations has been linked to suppression of carcinogen metabolic activation and/or increased detoxification through modulation of enzymes involved in phase I or II transformations. However, it is possible that such changes may also affect the efficacy or toxicity of concomitant medications since the same enzymes are responsible for drug metabolism. To ascertain whether such interactions might be a potential problem for resveratrol, the effect of once daily dosing on the activity of drug and carcinogen metabolizing enzymes, phase I cytochrome P450s plus the conjugating enzymes glutathione S-transferase (GST) and UDP-glucuronosyl transferase (UGT) 1A1, was investigated in healthy volunteers.[20] Subjects received resveratrol (1 g daily) for 4 weeks and when pre and post intervention activities were compared resveratrol was found to significantly inhibit the phenotypic indices of plasma CYP3A4, 2D6 and 2C9, while inducing 1A2. In addition, in subjects with low baseline values, intervention was associated with an induction of GST-π protein expression and UGT1A1 activity. Inhibition of CYP3A4 and 2C9 in particular, which metabolize a broad range of drugs, could lead to elevated plasma concentrations and increased likelihood of toxicity. The authors therefore acknowledged that although the enzyme modulation observed could be one of the mechanisms through which resveratrol inhibits carcinogenesis, doses of less than 1 g should be evaluated in future trials to minimize adverse drug interactions.[20]

A further demonstration of the capability of resveratrol to alter physiological parameters in humans was shown in a randomized double-blind placebo-controlled crossover study.[29] Single oral doses of resveratrol (250 or 500 mg) were associated with a dose-dependent pattern of higher cerebral blood flow in the prefrontal cortex during task performance, as indexed by total concentrations of hemoglobin. This is the first indication that resveratrol can alter cerebral blood flow variables in humans and supports additional research on its effects on brain function.

Conclusions

It is extremely encouraging for the future clinical development of resveratrol that doses of up to 5 g/day, taken for a month, are safe and reasonably well-tolerated. However, it seems likely that the occurrence of dose-related mild to moderate side effects, coupled with the ability of resveratrol to alter the activity of drug metabolizing enzymes, will limit the dosage employed in future studies to <1.0 g/day. In humans, resveratrol is efficiently absorbed after oral administration; however, rapid phase II metabolism drastically limits its plasma bioavailability, and this may also prove to be the case for internal tissues. In the absence of distribution data for specific human tissues, the levels attained in plasma should guide the choice of concentrations used in preclinical *in vitro* studies to more accurately reflect the clinical scenario. Consequently, concentrations employed should not surpass $\sim$1 μM in order to mimic the resveratrol levels attainable through intake of 1 g/day or less.[12] The presence of high concentrations of parent resveratrol in colorectal tissues, in excess of that required for activity *in vitro*, supports the colon as a target organ. The efficacy of resveratrol in other tissues may be largely dependent on whether its metabolites have significant activity or are able to regenerate resveratrol either locally or systemically. Alternatively, concentrations of resveratrol below the range typically studied *in vitro*, at least in cancer research, may be sufficient for activity. There is some evidence of biological activity at extremely low concentrations of resveratrol as well as biphasic dose–response relationships; these effects should be investigated further and the underlying mechanisms elucidated in order to help identify the optimum doses to be used in the next phase of clinical trials aimed at evaluating efficacy.[30]

Conflicts of interest

The authors declare no conflicts of interest.

References

1. http://clinicaltrials.gov/
2. Urpi-Sarda, M., O. Jauregui, R.M. Lamuela-Raventos, *et al.* 2005. Uptake of diet resveratrol into the human low-density lipoprotein. Identification and quantification of resveratrol metabolites by liquid chromatography coupled with tandem mass spectrometry. *Anal. Chem.* **77:** 3149–3155.
3. Vitaglione, P., S. Sforza, G. Galaverna, *et al.* 2005. Bioavailability of *trans*-resveratrol from red wine in humans. *Mol. Nutr. Food Res.* **49:** 495–504.
4. Burkon, A. & V. Somoza. 2008. Quantification of free and protein-bound trans-resveratrol metabolites and identification of *trans*-resveratrol-C/O-conjugated diglucuronides – two novel resveratrol metabolites in human plasma. *Mol. Nutr. Food Res.* **52:** 549–557.
5. Ortuño, J., M.I. Covas, M. Farre, *et al.* 2010. Matrix effects on the bioavailability of resveratrol in humans. *Food Chem.* **120:** 1123–1130.
6. la Porte, C., N. Voduc, G.J. Zhang, *et al.* 2010. Steady-State pharmacokinetics and tolerability of *trans*-resveratrol 2000 mg twice daily with food, quercetin and alcohol (ethanol) in healthy human subjects. *Clin. Pharmacokinet.* **49:** 449–454.
7. Zamora-Ros, R., M. Urpi-Sarda, R.M. Lamuela-Raventos, *et al.* 2006. Diagnostic performance of urinary resveratrol metabolites as a biomarker of moderate wine consumption. *Clin. Chem.* **52:** 1373–1380.
8. Goldberg, D.M., J. Yan & G.J. Soleas. 2003. Absorption of three wine-related polyphenols in three different matrices by healthy subjects. *Clin. Biochem.* **36:** 79–87.
9. Meng, X., P. Maliakal, H. Lu, *et al.* 2004. Urinary and plasma levels of resveratrol and quercetin in humans, mice, and rats after ingestion of pure compounds and grape juice. *J. Agric. Food Chem.* **52:** 935–942.
10. Walle, T., F. Hsieh, M.H. DeLegge, *et al.* 2004. High absorption but very low bioavailability of oral resveratrol in humans. *Drug Metab. Dispos.* **32:** 1377–1382.
11. Boocock, D.J., G.E.S. Faust, K.R. Patel, *et al.* 2007. Phase I dose escalation pharmacokinetic study in healthy volunteers of resveratrol, a potential cancer chemopreventive agent. *Cancer Epidemiol. Biomarkers Prev.* **16:** 1246–1252.
12. Brown, V., K. Patel, M. Viskaduraki, *et al.* 2010. Repeat dose study of the cancer chemopreventive agent resveratrol in healthy volunteers: safety, pharmacokinetics and effect on the insulin-like growth factor axis. *Cancer Res.* **70:** doi. 10.1158/0008-5472.CAN-10-2364.

13. Vitrac, X., A. Desmouliere, B. Brouillaud, *et al.* 2003. Distribution of [C-14]-*trans*-resveratrol, a cancer chemopreventive polyphenol, in mouse tissues after oral administration. *Life Sci.* **72:** 2219–2233.
14. Wenzel, E., T. Soldo, H. Erbersdobler, *et al.* 2005. Bioactivity and metabolism of *trans*-resveratrol orally administered to Wistar rats. *Mol. Nutr. Food Res.* **49:** 482–494.
15. Patel, K.R., V.A. Brown, D.J.L. Jones, *et al.* 2010. Clinical pharmacology of resveratrol and its metabolites in colorectal cancer patients. *Cancer Res.* **70:** 7392–7399.
16. Gescher, A.J. & W.P. Steward. 2003. Relationship between mechanisms, bioavailibility, and preclinical chemopreventive efficacy of resveratrol: a conundrum *Cancer Epidemiol. Biomarkers Prev.* **12:** 953–957.
17. Hoshino, J., E.J. Park, T.P. Kondratyuk, *et al.* 2010. Selective synthesis and biological evaluation of sulfate-conjugated resveratrol metabolites. *J. Med. Chem.* **53:** 5033–5043.
18. Calamini, B., K. Ratia, M.G. Malkowski, *et al.* 2010. Pleiotropic mechanisms facilitated by resveratrol and its metabolites. *Biochem. J.* **429:** 273–282.
19. Wang, L.X., A. Heredia, H. Song, *et al.* 2004. Resveratrol glucuronides as the metabolites of resveratrol in humans: characterization, synthesis, and anti-HIV activity. *J. Pharm. Sci.* **93:** 2448–2457.
20. Chow, H.H.S., L.L. Garland, C.H. Hsu, *et al.* 2010. Resveratrol modulates drug- and carcinogen-metabolizing enzymes in a healthy volunteer study. *Cancer Prev. Res.* **3:** 1168–1175.
21. Almeida, L., M. Vaz-da-Silva, A. Falcao, *et al.* 2009. Pharmacokinetic and safety profile of trans-resveratrol in a rising multiple-dose study in healthy volunteers. *Mol. Nutr. Food Res.* **53**(Suppl. 1): S7–S15.
22. Nunes, T., L. Almeida, J.F. Rocha, *et al.* 2009. Pharmacokinetics of trans-resveratrol following repeated administration in healthy elderly and young subjects. *J. Clin. Pharmacol.* **49:** 1477–1482.
23. Ibrahim, Y.H. & D. Yee. 2004. Insulin-like growth factor-1 and cancer risk. *Growth Horm. IGF Res.* **14:** 261–269.
24. Butt, A.J., S.M. Firth & R.C. Baxter. 1999. The IGF axis and programmed cell death. *Immunol. Cell Biol.* **77:** 256–262.
25. Lopez, T. & D. Hanahan. 2002. Elevated levels of IGF-1 receptor convey invasive and metastatic capability in a mouse model of pancreatic isle tumorigenesis. *Cancer Cell.* **1:** 339–353.
26. Samani, A.A., E. Chevet, L. Fallavollita, *et al.* 2004. Loss of tumorigenicity and metastatic potential in carcinoma cells expressing the extracellular domain of the type 1 insulin-like growth factor receptor. *Cancer Res.* **64:** 3380–3385.
27. Sandhu, M.S., D.B. Dunger & E.L. Giovannucci. 2002. Insulin, insulin-like growth factor-I (IGF-I), IGF binding proteins, their biologic interactions, and colorectal cancer. *J. Natl. Cancer Inst.* **94:** 972–802.
28. Renehan, A.G., M. Zwahlen, C. Minder, *et al.* 2004. Insulin-like growth factor (IGF)-1, IGF binding protein-3, and cancer risk: systematic review and meta-regression analysis. *Lancet* **363:** 1346–1353.
29. Kennedy, D.O., E.L. Whitman, J.L. Reay, *et al.* 2010. Effects of resveratrol on cerebral blood flow variables and cognitive performance in humans: a double-blind, placebo-controlled, crossover investigation. *Am. J. Clin. Nutr.* **91:** 1590–1597.
30. Scott, E.N., W.P. Steward, A.J. Gescher, *et al.* 2009. Dietary cancer chemoprevention agents – what doses should be used in clinical trials? *Cancer Prev. Res.* **2:** 525–530.

Erratum for Ann. N. Y. Acad. Sci. 1201: 1–7

Mei-Chen Lo,[1,2] Chin-I Lu,[2] Ming-Hong Chen,[3] Chun-Da Chen,[1,2,4] Horng-Mo Lee,[1,2,5] and Shu-Huei Kao.[1,2] 2010. Glycoxidative stress–induced mitophagy modulates mitochondrial fates. *Ann. N. Y. Acad. Sci.* **1201:** 1–7.

In the above article, the affiliations were incorrectly listed. The correct version is shown below.

[1]Graduate Institute of Medical Sciences, [2]School of Medical Laboratory Sciences and Biotechnology, College of Medicine, Taipei Medical University, Taipei, Taiwan. [3]Department of Pathology, Saint Paul's Hospital, Taoyuan, Taiwan. [4]Department of Laboratory Medicine, Taipei Medical University Municipal Wang-Fang Hospital, Taipei, Taiwan. [5]Institute of Pharmaceutical Sciences and Technology, Central Taiwan University of Science and Technology, Taichung, Taiwan

We apologize for this error.

doi: 10.1111/j.1749-6632.2010.05935.x

